INSTRUCTOR'S RESOURCE MANUAL

FOR STARR AND TAGGART'S

BIOLOGY

THE UNITY AND DIVERSITY OF LIFE

FIFTH EDITION

Susan M. Feldkamp

John Jackson
North Hennepin Community College

Larry Lewis
Millersville University

Jan A. Pechenik
Tufts University

Geraldine Ross
Bellevue Community College

Larry G. Sellers
Louisiana Tech University

Denny Smith
Clemson University

Jane B. Taylor
Northern Virginia Community College

Wadsworth Publishing Company
Belmont, California
A Division of Wadsworth, Inc.

Printed in the United States of America 49

1 2 3 4 5 6 7 8 9 10—93 92 91 90 89

ISBN 0-534-09191-1

CONTENTS

PREFACE

We believe that the material in this resource manual will help novice instructors establish good teaching practices and will also be useful to experienced professors. Because several instructors contributed to this manual, it represents a consensus as to what might be most suitable.

Chapters in the manual correspond to chapters in the fifth edition of *Biology: The Unity and Diversity of Life* by Starr and Taggart. In addition, there are three appendixes:

1. Suggestions for writing essays and term papers (these can be copied and handed out to students).
2. A descriptive list of some of the better software available for each major unit of the book and the addresses of suppliers.
3. The addresses for the film and video vendors listed in each chapter.

Each chapter of the manual has ten sections:

- **Revision Highlights:** Significant changes and improvements incorporated into the fifth edition are noted here.
- **Chapter Outline:** An outline of the headings within the textbook's chapters to give an overview of its contents.
- **Objectives:** General objectives that correspond to the main headings of each text's chapter.
- **Key Terms:** A complete listing of the boldface and italic terms in the chapter.
- **Lecture Outline:** A detailed, sequential outline of the chapter for instructors who may be preparing new lecture notes or updating their previous ones. The book's 400+ transparencies are referenced in this outline (TA refers to transparency acetates; TM refers to transparency masters).
- **Suggestions for Presenting the Material:** Helpful suggestions for presenting certain topics covered in the textbook. As for other sections, hints have been gathered from several classroom teachers.
- **Classroom and Laboratory Enrichment:** A unique and useful collection of visual demonstrations to help make the lecture a lively learning experience for your students; most of the ideas utilize common laboratory or household equipment.
- **Ideas for Classroom Discussion:** In contrast to questions at the end of each chapter in the text, these questions are written specifically to evoke a range of responses from your students in a group setting.
- **Term Paper Topics, Library Activities, and Special Projects:** For those instructors who want their students to "dig a little deeper" into the chapter content, a list of topics that require extra effort and research.
- **Films, Filmstrips, and Videos:** An annotated list of visual material to complement your lectures.

On behalf of all the contributors to this Instructor's Resource Manual, I hope our efforts will facilitate your presentation of the fascinating subject of biology.

Larry Sellers

1

ON THE UNITY AND DIVERSITY OF LIFE

Revision Highlights

New table summarizing the key characteristics of life, as outlined in this chapter (Table 1.1). Some passages tightened to strengthen concepts (pp. 5–6, for example). As for all chapters in the fifth edition, the Review Questions now have italic numbers indicating the pages on which students will find answers.

Chapter Outline

ORIGINS AND ORGANIZATION

UNITY IN BASIC LIFE PROCESSES

- **Metabolism**
- **Growth, Development, and Reproduction**
- **Homeostasis**
- **DNA: Storehouse of Constancy and Change**

DIVERSITY IN FORM AND FUNCTION

- **The Tropical Reef**
- **The Savanna**
- **A Definition of Diversity**

ENERGY FLOW AND THE CYCLING OF RESOURCES

PERSPECTIVE

Objectives

1. List features that distinguish living organisms from dead organisms. Then distinguish a dead organism from a rock.
2. Explain what is meant by the term *diversity*, and speculate about what caused the great diversity of life forms on Earth.
3. Describe the general pattern of energy flow through Earth's life forms, and explain how Earth's resources are used again and again (cycled).

Key Terms

organism
cell
adaptive potential
energy
subatomic particle
atom
molecule
organelle
tissue
organ
population
community
ecosystem
biosphere
energy transfer
photosynthesis
aerobic respiration
reproduction
metabolism
development
homeostasis
dynamic homeostasis
variations
inheritance

mutations
diversity
savanna
natural selection
ecology
producer
consumer
herbivores
carnivores
decomposers

Lecture Outline

I. Introduction
   A. What is life?
      1. The answer changes during our path to maturity.
      2. The degree of insight changes with experience and education.
      3. This book is biology revisited. . . . It will provide . . .
         a. Deeper understanding.
         b. A more organized level of understanding.
   B. To biologists, life reflects its ancient molecular origins and its degree of organization. It is . . .
      1. A way of capturing and using energy and materials.
      2. A commitment to a program of growth and development.
      3. A capacity for reproduction.
      4. An adjustment to changing conditions.
      5. This list only hints at a definition.
   C. Life cannot be understood in isolation from its history and adaptive potential.

II. Origins and Organization
   A. From bacteria to frog, organisms are more complex than nonliving rock.
   B. Organisms reproduce and most exhibit movement.
   C. Viruses have aspects of both the living and the nonliving.
   D. Liposomes are below the boundary of the living.
      1. They are small, water-filled sacs.
      2. They assemble spontaneously from lipid molecules.
      3. Like cell membranes, liposomes show potential for organization.
         a. Both can repair themselves if punctured.
         b. Both exclude some ions but let water enter freely.
   E. At the level of liposomes and viruses, the difference between the living and nonliving blurs.
   F. The structure and organization of the nonliving and the living world arise from the same fundamental properties of matter and energy.
   G. The "difference" between the living state and nonliving state lies in the degree to which energy is used and materials are organized.
   H. The first cells emerged by the evolution of complex systems of molecules.

III. Unity in Basic Life Processes **(Table 1.1, p. 3; TM 1)**
   A. Metabolism
      1. Underlying the assembling and tearing down of biological structures are energy transfers.
      2. Organisms must tap an energy source and transform it into forms they can use.
      3. Plants acquire energy from sunlight and transfer some of the energy into ATP.
      4. Energy stored in carbon-containing molecules may be released by the chemistry of energy-releasing programs.
      5. This capacity to acquire and use energy to assure survival is called metabolism.
   B. Growth, Development, and Reproduction
      1. Each organism proceeds through a series of changes in form and behavior.
      2. Each set of changes is characteristic for each type of organism.
   C. Homeostasis
      1. Homeostatic controls maintain the living state as internal and external conditions change.
         a. Food supply can change.
         b. Toxic substances must be avoided or eliminated.
         c. All environmental conditions are subject to change.

2. Homeostasis implies constancy. Internal conditions are maintained within tolerable ranges.
3. Dynamic homeostatic controls govern organism adjustments to directional changes in the internal and external environment.
   a. Dynamic homeostasis may be needed to complete development.
   b. At puberty in humans, hormones trigger maturation of reproductive structures.

D. DNA: Storehouse of Constancy and Change
1. What is responsible for inheritance?
2. Deoxyribonucleic acid, or DNA, is a storehouse of hereditary information.
3. Information in DNA must remain intact to assure faithful reproduction.
4. Mutations introduce variations on the basic plan.
   a. Most are harmful.
   b. Some may be harmless or beneficial.
5. The environment tests the combination of patterns expressed in each organism.

IV. Diversity in Form and Function

A. Introduction
1. The structure and organization of all organisms are similar.
2. Why are there so many kinds of organisms in each environment?
3. Consider two settings . . .

B. The Tropical Reef
1. The tropical reef has a vast array of interdependent animals and plants.
2. It includes corals, red algae, blue-green algae, fishes, sea stars, etc.

C. The Savanna
1. The savanna consists of grasslands punctuated by scattered shrubs and trees.
2. It includes ninety kinds of large plant-eating animals and other animals.

D. A Definition of Diversity
1. On the reef and savanna, diverse occupants have in common a specialization in who eats whom.
2. All organisms must be equipped to obtain a share of limited resources.
3. Diversity is the sum total of variations in form, functioning, and behavior that have accumulated in each species.
4. Some variant forms are adaptative and result in organisms that are more likely to survive.

V. Energy Flow and the Cycling of Resources (**Fig. 1.12, p. 18; TM 2**)

A. Most forms of life depend directly or indirectly on each other.

B. The dung beetle depends upon the droppings of elephants—an uncluttered environment results.

C. The interaction of organisms with each other and with their environment is the focus of ecology.

D. All life is linked by a one-way flow of energy from the sun.

VI. Perspective

A. There is a fundamental unity to all living organisms.
1. They have common origins.
2. All obey the laws that govern the organization of matter and energy.
3. All rely on metabolic and homeostatic processes.
4. All have the same molecular basis of inheritance.

B. There is great diversity in the appearance and behavior of the millions of species on earth.

C. Both the unity and the diversity are explained by the principle of evolution called natural selection.

D. A scientific principle is an idea whose validity is supported by repeated experiments and observations.

Suggestions for Presenting the Material

- Although Chapter 1 is a general introduction to biology and to this textbook, it will be viewed very differently by instructor and student. For the instructor, this chapter is a *review* rather than a *preview.* That means the instructor must take extra care not to "intimidate" the students during early lectures.
- A casual glance at the chapter contents will reveal terms unfamiliar to most students. These might include: *ATP, metabolism, homeostasis, DNA, ecology, herbivore,* and *carnivore.* The individual instructor will have to decide if these terms need explanation now or are to be deferred until later. Possibly this decision will depend on the time available.
- Obviously, it would be very easy to "lose" the attention and enthusiasm of newly enrolled students if *too much* is presented *too soon.*
- Figure 1.3 is an excellent "road map" and can be used throughout the course to guide the progression along the organizational ladder. It can also be used in the exercise listed in the "Enrichment" section below.
- Figure 1.12 is a diagram that also has relevance to future lectures. When introducing it here, you should stress the *flow* of *energy* and the *recycling* of *raw materials.*

Classroom and Laboratory Enrichment

- Bring several organisms into the classroom or lab. Ask your students to name characteristics that identify each item as living or nonliving (for some organisms, this may be difficult to do without specialized equipment, such as a microscope). Ask the students to identify equipment or experiments that would help to determine if an item is a living organism.
- Obtain an overhead transparency of levels of organization in nature (Figure 1.3). With the upper portions of the figure covered, ask your students to help you name each higher level as you ascend to the top of the figure.
- Select several 2 x 2 transparencies from your collection that show a representative variety of plants, animals, and decomposers. Ask students to characterize them as *producer, herbivore, carnivore,* or *decomposer.*
- Show the videotape "Life on Earth" by David Attenborough (available at retail outlets) as a general introduction to biological diversity.
- In the "Presentation" section above, Figure 1.3 was referred to as a "road map" for the text. With an overhead transparency of that figure in view, have the students mark the chapters in the "Contents in Brief" page of their text that amplify each level of organization.

Ideas for Classroom Discussion

- During your first lecture, ask students to name as many characteristics of living things as possible. While this may at first seem like an obvious and overly simple exercise, students will be surprised at some of the less obvious characteristics, such as homeostasis.
- How does our modern definition of "life" differ from the definition of life that a 17th-century biologist might have used?
- What is metabolism? What metabolic steps in humans are different from those found in green plants? What metabolic steps in humans are the same as those found in green plants?
- What are some examples of homeostasis? Why must living organisms be able to perform it?
- Present a list of ten random organisms (or, better yet, let your students do this). Identify ways in which all of the organisms are similar, then ways in which all of the organisms are different. How would you classify (that is, place into meaningful groups) these organisms?

- Why is it important for a species to be able to change? Wouldn't a species be more successful if it could be assured of remaining the same from one generation to the next?
- Name some organisms you might find in a grassy area nearby. Using arrows, arrange the organisms in a diagram depicting energy flow and the cycling of materials (for help, see Figure 1.12). What are some organisms that may be invisible to the eye but are essential for the recycling of nutrients during decomposition?
- An animal carcass infested with insect larvae is not an attractive sight. Yet it is a biological necessity. Explore the role of these and other "recyclers."
- Is there such a concept as the "balance of nature"?
- Humans are able to manipulate certain aspects of nature for their own benefit. However, it is often said "humans are the only animals that engineer their own destruction." Give examples to support this allegation.
- Death and decay are considered by religious fundamentalists as part of God's curse on mankind. What would the earth be like without these two processes?

Term Paper Topics, Library Activities, and Special Projects

- Are viruses alive? Trace the history of the discovery of viruses, and summarize modern viewpoints as to whether viruses are considered living organisms.
- Discover more about how the first cells are thought to have evolved. How do biologists "draw the line" between that which is living and that which is nonliving?
- Outline ways in which biologists attempted to explain inheritance of particular characteristics prior to the discovery of the role of DNA in heredity.
- Describe how any one of several modern scientific investigative tools (such as electron microscopy, radioactive labeling, gas chromatography, or gel electrophoresis) has made it possible to discover similarities and differences among living organisms.
- The pupal stage of insect metamorphosis is erroneously called the "resting stage." Actually there is a complete transformation of larval tissues to adult tissues. Consult several entomology and biochemistry texts to learn the current status of our knowledge concerning these transformations.
- The origin of life on this planet has always fascinated humankind. Several explanations have been advanced. Search for the principal ones that are still in contention today.
- The supply of easily obtainable energy sources is a matter of debate today. Some persons see a bleak future; others are optimistic. What are the issues that each of these camps sees?

Films, Filmstrips, and Videos

- *The Cell: Principles and Biogenesis.* MGHT, 16 minutes, 16 mm, color.
- *Coral Jungle.* DOUBLE, 1969, 23 minutes, 16 mm, color. A Jacques Cousteau film showing the community relationships of organisms inhabiting coral reefs.
- *Science Lab Safety, Part II.* FPSERV, 1975, 15 minutes, 16 mm, color. Explains safety procedures in using laboratory equipment and performing experiments in the biology laboratory.
- *Still Waters.* MGHT, 1969, 14 minutes, 16 mm, color. Shows the life cycle of fish in carefully photographed under- and above-water sequences accompanied by appropriate music.
- *The Time of Man: A Natural History of Our World.* BFA, 1970, 50 minutes, 16 mm, color. Examines the relationship between twentieth-century life and the environment. Stresses the importance of understanding and preserving the environment, which sustains man and all other forms of life.

2

METHODS AND ORGANIZING CONCEPTS IN BIOLOGY

Revision Highlights

This well-received chapter remains the same except for a more informative table introducing the five-kingdom classification system (p. 33).

Chapter Outline

ON SCIENTIFIC PRINCIPLES
- *Commentary*: Testing the Hypothesis Through Experiments

EMERGENCE OF EVOLUTIONARY THOUGHT
- Linnean System of Classification of Organisms
- Challenges to the Theory of Unchanging Life
- Lamarck's Theory of Evolution

EMERGENCE OF THE PRINCIPLE OF EVOLUTION
- Naturalist Inclinations of the Young Darwin
- Voyage of the *Beagle*
- Darwin and Wallace: The Theory Takes Form

AN EVOLUTIONARY VIEW OF DIVERSITY

PERSPECTIVE

SUMMARY

Objectives

1. List as many steps of the scientific approach to understanding a problem as you can.
2. Explain how people came to believe that the populations of organisms that inhabit Earth have changed through time.
3. Understand as well as you can the ideas and evidence that biologists use to explain how life might have changed through time.

Key Terms

principle	deduction	subjective	sampling error
hypothesis, -ses	suspended judgment	testable	significant
induction	theory	randomization	*Scala Naturae*

species	vestigial structures	artificial selection	Monera
binomial system of nomenclature	*Principles of Geology*	differential reproduction	Protista
genus, genera	uniformitarianism	phylogenetic system of classification	Plantae
specific epithet	*Essay on the Principle of Population*		Fungi
catastrophism	natural selection		Animalia
inheritance of acquired characteristics			

Lecture Outline

I. On Scientific Principles
   A. A principle, like evolution, is a way of explaining a major phenomenon of nature.
   B. Principles are derived from a scientific approach to understanding nature. This approach involves:
      1. Asking a question.
      2. Making a hypothesis (educated guess) using induction. That is, use all the information available to construct the general statement (hypothesis).
      3. Predicting the consequences if hypothesis is correct (deduction or "if-then" process).
      4. Testing deductions by experiments, observations, or models.
      5. Repeating tests to determine if results are consistent with hypothesis.
      6. Objectively noting results and drawing conclusions.
      7. Examining alternate hypotheses in the same manner.
   C. While methods of testing may differ, all hypotheses must be testable. **(Commentary art, p. 22; TM 3)**
   D. Science does not generate absolute truths.
   E. Science results in high probabilities that an idea is correct.
   F. A theory results when there is convincing evidence in support of the hypothesis.
   G. A principle results when the supporting evidence is overwhelming.
   H. Science is limited to questions that can be tested.
      1. Subjective questions cannot be addressed (e.g., Why do we exist?).
      2. All of human society must participate in moral, aesthetic, and other such judgments.
   I. Science becomes controversial when it explains an aspect of nature previously considered supernatural.
      1. Copernicus stated that the earth circled the sun—a heresy. Galileo supported this idea.
      2. The external world, not internal conviction, must be the testing ground for science.

II. Emergence of Evolutionary Thought
   A. Introduction
      1. Many ancient Greeks were committed to finding a natural explanation of the world.
      2. Aristotle started the study of biology and came to view nature as an organized "ladder of life."
      3. By the fourteenth century, this idea was transformed into a rigid view called the Great Chain of Being.
         a. This view held that each species had a fixed place in the divine order of things.
         b. Each species was thought to be unchanged since creation.
         c. This view became challenged when the global voyages of the sixteenth century uncovered many new species—often similar to European forms.
   B. Linnean System of Classification of Organisms
      1. The system was developed by Carl von Linné or Linnaeus, an eighteenth-century naturalist.
      2. It consists of a binomial system of nomenclature. **(Table 2.1, p. 24; TM 4)**
         a. Each scientific name has a first name or genus.
         b. The second name is the specific epithet.
         c. For example, a lobster is *Homarus americanus*.

d. The scheme was thought to mirror the patterns in the Great Chain of Being.
e. Higher levels of organization like family and order were added later.
f. The system reinforced the view that species are unchanging.

C. Challenges to the Theory of Unchanging Life
1. In the eighteenth and nineteenth centuries, evidence from comparative anatomy demonstrated that many animals share a common body pattern.
   a. For example, forelimbs in many animals have similar bones. **(Fig. 2.3, p. 24; TM 5)**
   b. But perhaps at creation, it was not necessary to have different body patterns for each species?
2. Some body parts seemed to have no function.
   a. For example, snakes have bones similar to the human pelvic girdle. **(Fig. 2.4, p. 25; TM 6)**
   b. Why should a "perfect" human backbone have tail bones?
3. The worldwide distribution of plants and animals did not seem consistent with the idea that there was one center of creation.
   a. Marsupials are uncommon elsewhere but extremely numerous in Australia.
   b. Cactus plants are found in North and South American deserts, but not in Australia or Asia.
   c. Why were many organisms restricted to one part of the world?
4. In the eighteenth century, Buffon suggested that the evidence did not favor a single center of creation. In addition, Buffon suggested that species became modified over time.
5. In the nineteenth century, Cuvier suggested that the abrupt changes in the fossil record in different rock strata reflected the concept of catastrophism.
   a. After each catastrophe, fewer species remain.
   b. This theory was never supported by evidence, but shows how prevailing beliefs can influence the explanations of scientists.

D. Lamarck's Theory of Evolution
1. Lamarck believed that simple forms evolved into more complex forms.
2. The cause of change was an internal drive toward perfection—up the Great Chain of Being.
3. This resulted in the inheritance of acquired characteristics, e.g., giraffe's neck.
4. Change resulted from environmental pressure and internal desires. Lamarck was criticized by his peers.
5. However, he correctly noted that species change over time and that the environment is an important factor.

III. Emergence of the Principle of Evolution

A. Naturalist Inclinations of the Young Darwin
1. A child in a wealthy family, young Darwin developed an interest in nature.
2. His grandfather developed an early theory of evolution.
3. At college Darwin first studied medicine, but left to go to the seminary at Cambridge.
4. At Cambridge, Professor Henslow arranged for Darwin to serve as the naturalist on a voyage of the H.M.S. *Beagle.*

B. Voyage of the *Beagle*
1. The trip to South America provided Darwin with many stops at islands, mountain ranges, and rivers.
2. Darwin read Lyell's *Principles of Geology* during the journey.
   a. This book stressed Hutton's concept of uniformitarianism—the notion that the forces that molded the shape of the earth are still at work.
   b. It implied that the earth was millions, not thousands, of years old.
   c. This was a staggering implication—there was enough time for evolution.

C. Darwin and Wallace: The Theory Takes Form
1. Clues from the Voyage
   a. Darwin wondered about the cause of diversity among animals.
   b. In Argentina, Darwin noted living armadillos near the fossil remains of a similar but extinct species. Perhaps the extinct species evolved into the present one?

c. Darwin noted that the differences between two populations of the same species increased as the distance between the populations increased.
d. Darwin noted that on the Galapagos Islands, each island had distinct species of finches. **(Fig. 2.8a, p. 29; TM 7)**
e. Perhaps all the species of finches evolved from one ancestral species?

2. Darwin's Deductions
   a. Darwin felt that one cannot see evolution happen because it occurs so slowly.
   b. Darwin was influenced by Malthus, who noted that populations tend to grow at a rate that exceeds the available resources.
   c. Darwin felt that some normal variant members of a species might be more likely to survive and reproduce.
   d. Darwin reasoned that "natural selection" of variants could be a mechanism for evolution.
3. The Theory of Natural Selection
   a. Darwin compiled evidence for evolution by using plant and animal breeding programs as a model for natural selection.
   b. Briefly, the theory of natural selection is . . .
      i. More offspring are born than can survive.
      ii. All members of a species differ in some trait.
      iii. Some of these traits are heritable and improve the chances of an individual surviving and leaving offspring.
      iv. Differential reproduction results in more individuals with such traits in subsequent generations.
      v. Natural selection is the result of differential reproduction.
   c. In 1858, Darwin received a paper from Wallace outlining the same concept.
   d. Both Darwin and Wallace presented papers to the Linnean Society.
   e. In 1859, Darwin published *On the Origin of Species by Means of Natural Selection.*
4. *On the Origin of Species*
   a. If evolution occurs, there should be evidence of one organism changing into a different type—a "missing link."
   b. No one had noted such forms and it would take almost seventy years before the theory of natural selection was accepted.

IV. An Evolutionary View of Diversity
   A. Evolutionary thought has resulted in a phylogenetic system of classification.
   B. In this text, all organisms belong to one of five kingdoms: Monera, Protista, Fungi, Plantae, and Animalia. **(Table 2.2, p. 33; TM 8)**

V. Perspective
   A. As in all science, Darwin and Wallace's principle remains open to test, open to revision.
   B. For example, is evolution always as gradual as envisioned by Darwin?

Suggestions for Presenting the Material

- Like the preceding chapter, this one is a collection of introductory topics. These include the *scientific method, evolution,* and *classification.*
- Because it focuses on derivation of scientific principles and thought, this chapter differs from most of the others in the book, which are more descriptive.
- Sometimes students think that methods of scientific investigation such as in the "Commentary" are used only by scientists. Show that this is not true by discussing the use of these methods in a routine investigation of "why won't the car start?" (see the "Enrichment" section below).
- Explain carefully the necessity for "control" groups in scientific investigations. Point out the difficulty of determining which groups of human patients will *not* receive a valuable drug (the controls) and who will receive a possibly life-saving medication.

- Having introduced the concept of deductive reasoning in science, the authors proceed to one of the great organizing principles in biology—evolution.
- Before proceeding to the men who proposed a changing biological world, point out that the prevailing thought 200 years ago was "fixity of species." Because of this belief that species did not change; it was incumbent on humans to classify all living things. Although Linnaeus believed in this "fixity," his system is nevertheless still very valid and useful.
- Although additional details will be added in Unit 6, the basic concept of biological evolution can, and should be, introduced early in the course. One way to present the historical development of evolutionary thought is to chronicle the contributions of persons such as Buffon, Lamarck, Lyell, Malthus, Wallace, and of course, Darwin.
- Students rarely hear about Darwin's life other than his famous journey. Present his biography before his theory to spark interest. Perhaps the videotape listed in the "Enrichment" section could be used. You can then proceed to *natural selection* by first explaining *artificial selection* (maybe using dogs rather than pigeons as Darwin did in his book).
- Table 2.2 is included in this chapter but is more relevant to Units 4 and 7.

Classroom and Laboratory Enrichment

- Show a film describing Charles Darwin's voyage on the HMS *Beagle* and his thoughts as he traveled.
- Present fossil evidence (or 35 mm transparencies, filmstrips, or films of fossils) showing how a group of related organisms or a single genus (for example, *Equus*) has evolved and changed through time.
- Give examples of several scientific names for local plants and animals that are well known to the students. Interpret the meanings of each Latin specific epithet.
- Show a phylogenetic tree of vertebrates (or any other group of organisms for which a phylogenetic tree is available) to demonstrate the phylogenetic system of classification. Present students with a set of diverse organisms; ask them how they would classify these organisms.
- Whales, like snakes, have pelvic girdles. Show an overhead diagram of this portion of the whale skeleton.
- Briefly list the steps of the scientific method in the wrong order. Ask the class to place them, one by one, in the correct order.
- Show how we use the scientific method in everyday problem solving as illustrated by this example:

Event	*Method Step*
a. Auto will not start	a. Observation
b. Battery dead	b. Hypothesis
Ignition problem	Hypothesis
Out of gas	Hypothesis
c. Turn on headlights	c. Experiment
Check spark at plug	Experiment
Check gas gauge	Experiment
Dip long stick into gas tank	Experiment
d. Headlights burn brightly (battery OK)	d. Analyze results
Strong ignition spark	Analyze results
Gauge says half tank but no gas on stick	Analyze results
e. Gas gauge is not accurate; car needs gas to run	e. Generalize; form principle

- Generate interest in Darwin's theory by bringing a copy of *Origin of Species* to class. Read selected chapter titles and portions of the text. Point out the lack of illustrations in the original edition.

Ideas for Classroom Discussion

- How is a principle different from a belief?
- What was Jean-Baptiste Lamarck's contribution to our modern understanding of evolutionary theory?
- Why is the term *scientific creationism* an oxymoron? Describe why this body of thought cannot be considered a science.
- What steps did Lamarck fail to perform before setting forth his hypothesis of inherited characteristics?
- How did the widespread discoveries of fossils in the 19th century help to support Darwin's views on evolution?
- Does belief in the principle of evolution exclude belief in religion? Why or why not?
- Why was extensive travel a key ingredient in the development of Darwin's evolutionary thought?
- What is artificial selection? How does it differ from natural selection?
- Can you think of any ideas commonly expressed today that are similar to Lamarck's understanding of evolution?
- How did the work of geologists such as Charles Lyell who were Darwin's contemporaries help Darwin to create his principle of evolution?
- Distinguish among independent, dependent, and controlled variables. Can you identify each if presented with an actual experimental design?
- Why is it difficult to obtain a control group when selecting volunteers to test a new anticancer drug?
- Those who wish to berate certain scientific principles sometimes say "it's only a theory." This statement is used by creationists when referring to evolution. Does the use of "theory" in biology mean the concept is in doubt? Explain using examples.
- If you asked the following question in a sidewalk survey, what do you think the responses would be? "Darwin wrote a very famous book on the origin of __________."
- Compare and contrast the principles of "uniformitarianism" and "catastrophism." Evaluate the physical evidence for each.

Term Paper Topics, Library Activities, and Special Projects

- Describe how a trip through the Grand Canyon with a paleontologist would reinforce our modern understanding of evolution.
- How do today's biologists reconcile their personal faith in an organized religion with their belief in evolution? Research the viewpoints of some famous scientists on this issue.
- Can we see evolution actually happening? Find examples of natural occurrences in the wild or experimental situations in the laboratory in which we can observe evolution occurring.
- Learn more about the discovery of fossils of *Archaeopteryx* and the reactions of the scientific community to them.
- Write a short biography of Linnaeus.
- Describe how Darwin's development of his principle of evolution was an example of the scientific method in action.
- From time to time, Lamarck's hypothesis of "inheritance of acquired characteristics" is revived as an explanation for certain events. Is the idea totally without merit, or could this be the explanation for the inheritance patterns in certain microorganisms? (See *Scientific American,* Nov. 1988, p. 34.)
- The great "catastrophe" that dominated Cuvier's thinking was the flood of Noah as recorded in the Book of Genesis. This is still the cornerstone of creationist thinking. Investigate creationist writings to see how this event is critical to their theories.

- Darwin's emerging ideas on natural selection were not welcomed by the *Beagle* captain. Investigate how this challenge to his views strengthened Darwin's hypotheses.
- It is ironic that Darwin and Wallace would arrive independently at so important a concept as natural selection. Investigate the path each took.

Films, Filmstrips, and Videos

- *Aristotle and the Scientific Method.* CORF, 1959, 14 minutes, 16 mm, color. Stresses Aristotle's contribution to the scientific method. Shows how he departed from Plato's ideas, made observations based on his own experiences, classified his data, performed experiments, and drew generalizations and principles in the fields of zoology and botany.
- *Classifying Plants and Animals.* CORF, 1961, 11 minutes, 16 mm, color.
- *Evolution and the Origin of Life.* CRMP (MGHT), 1972, 33 minutes, 16 mm, color. Simple demonstrations used to explain the theory of natural selection.
- *Natural Selection.* EBEC, 1963, 16 minutes, 16 mm, color.
- *Theories on the Origin of Life.* EBEC, 1969, 14 minutes, 16 mm, color. This presentation of four theories on the origin of life is designed to encourage students to discuss the evidence, to question, and to search for answers.

3

CHEMICAL FOUNDATIONS FOR CELLS

Revision Highlights

Minor text refinements, including clarification of polarity of water molecule (p. 41), and revised pH value for extracellular fluid of the human body (p. 44).

Chapter Outline

ORGANIZATION OF MATTER
- **Atoms and Ions**
- **Isotopes**
- **How Electrons Are Arranged in Atoms**

BONDS BETWEEN ATOMS
- **The Nature of Chemical Bonds**
- **Ionic Bonding**
- **Covalent Bonding**
- **Hydrogen Bonding**
- **Hydrophobic Interactions**
- **Bond Energies**

ACIDS, BASES, AND SALTS
- **Acids and Bases**
- **The pH Scale**
- **Buffers**
- **Dissolved Salts**

WATER MOLECULES AND CELL ORGANIZATION
- **Hydrogen Bonding in Liquid Water**
- **Solvent Properties of Water**
- **Water and the Organization Underlying the Living State**

SUMMARY

Objectives

1. Understand how protons, electrons, and neutrons are arranged into atoms and ions.
2. Explain how the distribution of electrons in an atom or ion determines the number and kinds of chemical bonds that can be formed.
3. Know the various types of chemical bonds, the circumstances under which each forms, and the relative strengths of each type.
4. Understand the essential chemistry of water and of some common substances dissolved in it.

Key Terms

element	proton	electron	atomic nucleus
compound	neutron	molecule	atomic number
atom	mixture	electric charge	mass number

atomic weight
ion
radioactive isotopes
isotopes
tracers
orbital
energy level
nonreactive
reactive
orbital model
shell model
chemical bond

energy relationship
formula
chemical equation
reactant
product
law of conservation of mass
mole
ionic bond
covalent bond
nonpolar covalent bond
polar covalent bond

hydrogen bond
hydrophilic
hydrophobic
bond energy
kilocalorie
hydrogen ion, H^+
hydroxide ion, OH^-
acid, acidic
base, basic
pH scale
buffers
salt

cohesion
stabilization
specific heat
heat of vaporization
temperature
evaporation
heat of fusion
solvent
lipid bilayer
solutes
spheres of hydration
dissolved

Lecture Outline

I. Introduction
   A. The interaction of sunlight with molecules in a plant is what channels energy to plants.
   B. Bacteria in our gut depend upon sugar molecules in our food for energy.
   C. All biological events begin with the organization and behavior of atoms and molecules.
   D. Hence, to understand life one must understand the structure and behavior of atoms and molecules.

II. Organization of Matter
   A. Introduction
      1. All the natural world contains one or more of about ninety elements. **(Table 3.1, p. 36; TM 9)**
      2. Elements cannot be decomposed into substances with different properties.
      3. A one- or two-letter symbol stands for each element.
      4. Compounds are formed from elements combined in fixed proportions.
      5. Mixtures are formed from elements present in varying proportions.
   B. Atoms and Ions
      1. An atom is the smallest unit of an element that retains the properties of the element.
      2. Atoms contain three major kinds of particles: protons, neutrons, and electrons.
      3. Atoms of each element have a distinct number of protons and electrons.
      4. Molecules are composed of at least two atoms bonded together.
      5. An atomic nucleus contains protons, (+ charge) and neutrons (no charge).
      6. Electrons (– charge) move around nucleus.
      7. Atomic number = number of protons in nucleus. Different for each element.
      8. Mass number = number of protons and neutrons in nucleus.
      9. An isolated atom has an equal number of protons and electrons, hence, net charge = zero.
      10. In nature, atoms may gain or lose an electron and end up with a + or – charge.
      11. A charged atom is an ion.
   C. Isotopes
      1. Atoms with the same number of protons but a different number of neutrons = isotopes. Carbon atoms may have a mass number of 12, 13, or 14, written ^{12}C, ^{13}C, and ^{14}C.
      2. Some radioactive isotopes have unstable nuclei and emit subatomic particles and energy.
         a. They decay or break down with an unvarying rate.
         b. They can be used to date rocks and fossils.
         c Some can be used as tracers to follow the path of an atom in a series of reactions or to diagnose disease.
   D. How Electrons Are Arranged in Atoms
      1. Energy Levels for Electrons **(Table 3.2, p. 38; TM 10)**
         a. Electrons are attracted to protons but are repelled by other electrons.
         b. Orbitals permit electrons to stay as close to the nucleus and as far from each other as possible.

c. Each orbital contains one or two electrons.
d. A hydrogen atom has a single electron in the 1*s* orbital—the lowest energy level.
e. If an atom has more than two electrons, they are located at higher energy levels. **(Fig. 3.2, p. 39; TM 11)**
f. The second energy level may have as many as eight electrons in four different orbitals.

2. Electron Excitation **(Fig. 3.3, p. 39; TM 12)**
a. When an atom absorbs energy, an electron may move into an orbital at a higher energy level.
b. In less than a second, the electron returns to the lowest available energy level and releases energy.
c. This is important for photosynthesis and other processes.
d. When excited, electrons move to one level or another, never in between.
e. Some atoms that have vacancies at the highest energy level tend to form bonds with atoms of the same or other elements; H, C, N, and O are examples.

III. Bonds Between Atoms

A. The Nature of Chemical Bonds
1. A chemical bond is a union between the electron structures of two or more atoms or ions.
2. Bonds are energy relationships—i.e., one atom may gain, give up, or share an electron.
3. Strong bonds form within molecules.
4. Weak bonds form between molecules or parts of molecules.

B. Ionic Bonding **(Fig. 3.5, p. 41; TM 13)**
1. Ionic bonding requires one atom to gain at least one electron and another to lose at least one electron.
2. This occurs when a positive and a negative ion are linked by the mutual attraction of opposite charges.

C. Covalent Bonding
1. Nonpolar and Polar Covalent Bonds
a. A covalent bond holds together two atoms that share a pair of electrons.
b. In a nonpolar covalent bond, atoms share electrons equally.
c. In a polar covalent bond, because atoms share the electron unequally, there is a slight difference in charge between the two poles of the bond. Water provides an example.
2. Polarity of a Water Molecule **(Fig. 3.6, p. 41; TM 14a)**

D. Hydrogen Bonding
1. Hydrogen bonding results when an electronegative atom interacts weakly with a hydrogen atom that is participating in a polar covalent bond. **(Fig. 3.7, p. 42; TM 14b)**
2. It imparts structure to large biological molecules and liquid water.

E. Hydrophobic Interactions
1. Polar substances are hydrophilic.
2. Nonpolar substances are hydrophobic.
3. After you shake a bottle of oil and water, hydrogen bonds form between the water molecules and push the oil molecules into droplets or a film.
4. Hydrophobic interactions help determine the architecture of membranes and other biological structures.

F. Bond Energies
1. Bond energies are measured in kilocalories per mole.
2. One kilocalorie is the amount of energy needed to raise 1,000 grams of water from 14.5°C to 15.5°C.
3. Covalent bonds = 80–110 kilocalories per mole.
4. Ionic bonds (in cell) and hydrogen bonds = 4–6 kilocalories per mole.

IV. Acids, Bases, and Salts

A. Acids and Bases
1. When a hydrogen atom leaves a molecule, its electron is stripped away, and a hydrogen ion or H^+ results.

2. An acid releases a hydrogen ion in solution.
3. A base accepts a hydrogen ion in solution.
4. The gain or loss of a hydrogen ion is reversible.

B. The pH Scale **(Fig. 3.8, p. 43; TM 15)**
1. The degree of acidity is measured by the pH scale.
2. The interior of each cell is not far from neutrality.

C. Buffers

Buffers in cells combine with or release H^+ to prevent drastic changes in pH.

D. Dissolved Salts
1. Salts are formed by the reaction between an acid and a base and dissociate into ions in water.
2. Consider $HCl + NaOH \rightarrow NaCl + H_2O$.

E. Water Molecules and Cell Organization
1. Water is 75%–85% of the weight of a cell.
2. The activities associated with living organisms require water.

F. Hydrogen Bonding in Liquid Water
1. Cohesion
   a. Cohesion is the capacity to resist rupturing under tension.
   b. It results from attractions between molecules (such as the hydrogen bonds in water).
   c. At the surface of a body of water (or a drop), hydrogen bonds pull on the molecules at the surface and create a high surface tension.
   d. Cohesion enables insects to walk on water and allows whole columns of water to be raised to the tops of trees.
2. Temperature Stabilization
   a. Because water has a high specific heat, water absorbs much energy before it increases greatly in temperature.
   b. Because water has a high heat of vaporization, water resists evaporation.
   c. Because water has a high heat of fusion, water resists changing from a liquid to a solid.
   d. All the above are due to hydrogen bonds and help stabilize the temperatures of cells.

G. Solvent Properties of Water **(In-text art, pp. 46, 47; TM 16)**
1. Because water is polar, it is a good solvent for ions and polar molecules.
2. Spheres of hydration form around ions and keep ions dissolved as solutes.

H. Water and the Organization Underlying the Living State

Water interacts with both polar and nonpolar molecules and thus contributes to the organization of the cell.

Suggestions for Presenting the Material

- There is no escaping the fact that Chapters 3 and 4 are "chemistry." And chemistry is intimidating—especially to nonscience majors. The material in the book is elementary and written in a lucid manner, but the quality of presentation is up to the individual instructor.
- Perhaps a quick survey of class members who have and have not had high school chemistry will aid in adjusting your level of presentation.
- One approach that might help your students in organizing this material is to write it in outline form on an overhead transparency. This may work especially well for this chapter because a large portion of the material consists of definitions.
- The use of ball-and-stick models (see the "Enrichment" section below) is very helpful. If the lecture room is large, you may have to "tour" the room with the models for better viewing.
- If students become discouraged, assure them that several of these topics will be reinforced in future chapters (hopefully before the next exam).
- The text gives careful attention to useful examples of isotopes, electron excitation, bonding, buffers, and water.

- Emphasize the text's definition of a chemical bond as ". . . not an object; it is an energy relationship."
- Using Figure 3.8 as your visual reference will help in explaining acid, base, and pH scale. Note particularly the pH values of common household products. Emphasize that acids and bases are not necessarily terms that describe *corrosive* substances!
- The properties of water are important to life on Earth. Describe the polarity of water molecules; then proceed to the influence that water molecules have on cells and cellular environments.

Classroom and Laboratory Enrichment

- Students often approach even basic chemistry with considerable trepidation, especially if they lack sufficient high school background in this area or they have been out of school for several years. Emphasize the biological significance of chemistry; stress that an elemental knowledge of chemistry is essential to understanding the structure and function of living things. Give students frequent opportunities to use new terms. Using overheads or diagrams, pause often and interject questions to gauge their level of understanding.
- Use as many models and diagrams as possible. If you wish to emphasize electron orbitals, use foam-and-stick models of the orbitals to make this concept seem clearer.
- Students frequently have trouble visualizing atoms and molecules as real entities. To help them get a clearer mental picture of some of the basic atoms and molecules, use ball-and-stick models that are very large and easy to see from the back of the room. (Be sure to relate these models to the other methods of diagramming atoms as shown in Figure 3.2.) These models will help students to understand the size relationships among molecules. Overhead transparencies of ball-and-stick diagrams will also help. Such models and diagrams will be especially useful when covering the larger carbon compounds discussed in the next chapter.
- Present sketches of a polar covalent molecule and a nonpolar covalent molecule. Ask students to identify which molecule is polar and which is nonpolar and to explain their choices.
- Ball-and-stick models are also useful for demonstrating the hydrogen bonding that occurs between water molecules and the latticework structure of ice.
- Fill a large jar with water, then add salad oil. Shake the bottle, then allow it to sit on the front desk. Ask students to explain what has happened. Add a few drops of methylene blue (a polar dye) and sudan III fat stain (a nonpolar dye) to the jar and shake. Students will note that the water layer is blue and the oil layer is red; ask them why this is so.
- Draw a pH scale on the board (or use an overhead transparency of Figure 3.8), and discuss pH values of familiar substances.
- If your class is small, demonstrate the use of a pH meter. For larger groups, pH paper can be used to give each student a chance to quickly determine the pH of some sample solutions.
- If you are teaching in a room with a "periodic table of the elements" hanging on the wall, point out the major elements, or use an overhead transparency to show the same items.
- Prepare a glass of iced tea (instant mix) with added sugar and lemon. Which ingredients are compounds? What are the components of the mixture?
- Bring a package of "buffered" and "regular" aspirin to class. Ask students to discover the difference(s) in ingredients.
- Using the names of the active ingredients on an antacid package, explain how they act as *buffers* to stomach acid.
- Show a filmstrip or slide/sound set on "atoms and molecules."

Ideas for Classroom Discussion

- Distinguish between: a compound and a mixture, an atom and a molecule.
- What chemicals are in the human body? Ask students to name as many as they can; help them complete their list.
- Discuss the role of electron excitation in photosynthesis.
- What is the difference between polar and nonpolar covalent bonds?
- Why do soft drinks have such a low pH? What ingredient is responsible for this low pH?
- What is acid precipitation? What chemical reaction is responsible for the mildly acidic pH of normal rainwater? What chemicals are responsible for acid precipitation?
- What is a calorie? What are we measuring when we determine the calorie content of different foods?
- What would happen to aquatic organisms living in temperate climates if water sank when it froze instead of floated?
- What is meant by the phrase *lipid bilayer*? Where would you find lipid bilayers inside a living organism?
- What is the difference between the composition of a *molecule* of a substance and an *atom* of that substance?
- If atoms are beyond the reach of visualization even by "super" electron microscopes, how then do we know so much about their structure?
- Water is the "universal solvent" for Earth. Do you know of any other compound that would serve as well, or better?
- Some pain relievers are advertised as "tribuffered." Is this a real advantage or just a sales gimmick?
- Television commercials portray the "acid stomach" as needing immediate R-O-L-A-I-D-S. Is the stomach *normally* acid? How do you know when there is too much acid down there?

Term Paper Topics, Library Activities, and Special Projects

- How are hydrophobic substances such as fats broken down in the human digestive tract? What chemicals are released by the body to assist with fat breakdown?
- Why are the cells lining the stomach able to withstand pH ranges between 1 and 3?
- How does the body measure blood pH? What are the homeostatic mechanisms that help the human body to regulate blood pH?
- Discuss strategies currently being considered by the United States and other nations to remedy acid precipitation. What suggestions would you make to help solve this problem?
- Describe some of the roles played by ions in the human body.
- Many elements have radioactive isotopes that are useful as tracers in biological systems. Show how $^{14}CO_2$ can be used to follow the fate of carbon as it is incorporated into carbohydrate.
- The structure of atoms can be deduced using nuclear magnetic resonance (NMR) and mass spectrometer machines. Report on the principles underlying the performance of each of these instruments.
- Using a pH meter, test the degree of acidity/alkalinity of common household products. If the substance is not a liquid, mix it with water according to package directions before testing.
- Most of the content of human blood is water. However, synthetic blood has been made and tested. What is the base in this fluid? Is it a feasible substitute? Report on its advantages and disadvantages.

Films, Filmstrips, and Videos

- *Atoms.* PBS, 1987, 1 hour, video, color. *Ring of Truth Series.* Philip Morrison leads a tour through the atom, elaborating on its properties and the quest for proof of its existence. New techniques in electron microscopy can visualize atoms, and Quantum Theory outlines behaviors within the atom.
- *Chemical Bonding.* Wards (MLA), 1965, 16 minutes, 16 mm, color. Describes chemical bonding in terms of the electrical interactions that bond two hydrogen atoms together. Shows the release of energy that occurs when hydrogen atoms combine on a platinum surface.
- *Chemical Systems of the Cell.* MGHT, 16 minutes, 16 mm, color.
- *Energy and Reaction.* MGHT, 1961, 15 minutes, 16 mm, color. The relationship between energy in its various manifestations and chemical reactions is shown. These concepts are then related to such basic ideas as the making and breaking of chemical bonds, activational energy, and the rate of a reaction.
- *The Molecular Theory of Matter.* EBEC, 1965, 11 minutes, 16 mm, color.
- *The Origin of the Elements.* CRMP (MGHT), 1973, 17 minutes, 16 mm, color. *Science Today Series.* Animation, graphics, and photographs from space explain the nature and origin of elements and isotopes.
- *The Physics and Chemistry of Water.* BFA, 1967, 21 minutes, 16 mm, color. A beautifully conceived film that demonstrates how the nature of the water molecule determines the physical and chemical properties of water. Also shows how life depends on some of the unusual characteristics of water.
- *The Structure of Atoms.* MGHT, 13 minutes, 16 mm, color.

4

CARBON COMPOUNDS IN CELLS

Revision Highlights

Minor text refinements.

Chapter Outline

THE ROLE OF CARBON IN CELL STRUCTURE AND FUNCTION
- Families of Small Organic Molecules
- Properties Conferred by Functional Groups
- Condensation and Hydrolysis

CARBOHYDRATES
- Monosaccharides
- Disaccharides
- Polysaccharides

LIPIDS
- Lipids With Fatty Acid Components
- Lipids Without Fatty Acid Components

PROTEINS
- Primary Structure: A String of Amino Acids
- Spatial Patterns of Protein Structure
- Protein Denaturation

NUCLEOTIDES AND NUCLEIC ACIDS

SUMMARY OF THE MAIN BIOLOGICAL MOLECULES

Objectives

1. Understand how small organic molecules can be assembled into large macromolecules by condensation. Understand how large macromolecules can be broken apart into their basic subunits by hydrolysis.
2. Memorize the functional groups presented and know the properties they confer when attached to other molecules.
3. Know the general structure of a monosaccharide with six carbon atoms, glycerol, a fatty acid, an amino acid, and a nucleotide.
4. Know the macromolecules into which these essential building blocks can be assembled by condensation.
5. Know where these carbon compounds tend to be located in cells or organelles and the activities in which they participate.

Key Terms

organic
inorganic
methyl group, —CH_3
ethyl group, —C_2CH_3
alcohols
phosphate group
condensation
carbohydrate

sugar	starch	oils	peptide bond
monosaccharides	cellulose	saturated	polypeptide chain
ribose	glycogen	unsaturated	primary structure
deoxyribose	chitin	phospholipid	secondary structure
glucose	lipids	waxes	tertiary structure
fructose	fatty acid	cutin	denaturation
disaccharide	glyceride	steroids	nucleotide
sucrose	monoglyceride	cholesterol	adenosine phosphates
lactose	diglyceride	protein	nucleotide coenzymes
maltose	triglyceride	amino acid	nucleic acids
polysaccharide	fats	R group	DNA

Lecture Outline

I. Introduction
  A. While our cells are 75%–85% water, when we walk we do not ooze forward like Jell-O.
  B. Oxygen, hydrogen, and carbon are the most abundant elements in our body.
    1. Much of the oxygen and hydrogen are in the form of water.
    2. Carbon is linked to hydrogen, oxygen, and other elements to form important structural elements in cells.

II. The Role of Carbon in Cell Structure and Function
  A. Introduction
    1. A carbon atom can form four covalent bonds with other carbon atoms or atoms of other elements.
    2. Many carbon atoms can be linked to form organic molecules.
    3. Simple inorganic molecules have no carbon chains or rings.
  B. Families of Small Organic Molecules
    1. These include compounds with at most twenty carbon atoms.
    2. They include simple sugars, fatty acids, amino acids, and nucleotides.
    3. They are used as an energy source or as building blocks for the synthesis of macromolecules.
    4. The main macromolecules are polysaccharides, lipids, proteins, and nucleic acids.
  C. Properties Conferred by Functional Groups **(Table 4.1, p. 50; TM 17)**
    1. Functional groups are atoms or groups of atoms covalently bonded to a carbon backbone.
    2. Functional groups convey distinct properties to the complete molecule.
  D. Carbon-Hydrogen Compounds
    1. Hydrocarbons can form linear chains, branching chains, or rings.
    2. Hydrocarbons are nonpolar and hence will not dissolve in water because they are hydrophobic.
  E. Carbon-Hydrogen-Oxygen Compounds
    1. These are called alcohols because they contain at least one hydroxyl group (—OH).
    2. Glycerol and sugars belong in this group.
    3. These compounds are hydrophilic because of hydrogen bonding of —OH groups.
  F. Condensation and Hydrolysis **(Fig. 4.2, p. 51; TM 18)**
    1. Small molecules can combine to form large ones because of special proteins called enzymes that can speed up a chemical reaction.
    2. In condensation an H^+ is removed from one molecule and an OH^- from another molecule such that water is formed.
    3. Polymers consisting of as many as millions of monomers can result.
    4. Hydrolysis, like condensation in reverse, breaks a covalent bond between two atoms by reacting with water.

III. Carbohydrates
  A. Introduction

1. Carbohydrates serve as an energy source or have structural roles.
2. They are monomers or polymers of sugar where each sugar has a composition of $(CH_2O)_n$.

B. Monosaccharides
1. The most common monosaccharides have a backbone of three, five, or six carbon atoms.
2. Those with five or six carbons tend to form rings in aqueous environments.
3. Glucose and fructose are both $C_6H_{12}O_6$, but they differ in structure and properties; they are structural isomers. **(Fig. 4.3, p. 52; TM 19)**

C. Disaccharides
1. A disaccharide results when two monosaccharides are covalently bonded.
2. Table sugar is sucrose and consists of glucose plus fructose.
3. Other important disaccharides are lactose (milk sugar) and maltose in germinating seeds.

D. Polysaccharides
1. A polysaccharide results when more than two monosaccharides are covalently bonded.
2. Starch and cellulose are both polymers of glucose that differ in the way the glucose units are arranged.
3 Glycogen is a branched polysaccharide found in fungi as well as in animal liver and muscle tissues. **(Fig. 4.7, p. 53; TM 20)**
4. Chitin is a modified polysaccharide containing nitrogen; hard external animal skeletons and firm fungal cell walls are chitin secretions.

IV. Lipids

A. Introduction
1. Lipids are largely hydrocarbon and hydrophobic.
2. They form the basic structure of membranes and have roles in energy metabolism.

B. Lipids With Fatty Acid Components
1. A fatty acid is a long hydrocarbon with a —COOH at one end.
2. Lipids with fatty acid components may be part of glycerides, phospholipids, and waxes.

C. Glycerides
1. Glycerides function as an energy source.
2. They may be mono-, di-, or triglyceride depending on the number of fatty acids attached to the glycerol.
3. Plants and animals store fat as triglycerides.
   a. Fats are solid because they are saturated (they have only single covalent bonds between carbons).
   b. Oils are liquids because they are unsaturated (they have at least one double bond between carbons).

D. Phospholipids
1. Phospholipids are found in cell membranes.
2. They have glycerol backbone, two fatty acids, and a phosphate group that is attached to hydrophilic groups.

E. Waxes
1. In waxes, fatty acids are linked to alcohols or carbon rings.
2. Waxes coat the surfaces of plants and animals and are used by bees as a structural component of hives.

F. Lipids Without Fatty Acid Components
1. Steroids have a backbone of four carbon rings.
2. Cholesterol is a component of cell membranes in animals and can be modified to form sex hormones.

V. Proteins

A. Introduction
1. Proteins contain N and usually S in addition to C, H, and O.
2. They are the most diverse class of molecules and function as enzymes in cell movements, in storage and transport, and as structural elements.

B. Primary Structure: A String of Amino Acids **(Fig. 4.11, p. 56; TM 21)**

1. The structurally diverse protein molecules are constructed of combinations of twenty different amino acids.
2. Each amino acid is composed of a carbon atom that is bonded to four parts: an amino group, a carboxyl group, a hydrogen atom, and an R group. **(Fig. 4.11, p. 56; TM 21)**
3. Amino acids are held together by peptide bonds to form a polypeptide chain.
4. The primary structure is the sequence of amino acids in the polypeptide chain.

C. Spatial Patterns of Protein Structure **(Fig. 4.14, p. 57; TM 22)**
1. Secondary structure is caused by hydrogen bonding and results in a helical or sheetlike structure.
2. Tertiary structure is caused by interactions among R groups and results in a complex three-dimensional shape. **(Fig. 4.15, p. 58; TA 1a)**
3. Quaternary structure is caused by interactions of at least two polypeptide chains, as in hemoglobin, for example. Collagen forms a triple helix and is fibrous. **(Fig. 4.16, p. 58; TA 1b)**

D. Protein Denaturation
1. A drastic change in the three-dimensional shape of a protein caused by high heat, chemical agents, or other means.
2. Function of biological molecule is lost upon denaturation.

VI. Nucleotides and Nucleic Acids **(Fig. 4.18 and in-text art, p. 59; TM 23)**

A. Each nucleotide has a five-carbon sugar (ribose or deoxyribose), a nitrogen-containing base, and a phosphate group.

B. Three kinds of nucleotide-based molecules:
1. Adenosine phosphates are chemical messengers, (cAMP) or energy carriers (ATP).
2. Nucleotide coenzymes transport hydrogen atoms and electrons (e.g., NAD^+ and FAD).
3. Nucleic acids are polymers of nucleotides.
   a. DNA contains the genetic instructions.
   b. RNA is concerned with protein synthesis.

Suggestions for Presenting the Material

- Although this chapter contains "chemistry" as did the previous chapter, it is less theoretical and contains more familiar terms such as *carbohydrate* and *protein.*
- It is valuable to point out that carbon, hydrogen, and oxygen are the principal atoms in the "molecules of life." Sulfur and nitrogen also participate in proteins and nucleotides.
- Your students will of course recognize these macromolecules as major food groups. You can capitalize on this to generate student interest.
- The extent to which each instructor requires the learning of molecular structure will vary. Perhaps you will ask your students to be able to draw the molecules; but recognition will usually suffice for the student who is a nonscience major.
- Make extensive use of the excellent overhead transparencies available for this chapter. Soon your students will be able to recognize these molecules on sight.
- Carbohydrates are easy to describe because they are built by assembling monomers into polymers. Lipids are a more diverse group and will need to be defined according to *solubility* rather than *common structural features.*
- Proteins are complex because of: (a) their number of amino acid subunits and (b) the levels of structure, that is, primary to quaternary. Use a string of beads and a "Slinky" to help here (see the "Enrichment" section below).
- You can preview the future lecture(s) on protein synthesis by stating "Amino acids are in a precisely defined sequence from one end of a protein to the other." How does the cell select from the twenty amino acid choices the proper one at the proper time?
- The *nucleotide* and *nucleic acid* section is obviously only a preview of more extensive information found in Chapters 15 and 16.

- Some instructors may also want to include two more of the molecules of life, namely *vitamins* and *minerals*. If so, these are discussed in Chapter 32 and Tables 32.4 and 32.5.
- Once again stress the importance of Table 4.2 as a useful summary of biological molecules.

Classroom and Laboratory Enrichment

- Use ball-and-stick models that are very large and easy to see to illustrate some of the basic carbon compounds. Overhead transparencies of ball-and-stick diagrams or "straight-stick" line drawings will also help students get a mental picture of each molecule.
- Use models or diagrams or transparencies to demonstrate the functional groups you wish to emphasize. Stress the importance of knowing several characteristic functional groups by identifying those functional groups present in diagrams or models of real molecules.
- To illustrate amino acid structure, draw a generalized amino acid stem (as shown in Figure 4.11 but with an empty spot at the R-group location) on an overhead transparency. Create different amino acids by changing the R groups, each sketched on a small piece of transparency material.
- Use models or overhead transparencies to show condensation and hydrolysis. Show an example (such as glucose and fructose combining to form sucrose), and ask students to state whether it is condensation or hydrolysis.
- Show a ball-and-stick diagram or three-dimensional model of any protein. An enzyme would be a good example; ball-and-stick diagrams of enzymes are readily available. Students will be amazed at the large size of proteins when compared to carbohydrates and lipids.
- Help students to understand nucleotide structure with models or diagrams. If students can get a good grasp of nucleotides now, they will have a better understanding of ATP and nucleic acids when these topics are covered in later chapters.
- Protein *primary structure* can be demonstrated by a string of beads or a Christmas tree garland. Individual beads can be colored with felt-tip markers for greater clarity and distinction. Secondary structure (alpha helix) is adequately illustrated by use of a "Slinky." You can even demonstrate tertiary structure by *carefully* folding a portion of the "expanded" Slinky.
- Students have been exposed to many words related to those in this chapter, whether in print or broadcast media. Use this opportunity to explain complex carbohydrates, polyunsaturates, cholesterol, fiber, high-fructose syrup, dextrose, and anabolic steroid.
- Select a variety of food products from your pantry and bring them to class. Ask students to check the ingredients list for forms of sugar. Can you find it in some very unlikely places, such as table salt?
- Show a filmstrip or slide/sound set on "atoms and molecules."

Ideas for Classroom Discussion

- How will your knowledge of carbohydrates, fatty acids, proteins, and nucleic acids help you in your study of biology? How will such information help you in personal health and diet issues?
- Compare the calorie contents of carbohydrates, lipids, and proteins.
- Why do alcohols dissolve in water?
- What is the difference between methyl alcohol and ethyl alcohol? How is each of these alcohols processed by the human body?
- What is a complex carbohydrate?
- Why don't animal cells contain cellulose? Can you think of at least one reason why cellulose in an animal cell could be considered a drawback?
- Why are saturated fatty acids solid at room temperature while unsaturated fatty acids are liquid?
- What is a steroid? What steroids are sometimes taken by athletes? What are the effects of these steroids on the body?

- What is the difference between a globular protein and a fibrous protein?
- Why is sugar (in various forms) so prevalent as an additive in our packaged food products?
- Which yields more energy: a pound of carbohydrate or a pound of fat?
- Cellulose and starch both consist of chains of glucose units. One is a useful source of energy to humans, the other is not. Identify which is which and why they differ.
- Where is glycogen stored in the human body? What regulates interconversions of glucose and glycogen?
- Television advertising implies that the ideal diet would include *zero* cholesterol. Is this feasible? Would it even be *desirable*?
- "The human body uses a lot of protein in its construction and function. Therefore, you should eat massive quantities to be even more healthy." Right or wrong? What could be some of the complications of a "high protein diet"?

Term Paper Topics, Library Activities, and Special Projects

- Describe the effects of alcohol on the human body.
- How are termites able to digest wood products?
- What is aspartame? How is it processed by the body? Describe studies that have been done regarding its safety as a food additive.
- What is dietary fiber? Describe its possible role as an anticarcinogen.
- Learn more about the recently synthesized artificial fats that can be used to replace fats in foods.
- Why do women have a higher percentage of body fat than men? Can you think of any adaptive value for this characteristic?
- Discuss the role of cholesterol in diet.
- Describe how scientists discovered the structure of hemoglobin.
- Search the "body building" magazines currently available for diet supplement advertising that might be misleading or outright false. Report on the distortions you find.
- Prepare a historical report on the cultivation and use of plant fibers (cellulose) from various sources in the construction of clothing.
- After searching for background information on the extent and variety of steroid use by athletes, interview persons who can give a "local and inside" perspective. Can you document any damage to heavy users?

Films, Filmstrips, and Videos

- *Carbon and Its Compounds.* CORF, 1971, 16 minutes, 16 mm, color. The atomic structure of carbon is detailed in animation and in models to clarify its unusual bonding behavior and to explain the wide differentiation in the carbon compounds.
- *Chemical Systems of the Cell.* MGHT, 16 minutes, 16 mm, color.
- *DNA: Blueprint of Life.* Wiley, 1968, 17 minutes, 16 mm, color. The DNA molecule, protein structure, and protein synthesis are explained using animated models. The film begins by indicating how DNA and chromosomes determine the phenotypes of organisms. The biochemical mechanisms of transcription, translation, and mutation are related to evolution by natural selection.
- *The Double Helix.* MGHT, 12 minutes, 16 mm, color.
- *Protein Structure and Function.* Wiley, 1972, 15 minutes, 16 mm, color. An excellent portrayal of the unity and diversity found in the proteins of life. Shows an enzyme catalyzing a hydrolytic reaction and emphasizes the importance that protein structure has for its function. Shows the active sites of trypsin and chymotrypsin and explains how they work.

5

CELL STRUCTURE AND FUNCTION: AN OVERVIEW

Revision Highlights

New illustration of cell size and units of measure (Figure 5.2), better comparison of image-forming abilities of different microscopes (same organism, at same magnification in all cases; Figure 5.4). New text and micrograph to introduce prokaryotic cells (Figure 5.7). Refined definitions of organelle functions (p. 68), new table on nuclear components, and better introduction to nuclear function. Simple text and art to start students thinking about "what is a chromosome." New text and overview illustration of cytomembrane system. Updated picture of cytoskeleton, emphasizing its newly defined roles in cell function and reproduction, and in shaping and organizing the developing embryo. New text and illustrations for flagella, cilia, centrioles, and basal bodies. Better introduction to cell junctions and extracellular matrix. Revised summary table comparing prokaryotic with eukaryotic cells.

Chapter Outline

GENERALIZED PICTURE OF THE CELL
- **Emergence of the Cell Theory**
- **Basic Aspects of Cell Structure and Function**
- **Cell Size and Cell Membranes**

PROKARYOTIC CELLS—THE BACTERIA

EUKARYOTIC CELLS
- **Function of Organelles**
- **Typical Components of Eukaryotic Cells**

THE NUCLEUS
- **Chromosomes**
- **Nucleolus**
- **Nuclear Envelope**

CYTOMEMBRANE SYSTEM
- **Endoplasmic Reticulum and Ribosomes**
- **Golgi Bodies**
- **Lysosomes**
- **Microbodies**

MITOCHONDRIA

SPECIALIZED PLANT ORGANELLES
- **Chloroplasts and Other Plastids**
- **Central Vacuoles**

THE CYTOSKELETON
- **Structure and Function of the Cytoskeleton**
- **Flagella and Cilia**
- **What Organizes the Cytoskeleton?**

CELL SURFACE SPECIALIZATIONS
- **Cell Walls**
- **The Extracellular Matrix and Cell Junctions**

SUMMARY

Objectives

1 Understand why cells generally fall into a predictable range of sizes.
2. Contrast the general features of prokaryotic and eukaryotic cells.
3. Describe the nucleus of eukaryotes with respect to structure and function.
4. Describe the organelles associated with the cytomembrane system, and tell the general function of each.
5. Contrast the structure and function of mitochondria and chloroplasts.
6. Describe the cytoskeleton of eukaryotes and distinguish it from the cytomembrane system.
7. List several surface structures of cells and tell how they help cells survive.

Key Terms

cell theory
plasma membrane
nucleus
cytoplasm
surface-to-volume ratio
surface area
compound light microscope
phase contrast microscope
resolution
transmission electron microscope
high-voltage electron microscope
scanning electron microscope
cell wall
ribosome
prokaryotic
eukaryotic
endoplasmic reticulum
Golgi bodies
lysosomes
microbodies
mitochondria
plastids
central vacuole
nucleus
nuclear envelope
cytomembrane system
endoplasmic reticulum
chromatin
chromosomes
Golgi bodies
nucleoprotein
nucleolus
exocytic vesicle
endocytic vesicles
lysosome
microbodies
peroxisomes
glyoxysomes
aerobic
mitochondrion, -dria
chloroplast
stroma
granum, grana
chromoplasts
amyloplasts
central vacuole
myosin
cytoskeleton
microtubules
tubulins
microfilament
actin
intermediate filaments
fluorescence microscopy
permanent
transient
flagella, flagellum
cilia, cilium
bacterial flagellum
microtubule organizing centers (MTOCs)
centrioles
basal bodies
cell walls
extracellular matrix
tight junctions
adhering junctions
gap junctions
plasmodesmata, plasmodesma

Lecture Outline

I. Generalized Picture of the Cells
   A. Emergence of the Cell Theory
      1. In the early seventeenth century, Galileo Galilei first used a microscope to observe a biological specimen (an insect eye).
      2. By mid-seventeenth century, Robert Hooke first used the term "cell" to refer to the holes in a slice of cork.
      3. Antony van Leeuwenhoek observed a bacterium with the use of a microscope.
      4. In the 1820s, Robert Brown found a structure in every plant cell and called it a nucleus.
      5. In 1839, Theodor Schwann noted cells in animal tissues.
      6. Matthias Schleiden concluded that all plant tissues are composed of cells.
      7. Schwann first stated the cell theory: All life is composed of cells and the cell is the basic unit of life.
      8. A decade later Rudolf Virchow added another part to the cell theory: All cells arise from preexisting cells.
   B. Basic Aspects of Cell Structure and Function
      1. All cells have a nucleus (nucleoid), cytoplasm, and a plasma membrane.
      2. A plasma membrane separates each cell from the environment, permits the flow of molecules across the membrane, and contains receptors that can affect the cell's activities.

3. A nucleus is membrane-bound and contains DNA. A bacterial cell's DNA is found in the nucleoid region that is not membrane-bound.
4. The cytoplasm contains membrane systems, particles, filaments (the cytoskeleton), and a semifluid matrix.

C. Cell Size and Cell Membranes **(Fig. 5.2, p. 63; TM 24)**

1. To see most cells, a microscope is required.
2. Light microscopes can explore details of 0.2 micrometer, and electron microscopes can can observe even smaller details. **(Fig. 5.4, p. 65; TA 2)**
3. The small size of cells permits efficient diffusion across the plasma membrane and within the cell.
4. As the surface area of a cell increases to the square of the diameter, the volume increases to the cube of the diameter. **(Fig. 5.5, p. 66; TM 25)**

II. Prokaryotic Cells—the Bacteria

A. Most bacteria have a cell wall that surrounds the plasma membrane.
B. DNA is concentrated in the nucleoid area.
C. Ribosomes, located in the cytoplasm and composed of RNA and protein, participate in protein synthesis.
D. Bacteria and cyanobacteria are prokaryotic because they do not have a nucleus.

III. Eukaryotic Cells

A. Function of Organelles

1. All eukaryotic cells contain organelles.
2. Organelles are membranous sacs or other compartmented portions of cytoplasm.
3. Organelles separate reactions in time (determining the proper sequence) and also separate incompatible chemical reactions.

B. Typical Components of Eukaryotic Cells

1. The nucleus controls access to DNA and permits easier packing of DNA during cell division.
2. The endoplasmic reticulum (ER) modifies proteins and is also involved with lipid synthesis.
3. Golgi bodies also modify proteins, sort and ship proteins, and play a role in the biology of lipids for secretion or internal use.
4. Lysosomes are involved with intracellular digestion.
5. Transport vesicles transport material between organelles and to and from the cell surface.
6. Ribosomes are "free" or attached to membranes.
7. The cytoskeleton determines cell shape and provides for motility.
8. Plastids are in photosynthetic cells and function in food production and storage.
9. Central vacuoles and a cell wall are found in many protistans, fungi, and plants.

IV. The Nucleus **(Fig. 5.12, p. 71; TA 3)**

A. Introduction

1. DNA contains instructions for building the proteins that determine cell structure and function.
2. The isolation of eukaryotic DNA in the nucleus functions to control access to DNA and to simplify packaging for cell division.
3. RNA brings the message of DNA to the cytoplasm, where the proteins are assembled at ribosomes.
4. Each nucleus is composed of chromosomes, nucleoplasm, a nucleolus, and a nuclear envelope.

B. Chromosomes

1. Eukaryotic DNA is associated with many different proteins, including:
   a. Enzymes needed for RNA assembly.
   b. Scaffolding that organizes DNA during cell division.
2. Sometimes DNA is threadlike, with proteins attached to it like beads on a chain; the word "chromatin" describes its appearance.

3. Chromosomes do not always look the same during the life of a cell. **(In-text art, p. 72; TM 26)**
   a. DNA may be unduplicated or duplicated.
   b. It may be condensed or uncondensed.

C. Nucleolus
The nucleolus is the portion of DNA that assembles parts of ribosomes.

D. Nuclear Envelope
1. The nuclear envelope includes two membranes with ribosomes attached to the outer surface and pores extending across the envelope.
2. It controls exchanges between the nucleus and the cytoplasm.

V. Cytomembrane System **(Fig. 5.14, p. 73; TA 4)**

A. Introduction
1. The cytomembrane system includes the endoplasmic reticulum, Golgi bodies, lysosomes, and vesicles.
2. It is a system because of the "traffic" between components. **(In-text art, p. 73; TM 27)**
3. Membranes modified or assembled in the ER and Golgi bodies may "bud" off as vesicles, move, and fuse with other membranes.

B. Endoplasmic Reticulum and Ribosomes
1. The membrane of ER consists of connected tubes and flattened sacs.
2. Smooth ER is free of ribosomes.
   a. Smooth ER is the site of lipid synthesis.
   b. It is abundant in cells that secrete steroids.
   c. In liver, it detoxifies drugs and potentially harmful metabolic by-products.
   d. The sarcoplasmic reticulum in skeletal muscle stores and releases calcium ions for muscle contraction.
3. Rough ER consists of stacked, flattened sacs and has many ribosomes attached.
   a. Rough ER produces proteins for secretion or for delivery to several organelles.
   b. It attaches polysaccharides to proteins. The polysaccharides are later modified in the Golgi bodies and determine the destination of the proteins.
   c. Rough ER is abundant in secretory cells and immature eggs.

C. Golgi Bodies **(Fig. 5.16, p. 75; TA 5)**
1. A Golgi body modifies, sorts, and packages proteins and lipids for specific destinations.
2. It resembles a stack of pancakes.
3. Vesicles with mailing tags break away from the topmost pancake.
4. Vesicles may be stored in secretory cells.

D. Lysosomes
1. Lysosomes are vesicles that bud from Golgi bodies.
2. They contain enzymes for intracellular digestion.
3. They act upon substances wrapped in membranes like cholesterol.
4. Lysosomes degrade worn-out organelles and destroy bacteria and foreign particles.

E. Microbodies
1. Microbodies are vesicles that bud from the ER.
2. Peroxisomes contain enzymes that degrade fatty acids and amino acids.
3. Another peroxisome enzyme degrades hydrogen peroxide to water or uses it to degrade alcohol.
4. Glyoxysomes in seeds convert fats and oils to carbohydrates needed for rapid, early growth.

VI. Mitochondria **(Fig. 5.18, p. 76; TA 6)**

A. Mitochondria use oxygen to extract energy (ATP) from carbohydrates.
B. They are bacteria-sized and are bound by a double-membrane system.
C. Their inner membrane is deeply folded.
D. They are abundant in cells with high energy demands.

VII. Specialized Plant Organelles

A. Chloroplasts and Other Plastids **(Fig. 5.19, p. 77; TA 7)**

1. Chloroplasts and other plastids are specialized for photosynthesis and storage.
   a. Chloroplasts contain photosynthetic pigments and have starch-storing capacity.
   b. Chromoplasts contain pigments and give color to flowers, fruits, and some roots.
   c. Amyloplasts have starch-storing capacity, but lack pigments.
2. Chloroplasts are oval or disk-shaped and have a double-membrane system.
   a. Parts of the inner membrane form stacked disks called grana.
   b. In the grana, sunlight is used to produce ATP.
   c. In the stroma, starch and proteins are synthesized.
   d. They appear green because of chlorophyll, their major pigment.
3. Chromoplasts store pigments that give color to petals, fruits, and roots (e.g., carrots).
4. Amyloplasts store starch grains (e.g., potato tubers).

B. Central Vacuole

1. Central vacuoles fill 50%–90% of cytoplasm in mature plant cells.
2. They store amino acids, sugars, and waste.
3. They increase cell size and surface area, and hence enhance absorption.

VIII. The Cytoskeleton

A. Structure and Function of the Cytoskeleton **(In-text art, p. 78; TM 28)**

1. Cell shape, internal organization, and motility are heritable features.
2. The protein filaments that form the cytoskeleton are specified by DNA.
3. The three classes of filaments are microtubules (tubulin), microfilaments (actin), and intermediate filaments (at least five types).
4. Cytoskeletal parts may be permanent (microtubules in flagella) or transient (spindle microtubules).

B. Flagella and Cilia **(Fig. 5.21, p. 79; TA 8)**

1. Flagella and cilia function to propel cells (such as sperm and trypanosomes) and fluids.
2. Both have the same 9 + 2 cross-sectional array of microtubules.
3. Dynein arms help adjacent doublets slide over each other to achieve movement.
4. Prokaryotic flagella are composed of flagellin and rotate like a propeller.

C. What Organizes the Cytoskeleton?

1. Microtubule organizing centers (MTOCs) may help to assemble and organize microtubules.
2. MTOCs may be associated with centrioles in animal cells.
3. Centrioles can act as a template to produce the basal bodies that give rise to cilia or flagella.
4. Centrioles with MTOCs govern the planes of division during development.

IX. Cell Surface Specializations

A. Cell Walls

1. Most are carbohydrate frameworks for mechanical support for bacteria, protistans, fungi, and plants.
2. They have spaces to permit movements of water and solutes.
3. The plants contain a "primary" wall of cellulose and may also have a multilayered "secondary" wall. **(Fig. 5.24, p. 81; TA 9)**

B. The Extracellular Matrix and Cell Junctions

1. In multicellular organisms cells must stick together, act in coordinated units, and exchange signals and nutrients.
2. The extracellular matrix holds cells together and influences metabolism and cell division.
   a. The extracellular matrix is composed of collagen, other fibrous proteins, glycoproteins, and polysaccharides that form the "ground substance."
   b. Nutrients, hormones, and other molecules must diffuse through the ground substance.
3. In plants, the middle lamella (pectin) cements primary walls of adjacent cells together.

4. Tight junctions in epithelial cells of animals prevent diffusion of molecules across the surface of the tissue.
5. Adhering junctions are like spot welds and hold cells in tissues together.
6. Gap junctions are channels that can permit diffusion of small molecules between adjacent cells.
7. Plasmodesmata are channels that permit diffusion between the cytoplasm of adjacent plant cells.
8. In multicellular organisms, coordination of cells requires linkage and communication between cells.

Suggestions for Presenting the Material

- For many readers, Chapter 5 represents the real entry into the realm of biology. Indeed, a discussion of the cell is fundamental to all future lectures.
- Because of the extent of knowledge concerning the cell, it is impossible to include all of it in one chapter. Therefore, Chapter 5 presents an overview that includes a fair amount of cell part description and a glimpse of functions, the details of which are explained in several successive chapters.
- Whether or not your lecture includes the historical sketch that opens the chapter, you should include: (a) the three elements of the cell theory, (b) the three basic cell components, and (c) some indication of cell size.
- This is also an excellent time to review the use of the word *theory* as explained in Chapter 2.
- A clear distinction between prokaryotic and eukaryotic cells should be made (see the "Enrichment" section below for visual aid suggestion). Additional details on prokaryotes (Monera) are contained in Chapter 39.
- As you begin the litany of cell organelles, use the overhead of Figure 1.3 (the "road map") to remind students of the progress they are making.
- Although the descriptions and diagrams of the cell organelles occupy only a small number of textbook pages, it is best to proceed carefully and deliberately. There is a dizzying array of nonfamiliar terms here.
- When describing each cell structure, a visual representation of some type should be constantly in view to the student. Each time a new cell part is introduced, Figure 5.8 or 5.9 should be shown for reference purposes.
- Stress the fact that several cell parts are so complex in function that greater detail will follow in future lectures. Examples are: nucleus and chromosomes in Unit 3; cytomembranes in Chapter 16; mitochondria in 9; chloroplasts in 8; and cell membranes in 6.
- Table 5.2 contains a wealth of information that is conformable to your needs. If you choose *not* to stress the difference between prokaryotic and eukaryotic cells with respect to structure (the two right-hand columns), then the table is reduced to one that lists cell structure and function (the two left-hand columns). It still remains a very useful table.

Classroom and Laboratory Enrichment

- Use sketches or models drawn to scale to demonstrate the size difference between prokaryotic and eukaryotic cells.
- Show an overhead transparency of a diagram or an electron micrograph of any cell. Ask if the cell is prokaryotic or eukaryotic. Is it an animal cell? A plant cell? Some other type of cell?
- Arrange for students to see an electron microscope and learn about specimen preparation and the operation of the microscope.
- Ask students to match "organelle" with "cellular task" at the board or on an overhead.

- If you present the historical sketch of the cell theory, include slides of the researchers you are discussing. These photos can usually be found in a variety of introductory biology texts or special texts on the history of biology.
- Construct a table (overhead or handout) listing side-by-side comparisons of prokaryotic and eukaryotic cells.
- Most departments possess some type of cell model. These are especially helpful in perception of the 3-D aspects of cell structure. They can also be useful in oral quizzing.
- If you have access to electron micrographs generated by your colleagues, bring some to class to pass around, or prepare 2 x 2 transparencies. Students will be impressed by the "home-grown" aspect of these micrographs.
- Show a filmstrip or slide/sound set that reviews cell structures and functions.

Ideas for Classroom Discussion

- List the tasks that a cell must do.
- What is the largest example of a single cell that you can think of?
- Why are there no unicellular creatures one foot in diameter?
- Discuss some of the methods by which single cells overcome surface-to-volume constraints (some examples: thin, flat cell shapes, invaginations to increase cell surface area, cellular extensions of the plasma membrane). Describe ways in which multicellular organisms solve this problem (some examples: thin or sheetlike body plans, transport systems).
- Distinguish between a nucleus and a nucleoid.
- Why must bacteria have ribosomes when they lack other organelles?
- How do photosynthetic prokaryotes perform photosynthesis without plastids?
- What is the difference between scanning electron microscopy and transmission electron microscopy?
- Where in your body would you find cells with high concentrations of mitochondria?
- Why do you think most plant cells have a central vacuole while animal cells lack this organelle?
- What is the significance of the word *theory* in reference to the basic properties of the cell?
- Does the cytoplasm have any functions of its own, or is it just a "filler" matrix in which other organelles float?
- Why is the term *nucleus* used to describe the center of an atom and the organelle at the center of the cell, when these are such different entities?
- In measurement of length, what are the largest cells (when mature) in the human body? What fundamental property of all cells is denied to these?
- What feature makes a eukaryotic cell a "true" cell?
- What organelle could be compared to the "control center" of an assembly line in a factory?
- Describe the interrelationship(s) of the individual members comprising the *cytomembrane system* (see Figure 5.14).
- Compare the functions of mitochondria and chloroplasts.

Term Paper Topics, Library Activities, and Special Projects

- Who first coined the term *organelle*? When did biologists discover that eukaryotic cells contained organelles?
- Learn more about plant tissue culture. What mechanisms govern cell differentiation in vitro?

- Discuss the development of electron microscopy. What are some of the advances in cell biology that electron microscopy has made possible? How are biological specimens prepared for examination with an electron microscope?
- How do antibiotics such as penicillin stop bacterial growth?
- Design a hypothetical cell that would function with maximum efficiency under conditions of extreme drought.
- Describe the function of liver cell smooth ER in the metabolism of drugs and alcohol. How might the liver cells of an alcoholic differ from those of a moderate imbiber or nondrinker?
- Prepare brief biographies of the researchers who are credited with early discoveries of cell structure and function.
- Search the library shelves for biology texts of twenty, thirty, forty, and fifty years ago. Find the diagrams of cell structure. Prepare a sequential composite of these, and compare each to the others and to your present text. What instrument made the difference in the drawings?
- Using a special dictionary of Latin and Greek root words, search for the literal meanings for each of the cell parts listed in Table 5.2. (Be careful with Golgi—it is a man's name!)

Films, Filmstrips, and Videos

- *The Cell: A Functioning Structure—Part I.* CRMP (MGHT), 1972, 32 minutes, 16 mm, color. David Suzuki explains many of the organelles of cells and presents a variety of microscopic photographs of *Paramecium.* Several microphotographic techniques show the function of microtubules, microfilaments, lysosomes, and other organelles.
- *The Cell: A Functioning Structure—Part II.* CRMP (MGHT), 1972, 32 minutes, 16 mm, color. Emphasizes and explains biochemical genetics. Not as useful a film as Part I with respect to the time it takes to show.
- *Cell Biology.* CORF, 1981, 17 minutes, 16 mm, color. *Biological Sciences Series.* Shows structure, function, and evolutionary history of cells.
- *The Living Cell: An Introduction.* EBEC, 1974, 20 minutes, 16 mm, color. *Physiology Series.* Uses graphics, animation, and photomicrography to explain the structure and function of living cells. Correlates development of cytology and instrumentation, particularly the microscope, and indicates theories about the origin of cells.
- *The Microscope.* MGHT, 11 minutes, 16 mm, color.
- *Protist Physiology.* MLA-Wards, 1977, 13 minutes, 16 mm, color. Shows techniques to stain living cells and prepare wet mounts of *Amoeba, Paramecium, Chilomonas, Euglena,* and others. Food capture and digestion are shown in *Amoeba* and *Paramecium;* the saprozoic lifestyle of *Chilomonas* and *Euglena* and the photosynthetic lifestyle of *Euglena* are examined.

6

MEMBRANE STRUCTURE AND FUNCTION

Revision Highlights

Updated picture of membrane proteins; major types (e.g., channel, transport, recognition, receptor proteins) are defined simply. Students should become familiar with the names and functions of these proteins, which play roles in ATP formation, neural function, kidney function, hormone action, embryonic development, and other topics described in later chapters. Updated illustration of fluid mosaic model of membrane structure. Better introduction to gradients in concentration, temperature, and electric charge. Updated text on active and passive transport mechanisms.

Chapter Outline

Objectives

1. Understand the essential structure and function of the cell membrane.
2. Know the forces that cause water and solutes to move across membranes passively (that is, without expending energy).
3. Understand which types of substances move by simple diffusion and which by bulk flow. Understand the importance of osmosis to all cells.

4. Know the mechanisms by which substances are moved across membranes against a concentration gradient.
5. Understand how material can be imported into or exported from a cell by being wrapped in membranes.

Key Terms

plasma membrane
internal cell membranes
phospholipid
lipid bilayer
glycoproteins
fluid mosaic model of membrane structure
channel protein
transport protein
electron-transfer protein
impermeability
surface receptors
hydrophobic barrier
concentration gradient
simple diffusion
bulk flow
osmosis
tonicity
isotonic
hypotonic
hypertonic
turgor pressure
water potential
membrane transport proteins
active transport
passive transport
facilitated diffusion
sodium-potassium pump
calcium pump
exocytosis
endocytosis
receptor-mediated endocytosis

Lecture Outline

I. Introduction
  A. Because the concentration of ions and other substances outside a cell may rapidly become too high or low, a mechanism is needed to selectively permit substances to enter or leave the cell.
  B. The plasma membrane—a surface of lipids, proteins, and some carbohydrate groups—achieves the goal of maintaining internal organization by regulating transport across the membrane.

II. Fluid Membranes in a Largely Fluid World
  A. The Lipid Bilayer
    1. The plasma membrane is a fluid layer that separates two aqueous environments.
    2. Properties of lipid molecules explain how a plasma membrane can seal itself when punctured with a fine needle.
    3. Phospholipids are the most abundant type of lipid in cell membranes.
    4. A phospholipid molecule is composed of a hydrophilic head and two hydrophobic tails. **(Fig. 6.1, p. 85; TM 29)**
    5. If phospholipid molecules are surrounded by water, the hydrophobic fatty acid tails cluster and a bilayer results; hydrophilic heads are at the outer faces of a two-layer sheet. **(Fig. 6.2, p. 87; TA 10)**
    6. Bilayers of phospholipids are the structural foundation for all cell membranes.
    7. Bilayers form to minimize the number of hydrophobic groups exposed to water.
    8. Within a bilayer phospholipids show quite a bit of movement; they diffuse sideways, spin, and flex their tails to prevent close packing and promote fluidity.
    9. Fluidity also results from short-tailed lipids and unsaturated tails (kink at double bonds).

  B. Membrane Proteins **(Fig. 6.2, p. 87; TA 10)**
    1. A variety of different proteins are embedded in the bilayer or positioned at its two surfaces.
    2. Most are glycoproteins with short polysaccharide chains extending into extracellular fluid; some are glycolipids.
    3. In 1972, S. J. Singer and G. Nicolson suggested the fluid mosaic model of membrane structure; some membranes differ—for example, plant cell membranes lack cholesterol.
      a. Lipids provide the basic membrane structure and, with their moving and packing produce the "fluid" nature.

b. Lipids render the membrane impermeable to water-soluble substances; "mosaic" refers to the lipid and protein membrane composition.
c. Membrane proteins serve as channels, transport proteins, electron transfer systems, recognition systems, and receptor systems.

4. Channel Proteins
   a. Embedded channel proteins permit water-soluble substances to cross the lipid bilayer.
   b. The channels extend through the protein bodies; some remain open but others have gates that close or open as needed (for example, Na^+ ions to excite nerve cells).
5. Transport proteins require energy to "pump" substances in specific directions (for example, the Na^+–K^+ ion pump).
6. Electron-transfer proteins accept electrons or H^+ ions from one molecule and carry them to another; cytochromes are examples.
7. Recognition proteins function in tissue formation and cell-to-cell interactions; their polysaccharide chains (and those of glycolipids) provide cues for adhesion of similar cell types.
8. Receptor proteins on membranes and within cells function as "switches" turned on by specific substances such as hormones; different cell types have different receptors.
   a. White blood cell receptors lock onto surface molecules of tissue invaders.
   b. Adjacent receptors open and close channel proteins.
   c. Somatotropin (hormone) receptors switch on enzyme production for cell growth and division.
   d. Receptor malfunctions can contribute to forms of cancer and diabetes.
   e. Membrane proteins can be imaged by freeze-fracturing and freeze-etching.

C. Summary of Membrane Structure and Function **(Fig. 6.2, p. 87; TA 10)**
1. Cell membranes are composed of lipids and proteins.
2. The hydrophilic heads of the lipids are at the outer faces of the phospholipid bilayer and the tails are sandwiched between them.
3. The lipid bilayer is the basic membrane structure and serves as a hydrophobic barrier between the cytoplasm and the extracellular fluid.
4. Membranes are fluid because of rapid movements and packing of lipid molecules.
5. Membrane proteins (embedded or surface) carry out most membrane functions.

III. Diffusion

A. Introduction
1. To understand membrane function, one must first understand the unassisted movements of water and solutes.

B. Gradients Defined
1. Concentration = the number of molecules in a given volume.
2. Molecules move from a region of high concentration to a region of lower concentration, that is, down the concentration gradient. This net movement is due to random collisions.
3. Gradients in pressure, temperature, or electric charge can also exist.

C. Simple Diffusion
1. Simple diffusion involves the random movement of molecules or ions down a concentration gradient.
2. The concentration gradient for each substance is independent of other substances present.
3. One can observe simple diffusion by placing a few drops of dye in water. **(Fig. 6.4, p. 89; TM 30)**
4. Diffusion rate depends on the extent of the concentration gradient, temperature, and molecular size.
5. Simple diffusion accounts for the greatest volume of substances that pass across cell membranes and accounts for most short-distance transport in cells.

D. Bulk Flow
1. Bulk flow is the tendency of different substances to move together due to a pressure gradient (an example is the circulatory system).

2. Bulk flow may enhance diffusion rates—in effect, it shrinks the distance that molecules must move.

IV. Osmosis

A. Osmosis Defined

Osmosis is the passive movement of water across a differentially permeable membrane in response to solute concentration gradients, a pressure gradient, or both.

B. Tonicity **(Fig. 6.5, p. 90; TM 31)**

1. Tonicity denotes the relative concentrations of solutes in two fluids (extracellular fluid and cytoplasmic fluid).
2. An isotonic fluid has the same concentration of solutes as the fluid in the cell.
3. A hypotonic fluid has a lower concentration of solutes than the fluid in the cell; cell lysis may occur.
4. A hypertonic fluid has a greater concentration of solutes than the fluid in the cell.
5. Cells may be adapted to live in hypotonic or hypertonic environments.

C. Water Potential

1. Land plants grow in hypotonic soil solutions.
2. Hence, water enters cells by osmosis and pushes against the cell wall, resulting in turgor pressure.
3. Turgor pressure can resist osmotic pressure, and the difference is the water potential.
4. Wilting may occur when the soil is dry or the solute concentration of the soils is too high.

V. Movement of Water and Solutes Across Cell Membranes

A. The Available Routes

1. Only small, nonpolar molecules readily diffuse across the bilayer (examples include carbon dioxide and oxygen).
2. Large, nonpolar molecules such as glucose cannot diffuse across the bilayer; neither do + or – ions, no matter how small.
   a. Some cross the membrane by active transport—a process that requires energy and may move a solute against a concentration gradient.
   b. Some cross the membrane through channel proteins by passive transport—a process that moves a solute with the concentration gradient.

B. Facilitated Diffusion **(Fig. 6.8, p. 93; TA 11)**

1. Facilitated diffusion assists solutes only in the direction that simple diffusion would take them.
2. Facilitated diffusion depends upon hydrophilic groups in channel proteins that bind with water-soluble molecules.
3. Binding triggers a change in the protein shape that results in the transport of the solute.

C. Active Transport **(Fig. 6.9, p. 93; TA 12)**

1. Transport proteins undergo a series of shape changes to move solutes rapidly into or out of cells.
2. Such shape changes require energy, usually as ATP.
3. Active transport pumps small ions such as Na^+–K^+ or Ca^{++}, small charged molecules, and large molecules against their concentration gradients.
4. It involves transport proteins that span the bilayer and are highly selective for what each one transports.

D. Exocytosis and Endocytosis

1. Exocytosis and endocytosis both involve the plasma membrane and transport vesicles.
2. Exocytosis moves substances out of a cell. **(Fig. 6.10a, p. 94; TM 32a)**
   a. Within the cytoplasm, secretory vesicles may first pinch off from Golgi membranes.
   b. These cytoplasmic vesicles fuse with plasma membrane and release their contents to the outside.
3. Endocytosis moves substances into a cell. **(Fig. 6.10b, p. 94; TM 32b)**
   a. Part of the plasma membrane encloses particles near the cell surface to form vesicles that move into the cytoplasm.

b. Many endocytic vesicles fuse with lysosomes (examples include the phagocytic *Amoeba* and white blood cells).
c. Endocytosis is called pinocytosis when a liquid is transported.
d. In receptor-mediated endocytosis, the region of the plasma membrane that participates forms coated pits; the pit is lined with surface receptors specific for the molecule being transported.
e. The pit sinks into the cytoplasm and forms an endocytic vesicle.

Suggestions for Presenting the Material

- This is the first of several chapters that will expand topics introduced in the cell overview chapter (Chapter 5).
- If you deferred the discussion of phospholipid function from Chapter 4 to the present chapter, now it can be resumed. Be careful to show that all of the phospholipid diagrams showing head and tail are characterizations of Figure 6.1.
- Don't assume students know, or remember, the definitions of *hydrophilic* and *hydrophobic.*
- Using the transparency of Figure 6.2, carefully distinguish between which portion of the membrane is the "fluid" (that is, lipid bilayer) and which is the "mosaic" (that is, proteins). Use demonstrations of a mosaic artwork (see the "Enrichment" section below) if you can locate such.
- Call attention to the concise "summary of cell membrane features" that is placed at the end of this section.
- The various methods by which molecules move, either through space or through membranes, can be confusing to students because of the subtle differences that distinguish each method. Perhaps you could begin with general, nonmembrane-associated phenomena such as diffusion and bulk flow. Then proceed to membrane-associated mechanisms such as osmosis, facilitated diffusion, active transport, and vesicle formation. (Note that "osmosis" is not listed under "membrane transport" in the text.)
- Note that each of the transport phenomena topics is accompanied by an illustration, which should be used to reduce the "abstract quality" of the mechanism.
- Notice also that there is a section in the chapter summary titled "Membrane Functions" that is a gleaning of five functions from throughout the chapter.

Classroom and Laboratory Enrichment

- Demonstrate the structure of the plasma membrane with a three-dimensional model or overhead transparencies.
- Use electron micrographs (in the form of overhead transparencies or 35 mm slides or passed around the room) to add to your description of plasma membrane structure.
- Discuss the terms *isotonic, hypotonic,* and *hypertonic* by showing students three sketches of semipermeable bags in beakers of distilled water (these can be drawn ahead of time on an overhead transparency if you wish). Vary the concentrations of the sugar solutions shown in the bags. Ask students what the direction of water movement would be in each case. Then ask what direction the sugar molecules will move in. Many students will believe that the sugar molecules will move across the membrane, even though they previously learned these sugar molecules are too large to cross the plasma membrane.
- To demonstrate that some molecules will pass through membranes and some will not, prepare two test tubes as follows: In one tube pour dilute Lugol's iodine solution until it is about nine-tenths full; in another tube pour 1 percent starch paste until it is about nine-tenths full. Cover the mouth of each test tube with a wet goldbeater's membrane; then secure it tightly by tying with thread or a tight rubber band. Invert the starch paste test tube into a beaker about one-half full of dilute

Lugol's solution; invert the test tube containing dilute Lugol's solution into a beaker containing a 1 percent starch paste. Ask students why the well-known blue-black color appears in the starch solution and not in the Lugol's solution. Students may decide that the starch requires digestion to a more soluble form before it can pass through the membrane.

- Set up two or more osmometer tubes (glass or plastic thistle tubes covered with a selectively permeable membrane) at the front of the room. Compare rates of osmosis by filling each tube with a colored sugar (you may use corn syrup) or salt solution (vary the concentrations) and placing the base of the tube into a beaker of distilled water.
- Living cells can be used to demonstrate the osmotic water passage through semipermeable membranes. Use an apple corer to remove a center cylinder of a raw white potato; leave about one-half inch of potato tissue at the bottom. Carefully pour a concentrated sucrose solution into the core; seal the opening with a one-hole rubber stopper through which a piece of glass tubing has been inserted. Pour melted paraffin around the stopper as a seal to avoid leakage. Place the potato in a beaker of water; use a clamp on a ringstand to support the glass tube.
- To show how molecules move from points of greater concentration to regions of lower concentration, pour red ink or dye along one side of a container of water. The liquid will diffuse through a liquid.
- In the classroom, demonstrate the loss of turgor pressure by flooding a small potted tomato plant with salt water at the beginning of the class period.
- Use models or overhead transparencies to show how channel proteins and transport proteins function during passive transport and active transport, respectively.
- In the lab, students can view plasmolysis under the microscope. Obtain a small *Elodea* leaf and place it in a small drop of distilled water on a microscope slide. Cover it with a cover slip. Now prepare a second slide, only this time mounting an *Elodea* leaf in a drop of 10% NaCl solution. Compare the cells of the second slide to those of the first slide.
- Watch hemolysis by diluting several drops of fresh blood with distilled water. Compare the resulting solution with one in which the same number of red blood cells were diluted with physiological saline (0.09% NaCl). The suspension that has undergone hemolysis will be clear (red blood cells have burst) while the solution made with physiological saline will be visibly cloudy (red blood cells remain intact in the solution).
- Open a bottle of perfume and place it on the desk early in the lecture period. Discuss principles involved later in the lecture.
- Demonstrate the diffusion of a liquid in a solid. Prepare three petri dishes, each filled with a layer of plain agar. Use a cork borer to carefully make three equidistant holes in the agar. Turn the dish over and number each hole. Put three or four drops of the following solutions in each of the appropriately numbered holes: (1) 0.02M potassium dichromate in the first hole, (2) 0.02M potassium permanganate, (3) 0.02M methylene blue. View results after at least one hour. Because the solutions have identical molarities, the rate of diffusion depends upon the molecular weight of each compound. Methylene blue (MW = 373) diffuses at the slowest rate, while potassium permanganate diffuses fastest (MW = 158). Potassium dichromate (MW = 294) diffuses at an intermediate rate.
- At least one day before lecture, dissolve gelatin in water and pour into a screw-top test tube; leave a small space between the top of the gelatin and the top edge of the test tube. Cool the tube in the refrigerator until the gelatin solidifies. At the beginning of the lecture, pour in a small amount of a bright-colored dye on top of the gelatin; replace and tighten the screw top. Allow the dye to diffuse through the gelatin until it reaches the bottom of the glass thread area of the test tube. At this time, turn the test tube to a horizontal position and place it so that students can observe progress of the dye through the gelatin. If one wishes to time the progress of the dye, begin timing when the tube is turned to the horizontal position.
- The diffusion of a gas through other gases (air) can be demonstrated easily as follows: Wet a circle of filter paper with phenolphthalein and insert it into the bottom of a large test tube. Next, invert the test tube over an open bottle of ammonium hydroxide. Ask students to explain the rather rapid color change of the filter paper to red. Set up a control with filter paper soaked in water.

- Demonstrate the diffusion of a gas in a gas using a glass tube at least one-half inch wide and eighteen inches in length. Plug one end of the tube with a cotton wad saturated with HCl (hydrochloric acid); plug the opposite end with a cotton wad saturated with NH_4OH (ammonium hydroxide). Label each end appropriately. Hydrochloric acid and ammonium hydroxide react together to produce ammonium chloride and water. Students will be able to see a ring of ammonium chloride that has formed closest to the HCl end of the tube; the ammonium ion (NH_4^+) has a smaller molecular weight than the chloride ion (Cl^-) and thus diffuses faster, meeting the Cl^- ion about two-thirds of the way down the tube.
- If you have an overhead or 2 x 2 transparency of the "sandwich" model of the plasma membrane as published in texts twenty years ago, show it to the students for comparison with the "fluid mosaic" model in the present text.
- Unless a student has an interest in art, he/she may not know what a "mosaic" is. If you can bring an example of such a piece of art, or at least a photo, it will aid your description of the fluid mosaic model.
- Bring a can of room deodorizer spray to class to demonstrate diffusion (in this case, a liquid in a gas) as a general example of how concentration gradients operate.

Ideas for Classroom Discussion

- Why do we use the fluid mosaic model to describe the plasma membrane?
- Why is the structure of the plasma membrane basically the same among organisms of all five kingdoms? Ask students to think about the common evolutionary origins of all life.
- What would happen to freshwater unicellular organisms if suddenly released in a saltwater environment?
- What are some organelles that contain internal compartmentalizations? How do internal compartments assist in the functioning of the organelle?
- Distinguish between diffusion and osmosis.
- Why do unicellular protists (such as *Paramecium*) not burst even though their cell interiors are hypertonic to their freshwater environments?
- What is physiological saline solution? (Hint: It is used to dilute samples of red blood cells in the laboratory.)
- How do exocytosis and endocytosis differ from passive transport and active transport?
- What is "fertilizer burn"? What can be done to correct it?
- Membrane models of several years ago had four layers. How many are present in the fluid mosaic model?
- Someone has said that a diagram of the fluid mosaic model resembles a scene from the North Atlantic. What they are referring to are "icebergs" representing ________________ (molecules) floating in a "sea" of ________________ (molecules).
- Based on your knowledge of membranes and solubility, which insecticide preparation would you expect to kill insects faster: one that is water formulated or petroleum-solvent formulated?
- What would be the result on blood cells of a substitution of pure water for physiological saline in an IV bottle?
- Observe Figures 6.8 and 6.9 in the textbook and distinguish between *facilitated diffusion* and *active transport.*
- In some cases, a 5 percent glucose solution is given intravenously to persons after surgery. If you were the doctor on such a case, would you order the glucose solution to be isotonic, hypertonic, or hypotonic to blood? Explain your decision.
- Discuss the precise meanings of the prefixes *hyper* and *hypo.* They are often confused by students and mental errors are compounded.

- Ask students what the prefixes *hyper* and *hypo* refer to in the text discussions. If they do not understand that these terms refer to solute concentration (not water), they will have difficulty with the concept.
- Plant cells have a rather rigid wall enclosing their plasma membranes; animal cells do not. Ask students to think about a comparison of consequences when plant and animal cells are placed in isotonic, hypertonic, and hypotonic solutions.

Term Paper Topics, Library Activities, and Special Projects

- What happens in arid climates when salts accumulate in cropland soils?
- How do cells "recognize" other cells of the same type during tissue formation?
- Learn more about the discovery and function of the sodium-potassium pump and the calcium pump.
- How are saltwater fish species able to cope with their extremely salty surroundings?
- From your library shelves, select a biology text written twenty years ago. Turn to the discussion of the ultrastructure of the plasma membrane. How does that model differ from your text's fluid mosaic one?
- Look up the composition of "physiological saline." Are there different varieties of this preparation for different animal species?
- The diffusion and transport phenomena discussed in the chapter are based on a property called Brownian movement that is demonstrated by all molecules, whether alive or not. What is the physical manifestation of this property, and what is the derivation of its name?

Films, Filmstrips, and Videos

- *Biological Membranes—Fundamental Characteristics.* WISSEN, 1979, 13 minutes, 16 mm, color. Covers general membrane structure and function, surface markers for intercellular identification and communication, and transmembrane transport.
- *Biological Membranes—Physical Models: Monolayer, Bilayer, Liposomes.* WISSEN, 1979, 17 minutes, 16 mm, color. Emphasizes the relationship among physical properties of water, membrane surfaces, and phospholipids. Uses models and animation to illustrate properties of natural membranes.
- *Chemical Properties of Water.* CORF, 1964, 14 minutes, 16 mm, color.
- *Diffusion and Osmosis.* EBEC, 1973, 14 minutes, 16 mm, color. Clever use of models and experimental setups to show that all life processes depend on the movement of molecules across the selectively permeable membranes of cells. Illustrates some practical applications of osmosis and asks students to explain how they work.
- *The Physics and Chemistry of Water.* BFA, 1967, 21 minutes, 16 mm, color. A beautifully conceived film that demonstrates how the nature of the water molecule determines the physical and chemical properties of water. Also shows how life depends on some of the unusual characteristics of water.
- *The Structure of Water.* MGHT, 14 minutes, 16 mm, color.

7

GROUND RULES OF METABOLISM

Revision Highlights

Simpler description of chemical equilibrium. New overview of metabolic pathways, with simple illustrations of linear and cyclic routes. Better diagram of enzyme-substrate interaction. Refined definitions and discussion of cofactors and energy carriers. Better introduction to controls over enzyme activity. ATP/ADP cycle now covered here instead of in Chapter 9. New, simple illustration of electron transport systems. (By covering the basics of metabolism in Chapter 7, the subsequent chapters on photosynthesis and aerobic respiration are kept uncluttered.)

Chapter Outline

THE NATURE OF ENERGY
- Two Laws Governing Energy Transformations
- Living Systems and the Second Law

METABOLIC REACTIONS: THEIR NATURE AND DIRECTION
- Energy Changes in Metabolic Reactions
- Reversible Reactions
- Metabolic Pathways

ENZYMES
- Enzymes Defined
- Enzyme Structure
- Enzyme Function
- Regulation of Enzyme Activity

COFACTORS

ATP: THE MAIN ENERGY CARRIER
- Structure and Function of ATP
- The ATP/ADP Cycle

ELECTRON TRANSPORT SYSTEMS

SUMMARY

Objectives

1. Know two laws that govern the way energy is transferred from one substance to another.
2. Provide an example of a metabolic pathway and explain what kinds of substances regulate activity of the pathway.
3. Tell exactly what enzymes do and how they do it.
4. Explain how a molecule can "carry" energy.

Key Terms

metabolism
first law of thermodynamics
second law of thermodynamics
spontaneous
entropy
exergonic reaction
endergonic reaction
dynamic equilibrium
metabolic pathways
degradative
biosynthetic
reactants
metabolites
enzymes
cofactors
energy carriers
end product
enzymes
substrate
controls
active site
enzyme-substrate
complex
induced-fit model
activation energy
regulatory enzymes
allosteric
feedback inhibition
cofactors
coenzymes
FAD
NAD^+
$NADP^+$
metal ions
ATP
cytochromes
ATP/ADP cycle
phosphorylation
electron transport systems
oxidized
reduced
oxidation-reduction reaction

Lecture Outline

I. Introduction
   A. Life has a need for energy.
   B. Metabolism is the controlled capacity to acquire and use energy.

II. The Nature of Energy
   A. Two Laws Governing Energy Transformations
      1. Energy is the capacity to do work.
      2. The first law of thermodynamics states that energy cannot be created or destroyed; it can only change form.
      3. The total energy content of a system and its surroundings remains constant.
      4. The second law of thermodynamics states that the spontaneous direction of energy flow is from high- to low-quality forms.
      5. The energy unavailable to do work is increasing.
      6. The entropy of an isolated system tends to increase.
   B. Living Systems and the Second Law
      1. Life obeys the laws of thermodynamics.
      2. Light energy intercepted by plants compensates for the energy lost during biological reactions and permits life to maintain its high degree of organization.

III. Metabolic Reactions: Their Nature and Direction
   A. Energy Changes in Metabolic Reactions **(In-text art, p. 99, TM 33)**
      1. Exergonic reactions result in products with less energy than the reactants had.
      2. Endergonic reactions result in products with more energy than the reactants had.
   B. Reversible Reactions
      1. Most reactions are reversible. **(Figs. 7.3, 7.4, p. 100; TM 34)**
      2. The greater the concentration of reactants, the faster the forward reaction.
      3. The greater the concentration of products, the faster the reverse reaction.
      4. All reversible reactions approach a state of dynamic equilibrium.
      5. Each reaction has an equilibrium constant, which is the ratio of reactant and product concentrations.
   C. Metabolic Pathways
      1. The amount of water per cell limits the number of compounds that can dissolve.
      2. Compounds must be in sufficient concentration to permit reactions but low enough to prevent disruptive side reactions.
      3. Metabolic pathways form a series of orderly reactions that operate under these constraints. Each step in a metabolic pathway is quickened with the help of a specific enzyme.

4. Metabolic pathway sequences may be linear or cyclic. **(Fig. 7.5, p. 100; TM 35)**
5. Major metabolic pathways are either degradative or biosynthetic.
   a. In degradative pathways, large molecules such as carbohydrates, lipids, and proteins are broken down to form products of lower energy. Released energy can be used for cellular work.
   b. In biosynthetic pathways, small molecules are assembled into large molecules (for example, simple sugars are assembled into complex carbohydrates).

IV. Enzymes

A. Introduction—Enzymes Defined **(Fig. 7.6, p. 101; TM 36)**
1. Enzymes are proteins that serve as catalysts.
2. Enzymes do not determine the direction of a reaction.
3. Enzymes are selective and act upon specific substrates.
4. Enzymes permit high reaction rates while concentrations of reactants are low.

B. Enzyme Structure
1. Active Sites
   a. An active site is a crevice that binds to a substrate due to weak bonds.
   b. The induced-fit model suggests that structural changes result in a complementary fit between the active site and substrate. **(Fig. 7.8, p. 102; TA 13a)**

C. Enzyme Function
1. Activation Energy
   a. Activation energy is the amount of energy needed for reactants to collide.
   b. It results in a transition state.
   c. Enzymes lower the activation energy. **(Fig. 7.9, p. 103; TA 13b)**
   d. Enzymes orient substrates in positions that promote reaction.
2. Effect of Substrate Concentrations
   a. Enzymes only increase the rate of a reaction.
   b. The net direction of a reaction is influenced by the concentrations of reactant and product molecules.
3. Effects of Temperature and pH **(Fig. 7.10, p. 103; TM 37)**
   a. The rate of enzyme activity increases with temperature until a maximum rate is reached.
   b. Greater temperatures result in a reaction rate decrease because high temperatures disrupt the weak bonds that determine the globular shape of the enzyme and can even denature the enzyme.
   c. Most enzymes are most effective at pH 7.
   d. An extreme in pH can also disrupt the weak bonds that determine the shape of the enzyme.
   e. Some enzymes can function at pH's other than 7.
   f. Enzymes function within a limited range of temperature and pH.
4. Regulation of Enzyme Activity
   a. Enzyme synthesis can be accelerated or slowed down.
   b. Reversible inhibitors may compete with a substrate for an active site.
   c. Irreversible inhibitors can bind with an enzyme and make catalysis impossible.
   d. In a metabolic pathway, the overall rate will be determined by the enzymes that catalyze the slowest reaction—regulatory enzymes.
   e. Allosteric enzymes are regulatory enzymes that contain at least one regulatory site that can reversibly bind to a substance that influences the enzyme activity.
   f. Feedback inhibition occurs when the end product of a metabolic pathway is able to reversibly inactivate an allosteric enzyme—this occurs when the concentration of product molecules is sufficient. **(Fig. 7.12, p. 105; TM 38)**
   g. Enzyme activity can be controlled by several mechanisms that can permit rapid adjustments in the rates of biosynthetic pathways.

V. Cofactors
   A. Introduction
      1. As enzymes catalyze the transfer of electrons, atoms, or functional groups, they may require the assistance of cofactors.
      2. Cofactors are nonprotein groups that bind to many enzymes and render them more reactive.
   B. Cofactor Types
      1. Several large organic molecules (coenzymes) and some metal ions serve as cofactors.
         a. Coenzyme FAD strips hydrogen atoms from substrates and becomes reduced to $FADH_2$.
         b. Coenzymes NAD^+ and $NADP^+$ are free-moving; each delivers a proton and two electrons from a substrate to other reaction sites (reduced to NADH and NADPH).
         c. Inorganic metal ions such as Fe^{++} also serve as cofactors when assisting membrane cytochrome proteins in their electron transfers.

VI. ATP: The Main Energy Carrier
   A. Introduction
      1. Cells cannot directly use chemical energy stored in carbohydrates and other organic molecules.
      2. Cells must first chemically transform the energy of organic molecules into the energy of ATP before that energy is available to accomplish cellular activities.
   B. Structure and Function of ATP
      1. ATP is composed of adenine, ribose, and three phosphate groups. **(Fig. 7.13, p. 105; TM 39a)**
      2. ATP donates at least one phosphate group to another molecule and thus transfers energy.
      3. ATP transfers energy to many different chemical reactions; almost all metabolic pathways directly or indirectly run on energy supplied by ATP.
   C. The ATP/ADP Cycle **(In-text art, p. 106; TM 39b)**
      1. Energy input links a phosphate group to ADP and ATP results.
      2. During a process known as phosphorylation, ATP can donate a phosphate group to a molecule and increase its store of energy; this primes the phosphorylated molecule for entering specific chemical reactions.
      3. The ATP/ADP cycle provides cells with a means of conserving energy and transferring it to specific reactions.
      4. ATP turnover rate is very rapid; each ATP gives up energy within sixty seconds of its formation.

VII. Electron Transport Systems **(Fig. 7.14, p. 106; TA 14)**
   A. ATP production depends on electron transport systems.
   B. Electron transport systems consist of membrane-bound enzymes and cofactors that operate in a highly organized sequence.
   C. Electrons are transferred stepwise through the components of electron-transport systems by oxidation-reduction reactions.
      1. The movement of electrons through electron transport systems generates usable forms of energy, which can be harnessed to do work.
      2. Such usable energy may be used to move hydrogen ions, which establishes gradients resulting in ATP production.

Suggestions for Presenting the Material

- Because students may be unfamiliar with the first and second laws of thermodynamics, it is important to distinguish clearly between the two laws; emphasize the central role of the sun in sustaining life on Earth.

- Acknowledge that this chapter speaks in *generalities* and defines terms that will be used to describe *specific* metabolic reactions in subsequent chapters.
- If you prefer to teach from specific examples, you may want to choose a specific metabolic pathway (of your own or from elsewhere in the book), draw it on an overhead transparency, and use it to explain the various terms found in the text.
- After presentation of the various capabilities of enzymes, students may think of them as "miracle workers." Remind the students that these are nonliving molecules—albeit, amazing ones. Also emphasize the limitations and vulnerability of enzymes, including causes and effects of denaturation.
- Because of their similarity, NAD and NADP need special emphasis to distinguish each in terms of structural and functional identity. Compare and contrast both terms with ATP.
- Instruct students to bring their texts to class and refer frequently to the excellent figures in it.
- This chapter provides a good opportunity to introduce the idea that oxidations release energy, reductions require energy, and NADH and NADPH contain more potential energy than their oxidized counterparts.

Classroom and Laboratory Enrichment

- The action of an enzyme (salivary amylase) can be easily demonstrated by the following procedure:
  a. Prepare a 6% starch solution in water and confirm its identity by a spot plate test with iodine solution (produces blue-black color).
  b. Collect saliva from a volunteer by having the person chew a small piece of Parafilm and expectorate into a test tube.
  c. Place diluted saliva and the starch solution in a test tube and mix.
  d. At suitable intervals, remove samples of the digestion mixture and test with iodine on the spot plate (lack of dark color indicates conversion of starch to maltose).
  e. Variations can include: heating the saliva to destroy the enzyme; adding acid or alkali; adding cyanide.
- The effect of ATP on a reaction can be demonstrated by use of bioluminescence kits available from biological supply houses.
- Show a videotape or slide/sound set depicting the role and function of enzymes.
- Show a film about energy transformations in cells.
- Demonstrate the two models of enzyme-substrate interactions in the following ways:
  a. *Rigid "lock and key" model:* Use preschool-size jigsaw puzzle pieces or giant-size Lego blocks.
  b. *Induced-fit model:* Use a flexible fabric or latex glove to show how the insertion of a hand (substrate) induces change in the shape of the glove (active site).
- The relationship between ATP, ADP, energy, enzymes, and phosphorylation may be illustrated by the use of a toy dart gun with rubber suction cup-tipped darts. It is helpful to have acetate transparencies of ADP and ATP structures that can be projected on a screen as the following demonstration is performed:
  a. Tell the students that the unloaded dart gun represents ADP and the dart represents inorganic phosphate (P). Show the structure of ADP on the screen.
  b. As you insert the dart into the gun, emphasize the need for the expenditure of energy to do this. Tell the students that the addition of P to ADP is, therefore, an endergonic reaction; it is also called a phosphorylation reaction. At this time show the structure of ATP on the screen. Also point out that the spring inside the dart gun is under much tension and as such has a great deal of potential energy. The same can be said for the P group that has been added to ADP.
  c. Next, demonstrate the hydrolysis of ATP. The trigger finger represents the necessary enzyme. Aim the gun at some vertical smooth surface (window or aquarium works well) and depress the trigger. Hopefully the dart will adhere to the surface (a substrate molecule being energized by phosphorylation). The reaction is thus exergonic and some of the energy has been transferred to the substrate molecule.

Ideas for Classroom Discussion

- The second law of thermodynamics is often used by creationists as the basis for an argument against evolution. According to their view, the second law depicts a world that must become more disordered. This is in contrast to evolution, which depicts a world that is becoming more ordered. Evaluate the creationists' argument. Is it valid?
- Pose several "what if" questions concerning enzyme action such as: What if the body temperature of humans mutated to 105°? Or body pH rose to 9.5?
- Preview the "inborn errors of metabolism" mentioned in Chapter 16 wherein a faulty enzyme blocks or diverts a metabolic pathway, as in phenylketonuria (PKU).
- Compare the energy "hill" that an enzyme must overcome to a ski slope. Does an enzyme act more like a chair lift or a bulldozer?
- Discuss what would happen to life on Earth if the flow of sunlight energy stopped.
- Name some of the different forms of energy. Are they interconvertible? Give some examples of interconversions. For example, trace the energy interconversions involved in cooking your breakfast on an electric stove. Begin with the dam over a hydroelectric plant.

Term Paper Topics, Library Activities, and Special Projects

- Research the basis for the first and second laws of thermodynamics. On the basis of what observations did scientists formulate these laws? What experiments can be done in the laboratory to confirm the laws?
- Explain the defect in the metabolic pathway that results in the condition known as phenylketonuria (PKU). How can this condition be treated? Is it curable?
- Construct clay models depicting an enzyme, a substrate, the enzyme-substrate complex, and induced fit.
- Explain the role of the B vitamins in human metabolism.
- Discuss some of the ways that coenzymes and inorganic cofactors participate in enzymatic reactions.

Films, Filmstrips, and Videos

- *The Energetics of Life.* Wiley, 1973, 23 minutes, 16 mm, color. Discusses and demonstrates the types and functions of energy used by living organisms. Kinetic and potential energy, thermodynamics, and chemical energy are explained. The structure and function of the ATP molecule, photosynthesis, and the metabolism of glucose are described in detail. Emphasizes the ability of the ATP molecule to harness solar energy and use it.
- *Energy and Reaction.* MGHT, 1961, 15 minutes, 16 mm, color. The relationship between energy in its various forms and chemical reactions is shown. Several different types of interesting chemical reactions that would be somewhat difficult to perform in the average laboratory are demonstrated. These demonstrations are then used to illustrate such ideas as the making and breaking of chemical bonds, the role of activation energy, and rate of reaction.
- *Energy Cycles in the Cell.* MGHT, 16 minutes, 16 mm, color.
- *Oxidation-Reduction.* MGHT, 1962, 10 minutes, 16 mm, color.

8

ENERGY-ACQUIRING PATHWAYS

Revision Highlights

New or refined illustrations; tighter text throughout; details verified by biochemists. Simple, early explanation of difference between sugar-phosphate intermediates of photosynthesis and the true end products (e.g., sucrose, starch). Photo of evidence of photosynthesis (oxygen bubbling from *Elodea*). A simple "bookkeeping" paragraph indicates where water molecules are formed and used in the reactions; this shows students where the "6 H_2O" of the summary equation comes from. Better text and illustrations of C4 pathway. Improved summary.

Chapter Outline

FROM SUNLIGHT TO CELLULAR WORK: PREVIEW OF THE MAIN PATHWAYS

PHOTOSYNTHESIS
- **Simplified Picture of Photosynthesis**
- **Chloroplast Structure and Function**

LIGHT-DEPENDENT REACTIONS
- **Light Absorption in Photosystems**
- **Two Pathways of Electron Transfer**
- **A Closer Look at ATP Formation**
- **Summary of the Light-Dependent Reactions**

LIGHT-INDEPENDENT REACTIONS
- **Carbon Dioxide Fixation and the Calvin-Benson Cycle**
- **Summary of the Light-Independent Reactions**
- **How Autotrophs Use Intermediates and Products of Photosynthesis**
- **C4 Plants**

CHEMOSYNTHESIS

SUMMARY

Objectives

1. Understand the main pathways by which energy from the sun or from specific chemical reactions enters organisms and passes from organism to organism and/or back into the environment.
2. Know the steps of the light-dependent and light-independent reactions. Know the raw materials needed to start each phase and know the products made by each phase.
3. Explain how autotrophs use the intermediates as well as the products of photosynthesis in their own metabolism.

Key Terms

autotrophic organism
photosynthetic autotroph
chemosynthetic autotroph
heterotrophic organism
photosynthesis
glycolysis
respiration
light-dependent reactions
light-independent reactions
intermediates
thylakoid membrane
stroma
granum, grana
pigments
photosystems
electron transport system
photophosphorylation
cyclic photophosphorylation
photosystem I
noncyclic photophosphorylation
photosystem II
photolysis
chemiosmotic theory
Calvin-Benson cycle
carbon dioxide fixation
PGAL
RuBP
chemosynthesis
ammonium ions, NH_4^+
photorespiration
C4 plants
mesophyll cells
bundle-sheath cells
C3 plants

Lecture Outline

I. Introduction
   A. A drought in the Midwest is linked to increased food prices in Los Angeles.
   B. Crop failures can lead to the death of a child in Ethiopia.
   C. The above events are both linked to how cells acquire and use energy.

II. From Sunlight to Cellular Work: Preview of the Main Pathways **(Fig. 8.1, p. 108; TA 15)**
   A. For all life based on organic compounds . . .
      1. Where does the carbon come from?
      2. Where does the energy come from to link carbon and other atoms into organic compounds?
      3. How does the energy in these organic compounds become available to do cellular work?
   B. Autotrophs use CO_2 from the air and water.
      1. Photosynthetic autotrophs (plant, protistan, and bacterial members) harness light energy.
      2. Chemosynthetic autotrophs (a few bacteria) extract energy from chemical reactions involving inorganic substances (such as sulfur compounds).
   C. Heterotrophs feed on autotrophs, each other, and organic wastes.
      1. Heterotrophs acquire carbon and energy from autotrophs.
      2. Heterotrophs include animal, protistan, and bacterial members.
   D. Thus, carbon enters the web of life through photosynthesis.
   E. Energy stored in photosynthetic reactions is released in glycolysis and aerobic respiration.

III. Photosynthesis
   A. Simplified Picture of Photosynthesis **(In-text art, p. 109; TM 40)**
      1. The light-dependent reactions convert light energy to ATP and NADPH.
      2. The light-independent reactions assemble sugars and other organic molecules using ATP and NADPH.
      3. Overall, for glucose formation . . . $12H_2O + 6CO_2 \xrightarrow{\text{sunlight}} 6O_2 + C_6H_{12}O_6 + 6H_2O$.
         a. Oxygen is obtained when water molecules are split.
         b. ATP and NADPH bridge the two sets of reactions.
   B. Chloroplast Structure and Function **(Fig. 8.2d, p. 110; TA 16)**
      1. Light-dependent reactions occur at the thylakoid membranes.
         a. The thylakoids are folded into grana and channels.
         b. The interior spaces of the grana and channels are continuous and are filled with H^+ needed during ATP synthesis.
      2. Carbohydrate formation occurs in the stroma.

IV. Light-Dependent Reactions
   A. Overview

1. Light energy is absorbed.
2. Electron and hydrogen transfers lead to ATP and NADPH formation.
3. Electrons are replaced in the system that first gives them up.

B. Light Absorption in Photosynthesis
1. Light-Trapping Pigments
a. A pigment molecule absorbs photons, which vary in energy as a function of wavelength.
b. The shorter the wavelength, the greater the energy in a photon.
c. Photosynthesis uses a narrow portion of the electromagnetic spectrum.
d. Chlorophylls absorb blue and red but transmit green.
e. Carotenoids absorb violet and blue but transmit yellow.
f. The wavelengths absorbed excite an electron in the pigment
2. Photosystems
a. A photosystem is a cluster of 200 to 300 light-absorbing pigments located in the thylakoid.
b. Over 90% of the pigments in a photosystem harvest energy.
c. The light energy absorbed excites an electron to a higher energy level.
d. The electron returns to a lower energy level, and the extra energy is released to hop from one pigment to another.
e. As energy is lost, the wavelength increases.
f. Only a few chlorophylls can act as a trap and absorb the longest wavelengths.
g. The energy absorbed excites an electron.
h. The excited electron is quickly transferred to an acceptor molecule in the thylakoid membrane. This is the first event in photosynthesis.

C. Two Pathways of Electron Transfer
1. Overview
a. Electrons expelled from a chlorophyll molecule go through one or two electron transport systems.
b. As the electron passes from one molecule to another in each system, ATP is produced by photophosphorylation.
2. The Cyclic Pathway **(Fig. 8.5, p. 113; TA 17)**
a. In cyclic photophosphorylation, electrons are first excited and then return to the original photosystem.
b. This photosystem is distinguished by one of its chlorophyll molecules, P700, which absorbs wavelengths of 700 nanometers.
c. The cyclic pathway is an ancient way to make ATP from ADP.
3. The Noncyclic Pathway **(Fig. 8.6, p. 115; TA 18)**
a. Noncyclic photophosphorylation transfers electrons through two photosystems and two electron transport systems (ETS).
b. ATP is made from ADP.
c. Electrons end up in NADPH—which is used in synthesis of organic compounds.
d. The noncyclic pathway begins at photosystem II, which contains chlorophyll P680.
e. Energy absorbed by photosystem II goes to an electron acceptor, then to an electron transport system, and then to photosystem I.
f. An electron in photosystem I is then excited to a higher energy level and is passed to an electron acceptor.
g. The electron acceptor transfers the electron to an ETS, and NADPH results ($2e^- + H^+ + NADP^+ \rightarrow NADPH$).
h. Thus, there is a one-way flow of electrons to NADPH.
i. Electrons lost by P680 are replaced with electrons from water.
j. Water is split into oxygen, hydrogen ions, and electrons = photolysis.
k. Oxygen, a by-product of the noncyclic pathway, contributed to the earth's early atmosphere and made aerobic respiration possible.

D. A Closer Look at ATP Formation **(In-text art, p. 114, TM 41)**
1. During photolysis, hydrogen ions accumulate inside the thylakoid region.
2. Electron transport systems also pump hydrogen ions inside the thylakoid compartment.

3. Thus, both electric and concentration gradients are established.
4. The energy in those gradients can be tapped to form ATP.
5. The hydrogen ions pass through a channel in an ATP synthase to phosphorylate ATP.
6. ATP formation through concentration and electric gradients across a membrane is known as chemiosmosis.

E. Summary of the Light-Dependent Reactions
1. Light absorption
Light absorbed by a photosystem is transferred to an electron acceptor, and then to a transport system.
2. Noncyclic Pathway
a. The noncyclic pathway is used by land plants to make ATP and NADPH.
b. It involves a one-way flow of electrons from photosystem II, to an electron transport system, to photosystem I, to a second electron transport system.
c. Electrons released during photolysis replace those lost in photosystem II.
d. During electron transport and photolysis, hydrogen ions are moved inside the thylakoid compartment.
e. Hydrogen ions flow down the gradient to achieve ATP synthesis.
f. The second electron transport system combines electrons, $NADP^+$, and H^+ to form NADPH.
3. Cyclic Pathway
a. Light excites electrons from photosystem I; these return to the same photosystem via an electron transport system.
b. Only ATP is produced.

V. Light-Independent Reactions **(Fig. 8.7, p. 116; TA 19)**

A. Introduction
1. ATP and NADPH are used for the synthesis part of photosynthesis.
2. These synthetic reactions do not require light, and normally occur at night.

B. Carbon Dioxide Fixation and the Calvin-Benson Cycle
1. The Calvin-Benson cycle is a cyclic pathway that occurs in the stroma and that requires ATP, NADPH, CO_2, RuBP (a five-carbon sugar), and enzymes.
2. Carbon is "captured," a sugar phosphate forms, and RuBP is reformed.
3. Consider the events associated with the capture of six CO_2 molecules.
a. CO_2 is captured by RuBP, and an unstable intermediate (six carbons) forms.
b. This breaks into two molecules of phosphoglycerate or PGA (a three-carbon sugar). This completes carbon dioxide fixation.
c. Next, PGA receives a phosphate group from ATP and then a H^+ and electrons from NADPH to form phosphoglyceraldehyde or PGAL.
d. For every twelve PGAL molecules, ten are used to make more RuBP.
e. The remaining two are used to make glucose-6-phosphate.
4. Thus, light energy, which excited electrons in the first stages of photosynthesis, is now stored as chemical energy in an organic compound. **(Fig. 8.8, p. 117; TA 20)**

C. Summary of the Light-Independent Reactions **(Fig. 8.7, p. 116; TA 19)**
1. CO_2 is fixed to RuBP, and an intermediate six-carbon compound breaks to form two PGA molecules (three carbons).
2. PGA is phosphorylated by ATP, and receives hydrogen ions and electrons from NADPH to form PGAL.
3. PGAL is rearranged to form RuBP and sugar phosphate.
4. Sugar phosphates can be used to form different carbohydrates.

D. How Autotrophs Use Intermediates and Products of Photosynthesis
1. In a corn plant, sugar phosphates are used for energy, and to make sucrose and starch.
2. Sucrose is transported from the leaves to all plant parts.
3. Starch is the main storage carbohydrate in leaves, stems, and roots.
4. Sucrose produced in the leaves of a potato is stored as starch in the tuber.
5. Some photosynthetic products can be used for lipid or protein synthesis.

E. C4 Plants

1. Plants in hot, dry environments close stomata and so not as much CO_2 is available.
2. In photorespiration, oxygen—not CO_2—is fixed to RuBP in the Calvin-Benson cycle. **(In-text art, p. 118; TM 42b)**
3. Because photorespiration forms one PGA + phosphoglycolate, which forms CO_2 + water, up to 50% of fixed carbon can be lost by photorespiration.
4. C4 plants, like corn, have a carbon-fixing system in mesophyll cells that precedes the Calvin-Benson cycle in bundle-sheath cells. **(Fig. 8.10, p. 119; TM 42a)**
   a. In mesophyll cells, CO_2 is added to phosphoenolpyruvate (PEP), forming oxaloacetate (a four-carbon compound).
   b. In bundle-sheath cells, oxaloacetate releases CO_2 to the Calvin-Benson cycle.
   c. Thus, C4 plants have a higher CO_2/O_2 ratio that prevents photorespiration but permits photosynthesis.
   d. C4 plants use more ATP, but can live with less water.

VI. Chemosynthesis
   A. The chemosynthetic autotrophs obtain energy from oxidation of inorganic substances, such as ammonium ions, iron, or sulfur compounds.
   B. Some nitrifying bacteria can oxidize NH_3 to NO_2 and NO_3.
   C. Nitrite and nitrate ions are more easily washed out of soil than are ammonium ions; nitrifying bacteria can lower soil fertility.

Suggestions for Presenting the Material

- Many students think that the only autotrophic organisms are those with green pigments, which are, of course, capable of using light for food manufacture; this may be their first introduction to *chemoautotrophs.*
- Students may have previously learned to refer to the two divisions of photosynthesis as the "light" and "dark" reactions. Starr and Taggart use the more accurate and currently acceptable terms *light-dependent* and *light-independent,* respectively. You may wish to explain why "dark reactions" is a poor designator and no longer used (the reactions can occur in both the dark and the light).
- ATP and NADPH were introduced in Chapter 7 as "coupling agents." Their use as bridges between the light-dependent and light-independent reactions is an excellent opportunity for reinforcement of these molecules as "bridges."
- The term "granum" dates from the days when microscopes revealed "grains" within the chloroplast. Now we realize that these grains consist of complex thylakoid membranes.
- Photosynthesis is much more complicated than the usual simple equation that a student may have seen in a previous course. Perhaps the best approach to presenting this topic is to follow the stepwise outline in "Summary of the Light-Dependent Reactions" and "Summary of the Light-Independent Reactions."
- Even though "cyclic photophosphorylation" is presented *prior* to the noncyclic pathway in the text, it should be noted that this is done in deference to its position as the first to evolve. It should be pointed out that most existing plants use the noncyclic pathway.
- The numbering of the photosystems (I and II) also is done with reference to evolution; when the noncyclic pathway is operating, the sunlight is *initially* absorbed by photosystem II.
- You may wish to note the similarities and differences between the chemiosmotic mechanism in chloroplasts and mitochondria (Chapter 9), or defer this comparison until mitochondria are discussed later.
- Although the diagram of the Calvin-Benson cycle (Figure 8.7) is intimidating, the most important features are the entry of *carbon dioxide* and the production of *glucose,* driven by *ATP* and *NADPH* from the light-dependent reactions.
- The emphasis on C3 and C4 plants can be moderated according to the interests of the instructor. Omitting it will not affect future discussions in the book.

- A discussion of the reason colored objects appear the color they do may facilitate the students' understanding of why, for example, green light is ineffective for photosynthesis.
- Show a diagram of the electromagnetic spectrum and discuss how the wavelength of the radiation is related to its energy content.

Classroom and Laboratory Enrichment

- You can demonstrate the production of oxygen by plants with the following:
  a. Place *Elodea* (an aquarium plant) in a bowl and expose to bright light.
  b. Invert a test tube over the plant and collect the bubbles.
  c. Remove the tube and immediately thrust a glowing wood splint into the tube.
  d. Result: The splint burns brightly in the high-oxygen air.
- Separate the pigments in green leaves by using paper chromotography. (Consult a botany laboratory manual for the correct procedure.)
- Many students have never seen the action of a prism in separating white light into its component colors; a demonstration would most likely be appreciated.
- If a greenhouse facility is readily accessible, a brief tour and explanation of the devices used to control light, water, air, heat, etc., is very instructive.
- Show a slide/sound set on photosynthesis.
- Provide a model of a chloroplast.
- Show an electron micrograph of a chloroplast and indicate where light-dependent reactions, light-independent reactions, and chemiosmosis occur.
- Cut out separated leaf pigments from a paper chromatogram and elute each pigment from the paper with a small amount of alcohol. Using a Spectronic 20, determine the absorption spectrum for each pigment and graph the results. (Consult a biology or botany laboratory manual for the procedure.)

Ideas for Classroom Discussion

- In what ways could the "greenhouse effect" hurt agriculture? In what ways could it possibly *help*, especially in Canada and the USSR?
- Suppose you could purchase light bulbs that emitted only certain wavelengths of visible light. What wavelengths would promote the most photosynthesis? The least?
- Assume you have supernatural powers and can stop and start the two sets of photosynthesis reactions. Will stopping the light-independent affect the light-dependent, or vice versa?
- What conflicting "needs" confront a plant living in a hot, dry environment? What is the frequent result in C3 plants? How is the problem avoided in C4 plants?
- What colors of the visible spectrum are absorbed by objects that are black? What about white objects?

Term Paper Topics, Library Activities, and Special Projects

- One of the hottest and driest summers (1988) in North America in recent years affected the production of food and grain crops. Prepare an analysis of the effect(s) of that summer's drought on future food availability and its cost to the consumer.
- Consult a biochemistry or advance plant physiology text to learn of a laboratory technique that would clearly indicate whether the oxygen produced by plants is derived from water or from carbon dioxide.

- Prepare a detailed diagram of the Calvin-Benson cycle showing the introduction of radioactively labeled carbon dioxide and its subsequent journey through several "turns" of the cycle.
- Of the total amount of sunlight energy impinging on a green plant, what percentage of the energy is actually converted into glucose?
- Compare the mechanisms of C_3 and C_4 photosynthesis and give examples of plants that fit into each category.
- Discuss why radiation with greater or lesser wavelengths than visible light is not generally used in biological processes.

Films, Filmstrips, and Videos

- *The Energetics of Life.* (See Chapter 7.)
- *Mechanism for Photosynthesis.* MGHT, 16 minutes, 16 mm, color.
- *Photosynthesis.* EBEC, 1967, 22 minutes, 16 mm, color.
- *Photosynthesis.* EBEC, 1982, 20 minutes, 16 mm, color. *Plant Life Series.* A look at photosynthesis: its importance in capturing energy and producing food at the base of food and energy pyramids, organelles and basic chemical processes involved, and research into developing ways to maximize its benefits.
- *Photosynthesis: Biochemical Process.* CORF, 1970, 17 minutes, 16 mm, color. Presents the basics of photosynthesis, including fundamental chemical steps, importance to life, and organelles involved. Animation is used to present the Calvin cycle.
- *Photosynthesis: Chemistry of Food-making.* CORF, 1964, 15 minutes, 16 mm, color. Although recommended for use in lower educational levels than college, the film effectively covers the fundamentals of photosynthesis. Traces the basic chemical steps involved in synthesizing glucose from water, carbon dioxide, and sunlight in the presence of ATP.

9

ENERGY-RELEASING PATHWAYS

Revision Highlights

New or refined illustrations; tighter text throughout; details verified by biochemists. Chapter still organized into two parts (overview of main degradative pathways, then details of aerobic respiration). Simpler picture of roles of oxidation-reduction and phosphorylation reactions in ATP formation. The overview illustration (Figure 9.2) now shows formation and fate of energy carriers NAD^+ and FAD; this makes the figure more useful for short courses that skip the details of respiration. NAD^+ regeneration now included in summary equations for fermentation routes. Better discussion of enzyme control of metabolism. Improved summary.

Chapter Outline

OVERVIEW OF THE MAIN ENERGY-RELEASING PATHWAYS
- **Glycolysis: First Stage of the Energy-Releasing Pathways**
- **Aerobic Respiration**
- **Anaerobic Electron Transport**
- **Fermentation Pathways**

A CLOSER LOOK AT AEROBIC RESPIRATION
- **First Stage: Glycolysis**
- **Second Stage: The Krebs Cycle**
- **Third Stage: Electron Transport Phosphorylation**
- **Glucose Energy Yield**

FUEL OR BUILDING BLOCKS? CONTROLS OVER CARBOHYDRATE METABOLISM

PERSPECTIVE

SUMMARY

Objectives

1. Understand what kinds of molecules can serve as food molecules.
2. Know the relationship of food molecules to glucose and thus to glycolysis.
3. Understand the fundamental differences between glycolysis + fermentation and glycolysis + aerobic respiration. Know the factors that determine whether an organism will carry on fermentation or aerobic respiration.
4. Know the raw materials and products of each of these processes: glycolysis, fermentation, the Krebs cycle, and electron transport phosphorylation.

Key Terms

phosphorylation
oxidation-reduction
glycolysis
pyruvate
aerobic respiration
aerobe
anaerobes
anaerobic electron transport
fermentation
alcoholic fermentation
lactate fermentation
Krebs cycle
electron transport phosphorylation
energy-requiring
energy-releasing
PGAL
substrate-level phosphorylation
ATP synthase system
chemiosmotic theory of ATP formation

Lecture Outline

I. Introduction
   A. Every organism uses organic compounds for energy and raw materials in the same way.
   B. Each one uses ATP as a prime energy carrier.

II. Overview of the Main Energy-Releasing Pathways **(Fig. 9.2, p. 123; TA 21)**
   A. Cells can use biological molecules as a fuel or as structural elements.
      1. Glucose is our main energy source.
      2. Energy is released from glucose through phosphorylation and oxidation-reduction reactions.
         a. Phosphorylation of glucose by two ATP is needed for cells to show a net return of two or thirty-six ATP per glucose.
         b. Oxidation-reduction: The difference in the number of ATP produced depends on how completely the glucose is oxidized by the pathways involved.
   B. Glycolysis: First Stage of the Energy-Releasing Pathways
      1. Glycolysis is a partial breakdown of glucose (a six-carbon sugar) that occurs in the cytoplasm.
      2. First, two phosphate groups are added from two ATP.
      3. Then, as the molecule is degraded, the H^+ and electrons are joined with NAD^+ to form NADH.
      4. Two pyruvates (three carbons each) are the end products.
      5. The net yield is two ATP.
   C. Aerobic Respiration **(Fig. 9.3, p. 124; TM 43)**
      1. Aerobic respiration dismantles pyruvate and NADH in mitochondria with oxygen as the final electron acceptor.
      2. H^+ and electrons are joined to coenzymes (NAD^+ and FAD), which donate them to an electron transport system.
      3. The electron transport system passes the electrons to oxygen, which combines with H^+ to form water.
      4. Many ATP are produced in the electron transport system. A total of thirty-six ATP per glucose result.
      5. Thus, the pyruvate is completely oxidized to CO_2 and water.
      6. Anaerobes do not use oxygen—it may even be toxic.
   D. Anaerobic Electron Transport
      Some microbes use an inorganic compound (e.g., sulfate) as a final electron acceptor.
   E. Fermentation Pathways
      1. Fermentation pathways are anaerobic routes where the final electron acceptor is a product of the glucose molecule being degraded.
      2. In alcoholic fermentation, pyruvate is broken down to ethanol and CO_2. **(Fig. 9.4, p. 125; TM 44a)**
      3. In lactate fermentation, pyruvate is converted into lactate. **(Fig. 9.4, p. 125; TM 44b)**
      4. Anaerobic electron transport or fermentation only has a net yield of two ATP from glycolysis—enough for many microbes.

III. A Closer Look at Aerobic Respiration

A. The Krebs cycle and a few other reactions degrade pyruvate, produce ATP, and release H^+ and electrons.

B. In electron transport phosphorylation, NAD^+ and FAD bring the H^+ and electrons to transport systems, which results in the formation of thirty-two ATP.

C. Enzymes catalyze each step.

D. First Stage: Glycolysis **(Fig. 9.5, p. 127; TA 22)**

1. The first steps are energy-requiring because phosphate groups from two ATP are added to glucose.
2. The 6-carbon intermediate splits into two 3-carbon PGAL (this initiates the energy-releasing steps of glycolysis).
3. Each PGAL loses two hydrogen atoms—one H^+ and both electrons join with NAD^+ to form NADH.
4. An inorganic phosphate group is added, and the intermediate becomes unstable.
5. Each of the two phosphate groups can be used to make an ATP by substrate-level phosphorylation.
6. The end products of glycolysis are ATP, NADH, and pyruvate.

E. Second Stage: The Krebs Cycle

1. The Krebs cycle occurs in the matrix of the mitochondrion. **(Fig. 9.6c, p. 128; TA 23)**
2. First, pyruvate is modified and attached to oxaloacetate for entry into the Krebs cycle. **(Fig. 9.7, p. 130; TA 24)**
3. Then, the hydrogen ions and electrons are removed and each of the three carbons leave as CO_2.
4. Each turn of the cycle produces an ATP, and there are two turns per glucose molecule.
5. The hydrogen ions and electrons are transferred to NAD^+ and FAD.

F. Third Stage: Electron Transport Phosphorylation

1. Electron transport phosphorylation occurs in the inner membrane of the mitochondrion. **(Fig. 9.6c, p. 129; TA 23)**
2. The electrons from the coenzymes pass down a membrane-bound electron transport chain. **(Fig. 9.8, p. 131; TM 46)**
3. Hydrogen ions accumulate in the outer compartment. **(In-text art, pp. 129, 131; TM 45)**
4. Hydrogen ions flow through an ATP synthase system to produce ATP.

G. Glucose Energy Yield

1. Each NADH forms three ATP.
2. Each FADH forms two ATP.
3. Complete breakdown of a glucose molecule is 39% efficient and yields thirty-six ATP.
4. Fermentation is only 2% efficient and yields only two ATP.

IV. Fuel or Building Blocks? Controls Over Carbohydrate Metabolism **(Fig. 9.10, p. 133; TM 47)**

A. Intermediates of glycolysis or the Krebs cycle can also be used for the synthesis of other molecules.

B. Cells can control the balance of synthetic and degradative pathways.

1. Phosphofructokinase is needed to add a second phosphate to fructose-6-phosphate group during glycolysis.
2. High levels of ATP inhibit this enzyme and slow down glycolysis.

V. Perspective

A. Consider how photosynthesis and aerobic respiration came to be linked.

B. Early cells produced ATP by a glycolysis-like pathway under anaerobic conditions. They used preformed molecules.

C. Later, by mutations degradative pathways were reversed to produce synthetic ones.

D. That is, molecules could be built from CO_2 and water, if enough energy was available.

E. The sun became the energy source, photosynthesis was invented, and O_2 accumulated in the atmosphere.

F. Mutants came to use oxygen as an electron acceptor.

G. Some of these mutants abandoned photosynthesis, and the forerunners of aerobic animals were born.
H. The aerobes produced CO_2—precisely what is needed for photosynthesis!
I. Thus, the flow of molecules and energy came full circle.
J. As long as sunlight flows into the web of life, life will be able to continue.
K. Life is a complex system of prolonging order that is sustained by energy transfusions.

Suggestions for Presenting the Material

- Assuming that photosynthesis has already been presented, Chapter 9 can be even more intimidating when the student views the several diagrams of complicated pathways. It would be comforting to your students if you could spend a few minutes presenting an *overview* that specifically relates Chapters 8 (energy acquiring) and 9 (energy releasing).
- The critical role of ATP must be emphasized. Distinguish clearly between the *transfer* of energy from carbohydrates to ATP and the *synthesis* of the ATP molecule.
- The material in the chapter is most easily and logically presented by skillful use of the figures on the overhead transparencies.
  a. Begin with Figure 9.2, which is really a flow chart of glycolysis (including fermentation), the Krebs cycle, and electron transport. Point out the entry molecules, exit molecules, and key intermediates as well as the total energy yield (36 ATP).
  b. For some instructors, Figure 9.2 may be of sufficient detail; however, if you choose, Figure 9.5 gives the individual steps of glycolysis; Figure 9.4 depicts fermentation; Figures 9.7 and 9.8 detail aerobic pathways.
- As you progress deeper into the pathway discussions, it is advisable to refer frequently to Figure 9.2 to maintain an overview.
- The chemiosmotic synthesis of ATP as depicted in Figure 9.6c and in the text is perhaps beyond the scope of some courses for nonscience majors. It may be omitted without detriment to the overall presentation.
- Summing up the total energy yield using Figure 9.9 can be enlightening or confusing. Notice that the energy yield (36 ATP) is the same as in Figure 9.2 but is presented on the right-hand side of the figure. If you make a point of telling the students that 1 NADH yields 3 ATP and 1 FADH yields 2 ATP, you should point out that *NADH from glycolysis* yields *only 2 ATP*.
- You can emphasize the roles of other foods (proteins and lipids) and their relationship to carbohydrates by following the arrows of Figure 9.10. This especially appeals to students interested in nutrition.
- A superb wrap-up to this unit on energy conversions is the "balance of nature" diagram (Figure 9.11) that traces the energy flow from the sun → plants → animals and the recycling of nutrient molecules.
- It should be stated that plants carry on aerobic respiration. Many students have the mistaken idea that plants only photosynthesize.
- Emphasize the processes taking place rather than require students to memorize the various reactions.
- Although the focus in this chapter is the generation of ATP, you may wish to explain to your students that another major function of respiration is the production of intermediates for biosynthetic reactions.

Classroom and Laboratory Enrichment

- Show a slide/sound set on cellular respiration.
- Demonstrate a computer simulation whereby basal metabolism rate (BMR) is calculated by measuring oxygen consumption.

- Select several persons who differ in physical stature and exercise conditioning. Allow them to exercise vigorously for several minutes; then determine heart rate and the length of time before breathing rate returns to normal (indicates extent of oxygen debt).
- Ask an exercise physiologist to talk to the class about the effect(s) of exercise on body metabolic rate.
- Show an electron micrograph of a mitochondrion and point out the matrix (inner compartment), cristae (inner membranes), and outer compartment. Relate this to the drawings in the text.
- If there is a brewery or winery nearby, arrange for a field trip. Brewmasters and winemakers generally are happy to conduct a tour through the facilities and explain the processes involved.

Ideas for Classroom Discussion

- Table wines, that is, those that have not been fortified, have an alcoholic content of about 10–12%. What factors could limit the production of alcohol during fermentation? Is it self-limiting, or do the vintners have to stop it with some additive?
- Your text lists two types of fermentation: one leads to alcohol, the other to lactate. Which occurs in yeasts, and why? Which pathway is reversible? What would be the consequences of nonreversible lactate formation in muscle cells?
- Your textbook says that the net energy yield from one molecule of glucose can be either *36* or *38* ATP. Textbooks commonly hedge on these figures; what is the explanation?
- Yeast is added to a mixture of malt, hops, and water to brew beer—a product in which alcohol and carbon dioxide are desirable! Why is yeast added to bread dough?
- Analyze the simple equation for cellular respiration shown in Figure 9.3 by telling exactly at what place in the aerobic metabolism of glucose each item in the equation is a participant.
- What is "metabolic water"?
- Why is fermentation necessary under anaerobic conditions? That is, why does the cell convert pyruvate to some fermentation product when it does not result in any additional ATP production?

Term Paper Topics, Library Activities, and Special Projects

- Rotenone is a fish poison and insecticide. Its mode of action is listed on container labels as "respiratory poison." Exactly where and how does it disrupt cellular respiration?
- Prepare a fermentation vat with grape juice and yeast (don't seal it!). Allow the process to proceed for a few days, then strain the fluid into a flask and distill it. What gas is produced during fermentation? What product distills over at 78.5°C?
- Certain flour beetles and clothes moths can live in environments where exogenous water is virtually unobtainable, yet they thrive. What mechanisms do they use for the synthesis and retention of water?
- Use diagrams to show how radioactive carbon 14 in glucose fed to rats could end up in body fat and proteins.
- Because ATP is the direct source of energy for body cells, why not bypass the lengthy digestion and cellular metabolism processes necessary for carbohydrate breakdown and eat ATP directly?
- Investigate the "set-point theory" of metabolism; discuss how it relates to people who are trying to lose or gain weight.
- Hydrogen cyanide is the lethal gas used in gas chambers. How does it cause death?

Films, Filmstrips, and Videos

- *The Energetics of Life.* (See Chapter 7.)
- *How the Body Uses Energy.* MGHT, 15 minutes, 16 mm, color.

10

CELL REPRODUCTION

Revision Highlights

Text and illustrations improved dramatically. Simpler overview of prokaryotic and eukaryotic division mechanisms. Simple, early explanation of inherently confusing terms (chromosome vs. chromatid, haploidy vs. diploidy, homologues). Students should carefully study pages 140–142, which are the conceptual foundation for the remaining chapters in the genetics unit. Much more accurate illustrations of mitosis and cell plate formation. Better summary.

Chapter Outline

OVERVIEW OF DIVISION MECHANISMS
- **Prokaryotic Fission**
- **The Eukaryotic Chromosome**
- **Mitosis, Meiosis, and the Chromosome Number**

WHERE MITOSIS OCCURS IN THE CELL CYCLE

STAGES OF MITOSIS
- **Prophase: Mitosis Begins**
- **Metaphase**
- **Anaphase**
- **Telophase**

CYTOKINESIS: DIVIDING UP THE CYTOPLASM

SUMMARY

Objectives

1. Understand the factors that cause cells to reproduce.
2. Know the differences that occur in prokaryotic and eukaryotic cell division.
3. Understand what is meant by *cell cycle* and be able to visualize where mitosis fits into the cell cycle.
4. Be able to describe each phase of mitosis.
5. Explain how the apportioning of cytoplasm to the daughter cells follows mitosis, which is a nuclear event.

Key Terms

reproduction	meiosis	centromere	gametes
life cycle	membrane growth	kinetochores	homologous chromosomes
prokaryotic fission	chromosome	somatic cells	haploid
mitosis	sister chromatids	germ cells	

diploid	prophase	anaphase	cleavage furrow
interphase	metaphase	telophase	spindle equator
mitosis	spindle apparatus	cytokinesis	cell plate formation

Lecture Outline

I. Introduction
   A. When cells asexually reproduce and divide in two, the daughter cells may be exact copies of the parental cells.
   B. Sexual reproduction introduces variation in offspring.
   C. Reproduction is part of the life cycle programmed by the DNA.
   D. For bacteria, their "cell cycle" is their life cycle.
   E. Multicellular organisms have a more complex life cycle, but a single cell still bridges each generation. **(In-text art, p. 139; TM 48a)**
   F. An understanding of reproduction requires one to know
      1. What elements are necessary for inheritance?
      2. How are they divided and distributed into daughter cells?
      3. What are the mechanisms of division?

II. Overview of Division Mechanisms
   A. Introduction
      1. Each daughter cell must receive DNA and some cytoplasm. **(In-text art, p. 139; TM 48b)**
      2. The cytoplasm contains enough machinery to start up on its own.
      3. Prokaryotic fission is used by bacterial cells.
      4. Mitosis and cytokinesis are used by eukaryotes for asexual reproduction and growth.
      5. Meiosis and cytokinesis are used by eukaryotes to form gametes.
   B. Prokaryotic Fission **(Fig. 10.1, p. 138; TA 25)**
      1. Bacteria have one circular DNA molecule, no nuclear envelope, and reproduce rapidly (thirty minutes or somewhat less).
      2. During initial growth, the DNA is first replicated.
      3. Both DNA molecules attach to the plasma membrane.
      4. Growth of the plasma membrane and wall separates the DNA molecules.
      5. Finally, new membrane and wall material separate the daughter cells by binary fission.
   C. The Eukaryotic Chromosome **(Fig. 10.2, p. 140; TA 26a)**
      1. Chromosomes are molecules of DNA complexed with proteins.
      2. Between divisions, each threadlike chromosome is duplicated, and a centromere joins sister chromatids. **(In-text art, pp. 140, 146; TA 26b, c)**
      3. Kinetochores attach to microtubules during division. **(In-text art, p. 140; TA 26b)**
   D. Mitosis, Meiosis, and the Chromosome Number
      1. Our body is composed of somatic cells and germ cells.
      2. Sexual reproduction begins with gamete production by meiosis and ends at fertilization.
      3. During development, mitosis maintains the chromosome number of the species (e.g., human somatic cells = 46). **(In-text art, p. 141; TM 49a)**
      4. Somatic cells contain pairs of homologous chromosomes.
      5. In meiosis, each diploid cell has its chromosome number halved, such that haploid gametes result. **(In-text art, p. 141; TM 49b)**

III. Where Mitosis Occurs in the Cell Cycle **(Fig. 10.4, p. 142; TM 50)**
   A. Only a small part of the cell cycle is spent in mitosis.
   B. About 90% of a cell's life occurs in interphase as the cytoplasmic components double and the DNA duplicates.
      1. G_1 may last from days to years.
      2. Adverse conditions may arrest amoebas in G_1 when they are deprived of a vital nutrient.

IV. Stages of Mitosis **(Fig. 10.5, p. 143; TA 27)**
   A. Prophase: Mitosis Begins **(Fig. 10.6, p. 144; TA 28)**
      1. The Prophase Chromosome
         Chromosomes become visible and condense into rodlike forms.
      2. The Microtubular Spindle **(Fig. 10.7, p. 146; TM 51)**
         a. The spindle apparatus forms to become the machine that will move chromosomes during mitosis.
         b. Two poles establish themselves due to microtubule organizing centers that may include a centriole pair.
   B. Metaphase **(Fig. 10.6, p. 145; TA 29)**
      1. Late in prophase, the nuclear envelope fragments such that by metaphase only fragments remain.
      2. The spindle penetrates the nuclear area, attaches to both kinetochores, and pulls equally toward opposite poles.
   C. Anaphase **(Fig. 10.6, p. 145; TA 29)**
      1. Sister chromatids separate and move toward opposite poles.
      2. Now each chromatid is an independent chromosome. **(In-text art, p. 146, TA 26c)**
   D. Telophase **(Fig. 10.6, p. 145; TA 29)**
      1. Telophase begins when chromosomes arrive at poles.
      2. The nuclear envelope forms from the fusion of small vesicles, and mitosis is complete.

V. Cytokinesis: Dividing Up the Cytoplasm **(Fig. 10.6, p. 145; TA 29)**
   A. Cytokinesis usually occurs from late anaphase through telophase.
   B. In animal cells, a cleavage furrow composed of contractile microfilaments pulls the plasma membrane inward.
   C. Plant cells form a cell plate that separates the two new cells. **(Fig. 10.10, p. 148; TA 30)**

Suggestions for Presenting the Material

- This is the first of two chapters concerning cell reproduction. The present one explains division in which the number of chromosomes remains the same in the identical daughter cells—mitosis. The next chapter explains a more complicated type of cell division in which the number of chromosomes is reduced during the production of cells destined to become gametes—meiosis.
- Remind students again of the differences between prokaryotic cells and eukaryotic cells. When students remember the differences that distinguish these two kinds of cells, it will be easier for them to understand why prokaryotes use a simple division technique like fission and eukaryotes use the more complex process of mitosis.
- Make sure that students are very familiar with terms, such as *centromere, chromatid, kinetochore,* and *homologues,* used to describe chromosomes and their parts.
- Emphasize the cell cycle, stressing that cells are not dividing all of the time. Students should be aware of the fact that the steps of cell division are part of a continuum. Our separation of the process into four stages is an artificial one, and it may be hard to say where one stage ends and the next begins when looking at a dividing cell. You may compare it to the showing of a "game tape" that athletes watch to see the errors committed in the big game.
- Students often have trouble following the number of chromosomes throughout the stages of mitosis. To help them, remind them that each chromosome has one centromere and it is not until centromeres divide in anaphase that "sister chromatids" are considered "chromosomes." Make certain that students understand where and when mitosis occurs in any organism.

Classroom and Laboratory Enrichment

- Show a film or film loop of time-lapse photography of cells undergoing actual cell division.
- Ask students (working individually or in small groups) to use chromosome simulation kits (available from biological supply houses) to demonstrate chromosome replication during the stages of mitosis. If kits are unavailable, make your own chromosomes using a pop-it bead (or bead and cord) for each chromatid and a magnet for each centromere.
- Demonstrate the phases of mitosis using either 35 mm slides or transparent colored chromosomes on the overhead projector.
- View ready-made squashes of mitotic material such as onion root tip or an overhead transparency of mitotic material using a microprojector. Ask students to estimate the length of each mitotic phase after counting the number of cells in each phase in several fields of view.
- Using purchased 2 x 2 transparencies or ones you have made yourself of Figure 10.6, project each stage of mitosis (not in correct sequence) and ask students to identify each phase (be sure to mask the name before you shoot the photo).
- Show a slide/sound set of mitosis, or better yet, a film showing animated and time-lapse photography of dividing cells.
- Perhaps the following analogy can help students visualize chromosomes during mitosis:

  2 chromatids = 2 matched socks
  1 centromere = 1 clothespin
  spindle fiber = clothesline

  (Hint: This analogy can be extended during meiosis to include two pairs of matched socks—a tetrad!)
- Prepare an overhead transparency of your own sketch, or use one from Chapter 15, to explain how one DNA molecule (that is, chromosome) becomes two identical DNA molecules (that is, daughter chromosomes).

Ideas for Classroom Discussion

- Bacteria divide by prokaryotic fission, and their cell cycle is much shorter than that of eukaryotes. Ask students to think of reasons why this might be so.
- Using a generation time of twenty minutes, calculate the size of a bacterial population that has arisen from a single bacterium growing under optimum conditions (for example, *Salmonella* in a bowl of unrefrigerated potato salad at a picnic on a warm summer day) for eight hours.
- Many of the drugs used in chemotherapy cause loss of hair in the individual being treated. Ask students if they can figure out why such drugs affect hair growth.
- Biologists used to believe that interphase was a "resting period" during the life cycle of the cell. Why did this appear to be so?
- Ask students how cell division in plant cells differs from that in animal cells.
- Loosely speaking, the process of one cell becoming two cells is referred to as mitosis, but to be completely accurate, what does mitosis specifically refer to?
- How can the mathematically impossible become the biologically possible—namely, a cell with 46 chromosomes splits to form two cells *each* with 46 chromosomes? This means 46 divided by 2 equals 46 plus 46???
- How can there be 46 chromosomes in a human cell at metaphase and also 46 chromosomes after the centromere splits in anaphase (see Figure 10.6)? Hint: Focus on the name change of chromatids to daughter chromosomes.
- What is there about the composition of an animal cell versus a plant cell that necessitates different methods of cleavage?

Term Paper Topics, Library Activities, and Special Projects

- Much progress in studying human disease has been made using the research technique of tissue culture. Describe techniques of tissue culture, explaining how cells can be induced to grow and divide in vitro.
- Explore diseases (such as cancer) that involve cell growth gone wrong. How do such diseases affect the mechanism of cell division? What drugs are used to halt runaway cell growth? How do these drugs work, and what are their side effects?
- Why do some cells of the human body (for example, epithelial cells) continue to divide yet other cells (for example, nerve cells) lose their ability to replicate once they are mature? Describe some of the latest research efforts to induce cell division in nerve cells.
- Colchicine is a chemical used to treat dividing plant cells to ensure that chromosomes of cells undergoing mitosis will be visible. How does colchicine achieve this effect? What is the natural source of colchicine?
- The 1956 edition of the high school biology text *Modern Biology* by Moon, Mann, and Otto was the last to state the human chromosome number as 48. This was not a misprint. What investigations resulted in assigning the correct number of 46 to humans?
- The preparation of a karyotype (picture of chromosomes) is a simple but multistep procedure. Provide a procedural outline for making such a preparation of chromosomes for publication in a book.
- There is some question as to exactly how chromosomes move to opposite ends of the cell in concert with the spindle fibers. Investigate the various theories, and report on their strengths and weaknesses.

Films, Filmstrips, and Videos

- *Cell Division: Mitosis and Meiosis*. CRMP (MGHT), 1974, 20 minutes, 16 mm, color. Excellent microphotography of mitotic and meiotic events that occur in living cells. Animation and time-lapse are used.
- *Cell Division: Mitosis and Meiosis*. MGHT, 1974, 20 minutes, 16 mm, color. Photomicroscopy and computer graphics combine to present the phases of mitosis and meiosis.
- *Mitosis*. EBEC, 1980, 13 minutes, 16 mm, color. *Heredity and Adaptive Change Series*. Introduction to cell division through time-lapse photomicrography and animation. Includes coverage of cloning and nuclear transplants.

11

A CLOSER LOOK AT MEIOSIS

Revision Highlights

Text and illustrations improved dramatically. New introduction compares asexual and sexual reproduction, and puts meiosis in the greater context of evolutionary change. Simple "refresher" paragraphs to define homologues, haploidy, diploidy, etc. There is now an extremely simple overview of the function of meiosis I and then meiosis II; details of both stages follow. Crossing over is introduced clearly and simply. New illustrations of meiosis I and II. New paragraph on importance of fertilization in contributing to phenotypic variation. New illustrations of spermatogenesis and oogenesis. Simpler introduction to plant life cycles. Much improved summary. New summary illustration comparing meiosis with mitosis.

Chapter Outline

ON ASEXUAL AND SEXUAL REPRODUCTION

OVERVIEW OF MEIOSIS
- **Think "Homologues"**
- **Overview of the Two Divisions**

STAGES OF MEIOSIS
- **Prophase I Activities**
- **Separating the Homologues**
- **Separating the Sister Chromatids**
- **More Gene Shufflings at Fertilization**

MEIOSIS AND THE LIFE CYCLES
- **Animal Life Cycles**
- **Plant Life Cycles**

SUMMARY
- **Key Features of Sexual Reproduction**
- **Stages of Meiosis**
- **Meiosis Compared With Mitosis**

Objectives

1. Contrast asexual and sexual types of reproduction that occur on the cellular and multicellular organism levels.
2. Understand the effect that meiosis has on chromosome number.
3. Describe the events that occur in each meiotic phase.
4. Compare mitosis and meiosis; cite similarities and differences.
5. Show where meiosis generally occurs in plant life cycles and contrast this with where it generally occurs in animal life cycles.

Key Terms

asexual reproduction
sexual reproduction
genes
homologous chromosomes
sister chromatids
prophase I
synapsis
crossing over
nonsister chromatids
alleles
interkinesis
meiosis I
genetic recombination
zygote
chiasma, chiasmata
metaphase I
anaphase I
metaphase II
fertilization
gametogenesis
zygote
spermatogenesis
sperm
oogenesis
egg
ovum
meiospores
sporophyte
gametophyte

Lecture Outline

I. On Asexual and Sexual Reproduction
   A. In asexual reproduction, one parent passes a duplicate of its genetic information to its offspring.
   B. In sexual reproduction with two parents, each parent contributes one gene for each trait.
   C. Mutations result in different alleles for each trait.
   D. Meiosis shuffles the alleles during gamete formation, and fertilization results in offspring with unique combinations of alleles.
   E. This variation is the testing ground for natural selection and is the basis for evolutionary change.

II. Overview of Meiosis
   A. Think "Homologues"
      1. Meiosis begins with diploid ($2n$) germ cells and results in haploid gametes.
      2. In $2n$ cells there are two chromosomes of each type = homologous chromosomes.
      3. Homologous chromosomes line up at meiosis.
      4. Sex chromosomes function as a pair of homologues even though their form differs.
      5. Meiosis produces gametes that have one of each pair of homologous chromosomes.
   B. Overview of the Two Divisions **(Fig. 11.3, p. 152; TA 31)**
      1. Meiosis consists of two consecutive divisions—meiosis I and II.
      2. During meiosis I, homologous chromosomes pair and cytokinesis follows.
      3. Each daughter cell receives a haploid number of duplicated chromosomes. **(In-text art, p. 152; TA 32a)**
      4. In meiosis II, the sister chromatids of each chromosome separate. **(In-text art, p. 152; TA 32b)**
      5. Cytokinesis follows and results in four haploid cells.

III. Stages of Meiosis
   A. Prophase I Activities
      1. In prophase I, homologous chromosomes pair by synapsis.
      2. Then, crossing over occurs—nonsister chromatids exchange parts. **(Fig. 11.4, p. 153; TA 33)**
      3. Because alleles of the same gene can vary, new combinations of genes in each chromosome can result.
      4. Crossing over leads to genetic recombination—that is, variations in the traits of the offspring result.
      5. After crossing over, as the nonsister chromatids begin to separate, they remain attached at chiasmata.
   B. Separating the Homologues **(Fig. 11.5 left, p. 154; TA 34)**
      1. During metaphase I, homologous chromosomes randomly align. **(In-text art, p. 154; TM 52a)**
      2. During anaphase I, homologous chromosomes separate. **(In-text art, p. 155; TM 52b)**

3. Two haploid cells with different mixes of maternal and paternal chromosomes result. **(Fig. 11.5 right, p. 155; TA 35)**
4. Telophase I and interkinesis are often fleeting stages.

C. Separating the Sister Chromatids
1. Meiosis separates each pair of sister chromatids at anaphase II.
2. Four haploid cells result at the close of meiosis.

D. More Gene Shufflings at Fertilization
1. Fertilization restores the diploid number in the zygote.
2. Because each gamete has a novel combination of alleles, each zygote is unique.

IV. Meiosis and the Life Cycles

A. Animal Life Cycles **(Fig. 11.7, p. 156; TM 53)**
1. Gametogenesis refers to both meiosis and gamete formation.
2. Males achieve gametogenesis by spermatogenesis. **(Fig. 11.8, p. 157; TA 36)**
   A $2n$ germ cell → primary spermatocyte—meiosis 1 → 2 secondary spermatocytes—meiosis II—cytokinesis → 4, $1n$ spermatids—develop into → sperm.
3. Females achieve gametogenesis by oogenesis, and this process differs from spermatogenesis in two ways. **(Fig. 11.9, p. 157; TA 37)**
   a. Cytoplasmic components accumulate in a primary oocyte.
   b. Meiosis results in one functional cell, or ovum, and the polar bodies do not form gametes.

B. Plant Life Cycles **(Fig. 11.10, p. 158; TM 54)**
1. Germ cells within plant tissues undergo meiosis and form haploid meiospores.
2. Each meiospore undergoes mitosis to become a haploid gametophyte.
3. Gametophytes produce cells haploid by mitosis that function as egg and sperm.
4. Fertilization results in a diploid sporophyte (e.g., a pine tree).

V. Meiosis Compared with Mitosis **(Fig. 11.11, pp. 160, 161; TA 38, 39)**

Suggestions for Presenting the Material

- This is a crucial subject area for beginning biology students. Students must have a good understanding of meiosis to comprehend the workings of inheritance explained in subsequent chapters.
- Before beginning this chapter, ask questions to make sure that students are well grounded in the events and purpose of mitosis. Because it was covered in the previous chapter, students should still recall the terms used to describe the parts of the cell involved in cell division.
- The events of meiosis can be confusing. Emphasize that meiosis makes it possible for organisms to undergo sexual reproduction. Remind students of the benefits of sexual reproduction; this helps them to understand why a process as complex as meiosis has evolved.
- Students often find it hard to understand when and how the chromosome number changes during meiosis, so be sure they understand that the two chromatids of one chromosome are each considered a chromosome after the centromere splits during meiotic anaphase II.
- Meiosis will be easier to grasp if students can become thoroughly acquainted with a typical animal life cycle (use the human life cycle as a familiar example) and a typical plant life cycle. Before finishing with this chapter, be sure to question the students about the events of meiosis and its consequences to the organism.

Classroom and Laboratory Enrichment

- Demonstrate the phases of meiosis using either transparent colored chromosomes on an overhead projector or 35 mm slides.

- Show students an overhead transparency of a karyotype of a normal man or woman to introduce the concept of homologous pairs.
- Compare a human karyotype to that of another organism.
- Ask students (working individually or in small groups) to use chromosome simulation kits (available from biological supply houses) to demonstrate chromosome replication and reduction of chromosome number during the stages of meiosis. If kits are unavailable, make your own chromosomes using a pop-it bead (or bead and cord) for each chromatid and a magnet for each centromere.
- Show overhead transparencies of adult plants or animals, and ask students to point out where meiosis occurs in each organism.
- Illustrate crossing over by using lengths of different colored string. Snip and tie the ends to create the products of a crossover (see Figure 11.4).
- Place Figure 11.5 on the screen and have students indicate the number of *human* chromosomes present at each stage (not how conveniently the *4* chromosomes in the diagram convert to *46* human ones). Use the centromere hint in the "Discussion" section below. Repeat the exercise with Figure 11.11.
- Prepare side-by-side comparisons of Figures 11.7 and 11.10 (generalized life cycles). Point out the subtle differences and the great similarity.
- If you can locate a segment of film depicting the union of gametes and subsequent cleavage, it will provide visual presentation of what mitosis and meiosis accomplish in a living cell.

Ideas for Classroom Discussion

- Do more advanced organisms have more chromosomes than primitive organisms? Review chromosome numbers of some common plants and animals.
- Why do cells undergoing mitosis require one set of divisions but cells undergoing meiosis need two sets of divisions?
- All but the most primitive species of organisms have the ability to reproduce sexually. Can you explain why sexual reproduction is considered a hallmark of evolutionary advancement? What are the advantages over asexual reproduction?
- Division of the cell cytoplasm is equal during spermatogenesis but unequal during oogenesis. Can you think of at least one reason why?
- The generalized life cycle of complex land plants is often described as "alternation of generations." Describe the meaning of this phrase.
- An old-fashioned name for meiosis II is "reduction division." Why?
- What would happen if meiosis did not halve the chromosome number?
- Ask students to compare an animal life cycle to a complex land-plant life cycle. How is the life cycle of a human different from that of a plant?
- Why does crossing over occur in prophase of *meiosis* and not *mitosis*?
- One of the meiotic series is very much like mitosis. Is it meiosis I or II?
- Does the reduction in chromosome number occur in meiosis I or II? Hint: To conveniently count the number of chromosomes (whether doubled as chromatids or newly formed daughter chromosomes) simply count the number of centromeres (or portions thereof) present in any particular stage.
- When do the processes of human spermatogenesis and oogenesis begin? Are they the same in males and females?
- What is the derivation of the prefix "chrom-" as used in describing the carriers of heredity?

Term Paper Topics, Library Activities, and Special Projects

- How are human karyotypes prepared? Discover the laboratory steps required in this procedure (briefly described in Figure 13.8).
- Trace the increasing dominance of the sporophyte in the life histories of land plants beginning with the bryophytes and ending with the angiosperms.
- Use the karyotypes of related species (for example, primates) to describe evolutionary relationships, if any, between the species.
- Meiosis precisely reduces the chromosome number so that union of several gametes restores the diploid number. How many extra or fewer chromosomes can a human body cell have and survive? Are there consequences? Does the same hold true for plants?
- What are the latest theories and evidence relative to the blockage of entry into the egg of all sperm subsequent to the first one?
- Select from the library shelves several biology texts from the past fifty years. Compare the formal definitions of "gene," and prepare a historical resumé of the changes.

Films, Filmstrips, and Videos

- *Cell Division: Mitosis and Meiosis.* CRMP (MGHT), 1974, 20 minutes, 16 mm, color. Excellent microphotography of mitotic and meiotic events that occur in living cells. Animation and time-lapse are used.
- *Meiosis.* (2nd ed.) EBEC, 1980, 15 minutes, 16 mm, color. The role of meiosis in sexual reproduction in both plants and animals is highlighted, as well as the direct implications of meiosis in producing genetic variations. Uses microscopic footage, concise animation, and artwork.
- *Meiosis: Sex Cell Formation.* EBEC, 1963, 16 minutes, 16 mm, color. Explains how reduction division occurs in the formation of sex cells. Shows various stages, using time-lapse, and explains how meiosis enables plants and animals to produce offspring with a wide variety of characteristics. Also explains the role of meiosis in natural selection.

12

OBSERVABLE PATTERNS OF INHERITANCE

Revision Highlights

Tighter writing (chapter is shorter by a page even though more illustrations added). Better definitions of gene pair and allele. New, simple illustrations of allelic segregation and of independent assortment. Details in dihybrid cross illustration are easier to track. Chapters 12, 13, 15, and others are deliberately written in a format that helps students understand what it means to think critically. This is just one of the many functions that an introductory textbook in biology must serve (see text Preface).

Chapter Outline

Objectives

1. Know Mendel's principles of dominance, segregation, and independent assortment.
2. Understand how to solve genetics problems that involve monohybrid and dihybrid crosses, as well as sex-linked inheritance.
3. Understand the variations that can occur in observable patterns of inheritance.

Key Terms

true-breeding strains	alleles	recessive	homologous chromosomes
cross-fertilization	homozygous	dominant allele	hybrids
genes	heterozygous	recessive allele	homozygote
locus	homozygous dominant	genotype	heterozygote
gene pair	dominant	phenotype	

monohybrid cross
homozygous genes
dihybrid cross
Mendelian principle of segregation
Punnett-square method
probability
testcross
dihybrid cross
Mendel's principle of independent assortment
incomplete dominance
codominance
multiple allele system
incomplete penetrance
cleft lip
variable expressivity
polydactyly
epistasis
melanin
albino
pleiotropy
sickle-cell anemia
discontinuous variation
continuous variation
quantitative inheritance

Lecture Outline

I. Mendel's Insights into the Patterns of Inheritance
   A. Introduction
      1. By the late nineteenth century, natural selection suggested that a population could evolve if members show variation in heritable traits. Variations that improved survival chances would be more common in each generation—in time, the population would change or evolve.
      2. However, many also believed that at fertilization, the traits of each parent blended like cream in coffee.
      3. Blending would produce uniform populations—and such populations could not evolve.
      4. Yet many observations did not fit blending. A white horse and black horse did not produce only gray ones.
      5. Near Vienna, a monk named Gregor Mendel first identified the rules governing inheritance.
      6. Mendel used experiments in plant breeding and a knowledge of mathematics to form his hypotheses.
   B. Mendel's Experimental Approach
      1. He cross-fertilized true-breeding garden pea plants having clearly contrasting traits (e.g., white vs. purple flowers).
   C. Some Terms Currently Used in Genetics **(Fig. 12.2, p. 163; TA 40)**
      1. We need to define some modern terms in order to appreciate Mendel's logic.
      2. Genes are instructions for producing a trait.
      3. Each gene has a locus on a chromosome.
      4. Diploid cells have two genes (a gene pair) for each trait—each on a homologous chromosome.
      5. If homozygous, both alleles are the same.
      6. If heterozygous, the alleles differ.
      7. When heterozygous, one allele is usually dominant (*A*)—the other is recessive (*a*).
      8. Thus, homozygous dominant = *AA*, homozygous recessive = *aa*, and heterozygous =*Aa*.
      9. Genotype is the sum of the genes, and phenotype is how the genes are expressed (what you observe).
   D. The Concept of Segregation
      1. Monohybrid crosses have two parents true-breeding for contrasting forms of a trait.
      2. One form of the trait disappears in the first generation, only to show up in the second generation.
      3. We now know that all members of the first generation are heterozygous because one parent could only produce an *A* gamete and the other could produce only an *a* gamete. **(Fig. 12.3, p. 164; TM 55)**
      4. The Mendelian principle of segregation states that $2n$ organisms inherit two genes per trait, and each gene segregates during meiosis such that each gamete will receive only one gene per trait.
   E. Probability: Predicting the Outcome of Crosses
      1. The numerical ratios of crosses suggested that genes do not blend.
      2. For example, the F_2 offspring showed a 3:1 phenotypic ratio. **(Fig. 12.5, p. 166; TM 56)**
      3. Mendel assumed that each sperm has an equal probability of fertilizing an egg. Consider the Punnett square . . . **(Fig. 12.6, p. 166; TM 57)**

4. Thus, each new plant has three chances in four of having at least one dominant allele.

F. Testcrosses **(Fig. 12.8, p. 167; TM 58)**
   1. To support his concept of segregation, Mendel crossed F_1 plants with homozygous recessive individuals.
   2. A 1:1 ratio of recessive and dominant phenotypes supports his hypothesis.

G. The Concept of Independent Assortment **(Fig. 12.9, p. 168; TA 41)**
   1. Mendel crossed plants with contrasting forms of two traits—a dihybrid cross.
   2. Mendel correctly predicted that all F_1 plants would show both dominant alleles (they would have purple flowers and would be tall). **(Fig. 12.10, p. 169; TA 42)**
   3. But if you cross the F_1 plants, would the genes for color and height travel together or independently?
   4. We now know that genes located on *non*homologous chromosomes segregate independently of each other.
   5. Hence, Mendel observed a 9:3:3:1 phenotypic ratio.
   6. If parents differ in only 10 gene pairs, 60,000 genotypes are possible.
   7. In 1865, Mendel reported that hereditary material retains its identity through the generations, but his report had no impact.

II. Variations on Mendel's Themes

A. Dominance Relations
   1. In incomplete dominance, a dominant allele cannot mask the expression of another.
   2. In codominance, both alleles are expressed in heterozygotes (e.g., the ABO blood group—also a multiple allele system).
   3. In incomplete penetrance, dominant alleles are expressed in some individuals but not in others.
   4. In variable expressivity, the phenotype expressed by a dominant allele varies with the individual (e.g., polydactyly). **(Fig. 12.12, p. 171; TM 59)**

B. Interactions Between Different Gene Pairs
   1. Genes often interact with each other to produce the phenotype.
   2. Two gene pairs may *cooperate* to produce a unique phenotype (e.g., comb shape in chickens).
   3. In epistasis, one gene pair masks another (e.g., albino rabbits may result when they are homozygous recessive for a gene that codes for tyrosinase, which is needed for melanin production).

C. Multiple Effects of Single Genes
   1. Pleiotropy occurs when a single gene affects an unrelated aspect of the phenotype.
   2. The gene for sickle-cell anemia codes for a variant form of hemoglobin. The altered hemoglobin in turn affects the shape of the red blood cells, which clump together and block capillaries. Impaired gas flow damages tissues.

D. Environmental Effects on Phenotype
   1. Temperature-sensitive enzymes may result in Siamese cats' coloration.
   2. Water buttercup leaves that grow submerged are finely divided compared with leaves growing in the air.

E. Continuous and Discontinuous Variation **(Fig. 12.18a, p. 174; TM 60)**
   1. Mendel's traits show discontinuous variation because they belonged to one or more clear classes.
   2. Most traits are not qualitative but show continuous variation and are transmitted by quantitative inheritance.
   3. Quantitative inheritance arises from the combined influence of three or more gene pairs.

Suggestions for Presenting the Material

- Students will be curious and very interested in genetics. Start first with the simple examples of Mendel's monohybrid and dihybrid crosses before fielding questions on human traits such as height or eye color. Emphasize the remarkable nature of Mendel's work; remind the students that he knew nothing of chromosomes and their behavior as described in Chapters 10 and 11 and that the term *gene* did not exist.

- Use Mendel's experiments and his conclusions as real-life examples of the scientific method at work. Ask questions to make sure students understand monohybrid and dihybrid crosses and testcrosses. Use the information in the illustration of genetic terms (Figure 12.2), or use the figure itself as a sketch or overhead transparency, to ensure that students know the meaning of homologous chromosomes, gene locus, alleles, and gene pairs.
- Students should be able to relate the events of meiosis to the concepts of segregation and independent assortment; if their understanding of meiosis is weak, they will have trouble doing this.
- Beginning with this chapter, students will be quick to ask questions about human traits, many of which are governed by mechanisms more complex than those postulated by Mendel. Answer questions in this area during (or after) the discussions of variations on Mendel's themes presented in the second half of this chapter.

Classroom and Laboratory Enrichment

- Ask groups of students to conduct coin tosses. Demonstrate the importance of large sample size by having the students vary the number of tosses before calculating variation from expected ratios.
- Distribute PTC tasting paper to your students, and calculate the number of tasters and nontasters in the classroom. Or use easily seen physical traits such as tongue rollers versus nontongue rollers or attached earlobes versus unattached earlobes to demonstrate traits governed by simple dominance/recessiveness.
- Prepare a biographical sketch of Mendel, including his education and practice as a clergyman. Enliven your presentation with as many slides of photos as you can find.
- Hand out a partially completed pedigree, and show students how to assign squares and circles for their family. Then ask them to select a trait and complete the pedigree after surveying the family members for presence/absence of the trait.
- Select a portion of the class to reenact the photo in Figure 12.18b. If the quantity of students chosen does not provide a bell-shaped curve, use this as an illustration of how the greater number of trials/subjects/experiments tends to increase probability.
- Prepare 2 x 2 transparencies of the examples of "variations on Mendel's themes." This will remove some of the "abstractness" of these topics.
- Show a slide/sound set on genetics.

Ideas for Classroom Discussion

- Describe the behavior of one trait with regard to its inheritance in a particular cross; then ask students to identify the genetic mechanism at work (simple dominance/recessiveness, incomplete dominance, codominance, incomplete penetrance, variable expressivity, epistasis, pleiotrophy, continuous variation).
- List some human traits that you would guess are each governed by a single gene.
- Give several reasons why Mendel's pea plants were a good choice for an experimental organism in genetics. Give an example of an organism that would be a poor choice for genetic research and explain your choice.
- Describe several different crosses using organisms such as Mendel's pea plants. Then ask students to calculate phenotypic and genotypic ratios for each cross.
- Discuss the significance of Mendel's use of mathematical analysis in his research.
- Why do you think Mendel was not immediately recognized as the discoverer of a new area of biology—genetics?

- What conclusions would Mendel have come to if he had chosen *snapdragons* instead of *peas* for his study material?
- Why are the traits of (a) human skin color and (b) human height not suitable for explaining the concept of simple dominance?
- There are four possible blood types in the ABO system. But how many *different* alleles are in the human population for this marker?
- What is the subtle difference between *incomplete dominance* and *codominance*?
- What is the significance of using upper- and lowercase versions of the same letter (for example, A and a) for the dominant and recessive trait, respectively, rather than a capital A for dominant and a B (or b) for recessive?

Term Paper Topics, Library Activities, and Special Projects

- What organisms are used most frequently in modern genetic research, and why?
- Describe the legal role now played by blood type evidence in paternity cases. Are other aspects of blood genetics (besides the ABO series) now used in deciding such cases?
- How do you think Charles Darwin's writings on his theory of evolution might have changed had he known of Mendel's work?
- Describe examples of how modern knowledge of genetics has led to improved agricultural strains of plants and animals.
- Search for details of Mendel's life and work. Seek answers to the allegations that his results may have been "too good."
- When studying genetics, it is easy to discover variations in plants and animals that result in organisms that even though related are very different in appearance. How does a researcher prove whether or not the variants are/are not the "same species"?
- Prepare an update on the extent of sickle-cell anemia in the United States and the world. Include in your report the consequences to those persons who are recessive and those who are carriers.
- Sickle-cell anemia results from an abnormal hemoglobin in which valine is substituted for glutamate. Is this substitution random, or does it occur under the direction of the molecules that direct protein synthesis? Explain the mechanism.

Films, Filmstrips, and Videos

- *Genetics and Plant Breeding.* BFA, 1970, 17 minutes, 16 mm, color. This film shows the work of Mendel and the techniques of modern geneticists involved in increasing plant productivity. The basic laws of Mendelian inheritance are shown simply and clearly.
- *Gregor Mendel.* UCEMC, 1973, 24 minutes, 16 mm/video, color. *Great Scientists Speak Again Series.* Richard Eakin plays Mendel to explain the man's work and philosophy. The logic of Mendel's approach, independent assortment, and principles of segregation are explained.

13

CHROMOSOMAL THEORY OF INHERITANCE

Revision Highlights

Better coverage of sex determination, including reference to mammalian TDF gene coding for the protein that dictates gender. Clarification of sex-linked vs. X-linked genes. Cleaner description of Morgan's experiments and reasoning. *Drosophila* diagrams redrawn so it is easier to track the eye color phenotypes in the crosses. New illustration of effect of crossing over on gene linkage. Simpler text on chromosome banding. New micrograph of translocation involving a human chromosome. Cri-du-chat used as example of chromosomal deletion. New generalized illustration on nondisjunction.

Chapter Outline

RETURN OF THE PEA PLANT

THE CHROMOSOMAL THEORY

CLUES FROM THE INHERITANCE OF SEX
- **Sex Determination**
- **Sex-Linked Genes**
- **Morgan's Studies of Inheritance**

LINKAGE

CROSSING OVER AND LINKAGE MAPPING OF CHROMOSOMES

CHANGES IN CHROMOSOME STRUCTURE
- **Chromosome Banding and Karyotypes**
- **Deletions, Duplications, and Other Structural Rearrangements**

CHANGES IN CHROMOSOME NUMBER
- **Missing or Extra Chromosomes**
- **Polyploidy**

SUMMARY

Objectives

1. Describe the chromosomal theory of inheritance and explain how the theory helps to account for events that compose mitosis and meiosis.
2. Name some ordinary and extraordinary chromosomal events that can create new phenotypes (outward appearances).
3. Know how fruit fly experiments have helped us understand chromosomal behavior.
4. Understand how changes in chromosome structure and number can affect the outward appearance of organisms.

Key Terms

chromosome
chromosomal theory of inheritance
genes
homologues
diploid
$2n$
haploid
assort independently
crossing over
chromosomal aberrations
autosomes
sex chromosomes
X chromosome
Y chromosome
gender
sex-determining genes
X-linked
Y-linked
sex-linked
white-eyed male
wild-type
marker
linkage
linkage mapping
polytene chromosome
G-banding
karyotype
duplication
deletion
inversion
translocation
nondisjunction
aneuploidy
polyploidy
triploid
autopolyploidy
allopolyploidy

Lecture Outline

I. Return of the Pea Plant
   A. By 1882, Flemming observed threadlike chromosomes and discovered mitosis.
   B. By 1887, Weismann suggested that meiosis halves the number of chromosomes when gametes are made.
   C. By 1900, Mendel's work was finally appreciated; that is, $2n$ cells have two units for each trait, and the units segregate during gamete formation.
   D. Later, it would be clearly demonstrated that the "units" lie on chromosomes.

II. The Chromosomal Theory **(Fig. 13.1, p. 177; TA 43)**
   The chromosomal theory of inheritance states that
   A. Genes lie upon chromosomes in a linear array.
   B. Diploid cells have two chromosomes of each type, called homologues.
   C. The two sex chromosomes pair as homologues during meiosis.
   D. After pairing in prophase I, each chromosome segregates and the diploid number is reduced to the haploid number.
   E. Chromosomes assort independently at meiosis.
   F. Genes on the same chromosomes tend to stay together, but crossing over can lead to genetic recombination.
   G. Chromosomal aberrations occur.
   H. Independent assortment, crossing over, and aberrations play roles in evolution.

III. Clues from the Inheritance of Sex
   A. Sex Determination
      1. By the early 1900s, most chromosomes were recognized as autosomes, and one or two were called sex chromosomes.
      2. In many animals like fruit flies and humans, the males = XY and the females = XX. **(Fig. 13.2, p. 179; TM 61)**
      3. Gender is determined by which sex chromosome is carried in the sperm—X or Y. **(Fig. 13.3, p. 179; TM 62)**
      4. The Y chromosome carries a gene that codes for testis determining factor, and a few other genes.
   B. Sex-Linked Genes
      1. The X chromosome may carry as many as 200 sex-linked genes; many of them deal with nonsexual traits.
      2. Sex-linked genes may be X-linked or Y-linked.
   C. Morgan's Studies of Inheritance
      1. In the early 1900s, Morgan studied the fruit fly, *Drosphila melanogaster.*

2. Because a fruit fly can breed after two weeks, almost thirty generations a year could be studied.
3. In 1910, a white-eyed male was noted that differed from the wild type that had brick-red eyes.
4. The phenotypic outcomes of reciprocal crosses differed. **(Fig. 13.4a, p. 180; TM 63)**
5. Morgan thought that the gene for eye color must lie on the X chromosome and be sex-linked.

IV. Linkage

A. Morgan reasoned that if many genes were on the same chromosome, they would be physically linked during meiosis and end up in the same gamete.

B. Morgan found four groups of linked genes that corresponded to the haploid number of chromosomes.

C. Linkage is the tendency of genes located on the same chromosome to be transmitted together in inheritance. **(Fig. 13.5, p. 181; TM 64)**

V. Crossing Over and Linkage Mapping of Chromosomes

A. Crossing over is an exchange of parts of homologous chromosomes. **(Fig. 13.6, p. 182; TM 65)**

B. Novel combinations of parental traits result in offspring.

C. Crossing over can disrupt gene linkages at any point on the chromosome.

D. The probability of crossing over separating two genes on a chromosome is proportional to the distance between them.

E. Thus, genes *A* and *B* would be disrupted twice as often as *C* and *D* if genes *A* and *B* are located twice as far apart on the chromosome as genes *C* and *D*.

F. The farther apart two genes on a chromosome are, the greater the frequency of crossing over.

G. The relative frequency of crossing over between different genes can be used for linkage mapping.

VI. Changes in Chromosome Structure

A. Chromosome Banding and Karyotypes

1. Stained chromosomes have banding patterns that can be used to help identify individual chromosomes. **(Fig. 13.8, p. 183; TM 66)**

B. Deletions, Duplications, and Other Structural Rearrangements

1. Chromosomes can undergo abnormal structural rearrangements that change the normal banding pattern.
2. A change in phenotype can result and be correlated with the change in banding pattern.
3. A deletion is the loss of a chromosome part.
   a. A deletion can be caused by a terminal segment breaking off, or because of breaks due to viruses, chemicals, or irradiation.
   b. It can cause problems like mental retardation.
4. Duplication occurs when a gene sequence is in excess of the normal amount. Pairing of homologues at prophase I may be affected. **(In-text art, p. 184; TM 67a)**
5. An inversion alters the position and sequence of the genes; gene order is reversed. **(In-text art, p. 184; TM 67b)**
6. A translocation occurs when a part of one chromosome is transferred to a nonhomologous chromosome.
   a. It is seen in some forms of cancer.
   b. It may result in the fusion of two nonhomologous chromosomes and reduce the chromosome number.

VII. Changes in Chromosome Number

A. Nondisjunction at anaphase I or anaphase II frequently results in a change in chromosome number. **(Fig. 13.11, p. 185; TA 44, 45)**

B. Missing or Extra Chromosomes

1. Nondisjunctions can result in aneuploidy—one extra or one chromosome less than normal.
2. An extra chromosome results in trisomy.
3. A missing chromosome results in monosomy—which is often lethal in humans.

C. Polyploidy
1. Polyploidy means having a multiple of the parental chromosome number; it occurs in about half of flowering plants.
2. Complete nondisjunction can result in a triploid or tetraploid zygote; this is called autopolyploidy.
3. Triploid plants are usually sterile.
4. Polyploidy can also arise when the DNA duplicates, but the cell does not divide.
5. Triploid bananas, potatoes, etc., are of great commercial value, since they are larger and hardier than diploid forms.
6. Autopolyploidy can be induced with colchicine.
7. Related plant species can sexually reproduce, and if at least one gamete is diploid, allopolyploidy results. **(Fig. 13.13, p. 187; TM 68)**

Suggestions for Presenting the Material

- Students should be well grounded in their understanding of chromosomal structure before attempting to tackle the material in this chapter. The use of sketches, diagrams, and overhead transparencies will greatly assist in making this material as clear as possible.
- They also must understand the events of meiosis or they will have difficulty comprehending crossing over and changes in chromosome number resulting from nondisjunction. Emphasize the information in Figure 13.1 (an overview of some key aspects of chromosome structure and function) by either sketching the chromosomes as shown or using the figure as an overhead transparency.
- Remind students that crossing over and genetic recombination create variability among sexually reproducing organisms; encourage students to think about the role this plays in evolution.
- To reduce the difficulty that students often have when learning about sex-linked genes, remind them that more precise terms for genes on either sex chromosome are X-linked or Y-linked. Describe the steps of Thomas Hunt Morgan's work to show how sex-linkage was discovered. Ask students to solve the genetics problems that deal with sex-linked genes at the end of the chapter. To assess how well students understand this material, work on as many of these problems together in class as time allows.
- Linkage is another concept that is sometimes confusing. Use as many drawings and diagrams as possible to show students what linkage means and how genes can be separated by crossing over. Include, if possible, gene maps of human chromosomes; students will be fascinated by this.
- Explain how karyotypes are prepared before showing a human karyotype; otherwise students might think that human chromosomes naturally occur in pairs as shown.

Classroom and Laboratory Enrichment

- Ask students (working individually or in small groups) to use chromosome simulation kits (available from biological supply houses) to review chromosome structure, homologous pairing, crossing over, and independent assortment during gamete formation. If kits are unavailable, make your own chromosomes using a pop-it bead (or bead and cord) for each chromatid and a magnet for each centromere.
- Show karyotypes of males and females of different species without revealing the sex of the individual. Ask students to identify the sex.
- Discuss gene mapping in humans using an overhead transparency showing some of the known locations of particular genes.

- Use large, transparent chromosomes with labeled gene sequences to demonstrate deletions, duplications, inversions, and translocations on the overhead projector.
- If you have not done it previously, you may wish to demonstrate crossing over by using the colored string as explained in the "Enrichment" section for Chapter 11 (this manual).
- To dramatize independent assortment, use the following simulation of a cocktail party. Have several students pair off and line up as couples. Next, allow all participants to intermingle, but ask a few couples to switch partners and form new couples. (Explanatory comments: Those couples that did not switch but remained together represent genes linked together on the same chromosome; they travel together. Those that did switch were not linked and could independently assort themselves.)
- Ask a local health unit or testing lab if you can copy (anonymously of course) some karyotypes that show chromosomal defects. Show transparencies of these to the class, and ask if the students can spot the defect before it is revealed to them.
- Prepare a hand-drawn transparency to relate the following:

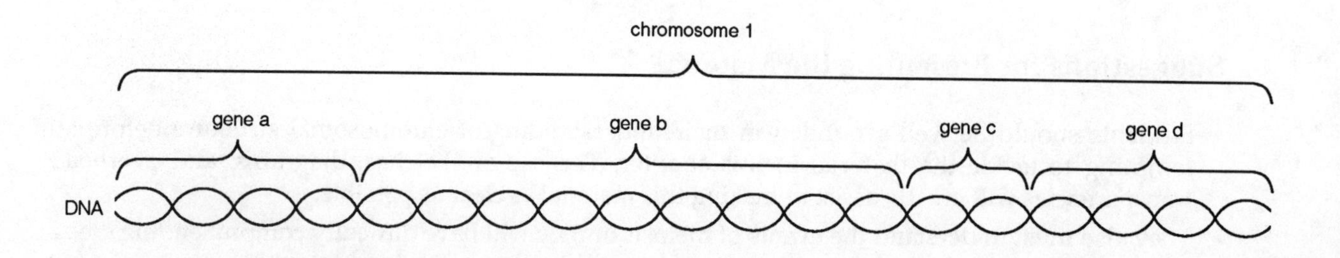

Ideas for Classroom Discussion

- If it becomes possible to easily and inexpensively choose the sex of your child, how will this change the male:female ratio among newborns? Do you think it is ethically correct to select the sex of your children?
- Do you think that any of the traits Mendel followed in the garden pea were linked? Why or why not?
- If male and female offspring occur at a ratio of approximately 50:50, why do some couples have only boys or only girls?
- What is the distinction between the terms *gene* and *allele*?
- Why do individual chromosomes present at the conclusion of meiosis not have the same genetic constituency as they did before meiosis?
- If one sex of offspring tends to exhibit a trait more frequently than the other sex, this is an indication of what?
- What is the derivation of the numerals designated as "map units" in Figure 13.7?
- Why is polyploidy more common among plants than among animals?
- What is the physical relationship of *genes* to *chromosomes* to *DNA*?
- What is the difference between a *translocation* of chromosomal segments and *crossing over*?

Term Paper Topics, Library Activities, and Special Projects

- How do the sex chromosomes actually determine the sex of the individual? What are some of the characteristics governed by genes of the X and Y chromosomes?
- What human genes have been mapped?

- Because the sex of an individual is determined by the sperm at the moment of conception, is it possible to select the sex of a child by separating sperm with X chromosomes from those with Y chromosomes? Discuss recent experimental techniques that attempt to do this.
- How are human karyotypes prepared? Discover the laboratory steps required in this procedure (briefly described in Figure 13.8).
- Find specific examples of how irradiation, chemical action, or viral attack can cause chromosome breakage.
- Are any human traits Y-linked?
- Describe how (and why) artificially induced changes in chromosome number have been used to create new varieties of fruits and vegetables.
- As indicated above, the predicted ratio of newborns should show a 50:50 sex ratio. Ask a local hospital to provide statistics on the sex ratio in order to confirm or deny this prediction.
- Report on the progress being made on the human chromosome mapping project.
- Prepare a protocol for rearing *Drosophila* in the lab and making simple crosses.

Films, Filmstrips, and Videos

- *The Chromosomes: General Considerations.* MIFE, 1976, 14 minutes, 16 mm, color. Demonstrates the preparation of normal human karyotypes and discusses the nomenclature and normal variations. Clinical findings and cytological observations are correlated, and autosomal aneuploidy is discussed in detail. Structural aberrations such as translocations, deletions, and inversions are also presented.
- *The History of Genetics: The Physical Basis of Inheritance.* MIFE, 1976, 19 minutes, color. From Mendel to Müller, the history of genetics is outlined. The contributions of Garrod, Bateson, Sutton, Boveri, Hardy and Weinberg, Johanssen, Morgan, Bridges, and Landsteiner are demonstrated. The processes and significance of mitosis and meiosis are described and discussed. Verbal level high.

14

HUMAN GENETICS

Revision Highlights

More straightforward organization. Clearly distinguishes between disease, disorder, and abnormality at the outset. New table on human genetic disorders. New illustrations of autosomal recessive inheritance. New coverage and new illustration of autosomal dominant inheritance, using achondroplasia and Huntington's disorder as examples. Frequency of Huntington's in South Africa cited as example of founder effect. Photos of children with Down syndrome, emphasizing that affected individuals are no less human than the rest of us. New section on RFLPs and genetic fingerprinting.

Chapter Outline

DISORDERS ARISING FROM GENE MUTATIONS
- **Autosomal Recessive Inheritance**
- **Autosomal Dominant Inheritance**
- **X-Linked Recessive Inheritance**

DISORDERS ARISING FROM CHANGES IN CHROMOSOME NUMBER
- **Down Syndrome**
- **Turner Syndrome**
- **Klinefelter Syndrome**
- **XYY Condition**

PROSPECTS AND PROBLEMS IN HUMAN GENETICS
- **Treatments for Phenotypic Defects**
- **Genetic Screening**
- **Genetic Counseling and Prenatal Diagnosis**
- **RFLPs**

SUMMARY

Objectives

1. Distinguish *autosomal recessive inheritance* from *sex-linked recessive inheritance.*
2. Give two examples of each of the above types of inheritance.
3. Explain how changes in chromosomal number can occur and present an example of such a change.
4. List three examples of phenotypic defects and describe how each can be treated.
5. Explain how knowing about modern methods of genetic screening can minimize potentially tragic events.

Key Terms

phenotypic treatment
genotypic cure
abnormal
disorder
disease
galactosemia
autosomal recessive inheritance
autosomal dominant inheritance
achondroplasia

Huntington's disorder	trisomy 21	sex chromosome abnormality	cleft lip
hemophilia A	Down syndrome	Klinefelter syndrome	genetic screening
X-linked recessive inheritance	Turner syndrome	XYY condition	amniocentesis
			RFLPs

Lecture Outline

I. Introduction
   A. Human genetics is difficult to study.
      1. We live under variable conditions in diverse environments.
      2. Humans mate by chance and may choose to reproduce.
      3. Humans live as long as those who study them.
      4. The small family size characteristic of human beings is not good for statistical studies.
   B. One can study pedigrees.
   C. More than 3,500 single-gene abnormalities have been recorded thus far.
      1. There are few phenotypic treatments, which involve drugs or diet modification.
      2. Some day, genotypic cures may be possible.
   D. Remember that "abnormal" does not always mean "disorder"—many "abnormalities " are subjective.

II. Disorders Arising from Gene Mutations
   A. Autosomal Recessive Inheritance **(Fig. 14.1, p. 190; TA 46)**
      1. A single gene mutation causes galactosemia—lactose cannot be metabolized (about 1 in 100,000 newborns suffers from this).
      2. Galactosemia is an example of autosomal recessive inheritance and can be lethal if untreated. **(In-text art, p. 190; TM 69)**
   B. Autosomal Dominant Inheritance **(Fig. 14.2, p. 191; TA 47)**
      1. Even if a dominant allele is selected against, it can reappear because of mutation.
      2. Some dominant alleles don't affect reproduction or are expressed after reproductive age.
      3. Achondroplasia is a type of dwarfism caused by an autosomal dominant allele.
      4. Huntington's disorder is often expressed after age forty and is lethal.
   C. X-Linked Recessive Inheritance **(Fig. 14.3, p. 192; TA 48)**
      1. One mutation gives rise to hemophilia A—the affected do not produce enough clotting factor VIII.
      2. The gene is on the X chromosome and is expressed in males and homozygous recessive females.

III. Disorders Arising from Changes in Chromosome Number **(Fig. 14.7, p. 194; TA 49)**
   A. In addition to gene mutations, nondisjunction will cause disorder in 1 of 200 liveborns.
   B. Down Syndrome
      1. Down syndrome results from trisomy 21—1 in 1,000 liveborns in North America are affected.
      2. Most children with Down syndrome show mental retardation, and 40% have heart defects.
      3. Down syndrome occurs more frequently in children born to women over 35 and men past 55.
   C. Turner Syndrome
      1. Turner syndrome involves females with only one X chromosome (XO) and is caused by nondisjunction; it affects 1 in 5,000 liveborns.
      2. Affected individuals are sterile with other phenotypic problems—e.g., they may have a defective aorta or kidney, and they may age prematurely.
   D. Klinefelter Syndrome
      Klinefelter syndrome leads to XXY males, who are sterile and mentally retarded.

E. XYY Condition

The XYY condition manifests itself in mostly normal to mildly retarded males, who are often convicted of nonviolent crimes.

IV. Prospects and Problems in Human Genetics

A. Introduction

1. 1% of newborns have a problem linked to a chromosomal aberration.
2. 1%–3% have problems linked to mutant genes.
3. 10%–25% of children in hospitals are treated for genetic problems.

B. Treatments for Phenotypic Defects

1. Diet Modification
   a. Phenylketonuria (PKU) is treated by a diet with limited phenylalanine.
2. Environmental Adjustments
   a. Albinos can avoid sunlight.
   b. People with sickle-cell anemia can avoid strenuous activity.
3. Surgical Correction
   a. Cleft lips or palates can be corrected in terms of appearance and function.
4. Chemotherapy
   a. People with Wilson's disorder deposit copper in tissues—a drug can bind with copper and flush it from the body.
5. Genetic Screening
   a. Newborns are screened for PKU.
6. Genetic Counseling and Prenatal Diagnosis
   a. This begins with accurate diagnosis of disease, including construction of a family pedigree.
   b. Predictions of producing affected children can be made in some cases.
   c. Amniocentesis can diagnose more than 100 disorders early in pregnancy. **(Fig. 14.8, p. 196; TM 70)**
   d. Chorionic villi sampling can be used as early as the eighth week of pregnancy.

C. RFLPs

1. Detection and treatment of genetic diseases will be easier once all human chromosomes have been mapped.
2. The mapping is being achieved through a technique in recombinant DNA technology called "restriction fragment length polymorphisms" (RFLPs).
   a. Restriction enzymes cut DNA at specific sites and a set of fragments is generated.
   b. Mutant alleles have unique restriction sites.
   c. This technique can be used to identify criminals from blood or semen.

Suggestions for Presenting the Material

- Students need practice to learn how the different types of inheritance (autosomal recessive, autosomal dominant, and X-linked recessive) actually influence the inheritance of a trait in real-life examples. Review, if necessary, basic genetic terms such as *homozygous, heterozygous, dominant,* and *recessive.* To see how well students understand these types of inheritance, begin by working through some simple examples (as shown in Figures 14.1, 14.2, and 14.3) of autosomal recessive inheritance, autosomal dominant inheritance, and X-linked recessive inheritance at the blackboard; ask students to predict the possible phenotypic outcomes in each example.
- During lectures, use the genetics problems at the end of the chapter as they apply, working through one or two examples at the blackboard with your class as a whole and then asking students to complete the rest in class (possibly as part of a quiz) or on their own time. Students will enjoy the puzzle-solving aspects of pedigree analysis while at the same time measuring their level of understanding of the different types of inheritance.

- Many of the genetic disorders and abnormalities mentioned in this chapter are ones whose names students have heard but whose mechanisms of inheritance were unknown to them before reading this chapter. To lend more meaning to the conditions described here, ask students to think about the social and ethical problems associated with some of the diseases mentioned in this chapter.

Classroom and Laboratory Enrichment

- Prepare unlabeled overhead transparencies of normal individuals and individuals with chromosomal abnormalities such as Turner's syndrome, Klinefelter's syndrome, or Down syndrome. Ask students if the karyotype appears normal; if not, what is wrong?
- Ask a genetic counselor to speak to your class about his/her job.
- Draw a pedigree for an unnamed genetic condition. Ask students if the disorder is autosomal dominant, autosomal recessive, sex-linked dominant, or sex-linked recessive.
- Obtain from your local health unit several brochures that explain the various genetic problems that can be inherited. Make projection transparencies and show these to the class for their evaluation and information.
- From the same source mentioned above you may be able to obtain a list of those genetic "diseases" for which there is mandatory testing (usually of newborns) in your state. What voluntary testing programs are available?

Ideas for Classroom Discussion

- Discuss the risks and benefits of amniocentesis. Would you elect to undergo this procedure (or urge your spouse to do so) if you had a history of genetic abnormalities in your family or you or your spouse were over 35? Why or why not?
- Why do harmful genes remain in the human population?
- Can you think of some of the ethical questions involved in performing genetic research on humans? As our society learns more about genetic diseases, do you think couples who plan to marry and have children should be forced to have genetic counseling or undergo amniocentesis?
- Your text refers to *phenotypic* cures for genetic diseases. What are the hindrances to the perfecting of *genotypic* cures?
- Why is hemophilia more threatening to the life of a female victim than to a male victim?
- Why do so many people insist that girls cannot be red-green color blind?
- Do you have any idea why the incidence of Down syndrome increases with the mother's age?
- What is the difference between *genetic screening* and *genetic counseling*?

Term Paper Topics, Library Activities, and Special Projects

- Construct a pedigree for your own family using a trait governed by simple Mendelian inheritance.
- Describe the history of the discovery and treatment of victims of any one of the human genetic abnormalities or disorders described in Table 14.1.
- Discover information about some of the tests now available for some genetic diseases (for example, Huntington's disorder) to determine if one is a carrier for that disease. What are some of the ethical questions raised by such tests?
- Learn more about hemophilia: Discuss its history, its role in the downfall of the Russian monarchy, and modern treatments for the disease.
- Contact your local health unit for statistics on the incidence of Down syndrome and other genetic disorders whose causes are known.

- Research the current treatment for hemophilia. Is there a difference in the regimen for males and females?
- Consult a book on medical ethics to learn of the objections that have been raised to genetic screening programs.

Films, Filmstrips, and Videos

- *Chromosomal Abnormalities—Autosomes.* INFORM, 1972, 21 minutes, 16 mm, color. *Medical Genetics Series.* Analyzes Down Syndrome, trisomy 13 and 18, and the probability of defective offspring from these and other chromosomal abnormalities.
- *Chromosomal Abnormalities—Sex Chromosomes.* INFORM, 1972, 25 minutes, 16 mm, color. *Medical Genetics Series.* Sex chromosome defects are illustrated with photomicroscopy, graphics, and animation, including the Barr Body, Lyon Hypothesis, Klinefelter Syndrome, Turner Syndrome, hermaphroditism, and nondisjunction of sex chromosomes.
- *Chromosomal Abnormalities: The Sex Chromosomes.* MFIORH, 25 minutes, 16 mm, color. Turner's, Klinefelter's, superfemale, XYY syndromes are discussed. A patient with Lesch-Nyhan syndrome is shown.
- *The Chromosomes: General Considerations.* MIFE, 1976, 14 minutes, 16 mm, color. Demonstrates the preparation of normal human karyotypes and discusses the nomenclature and normal variations. Clinical findings and cytological observations are correlated, and autosomal aneuploidy is discussed in detail. Structural aberrations such as translocations, deletions, and inversions are also presented. High verbal level.
- *Enzyme Defects and DNA.* MIFE. 1976, 15 minutes, 16 mm, color. The one-gene/one-polypeptide concept is emphasized. This film is presented on a high verbal level. The genetic basis and clinical descriptions of PKU, galactosemia, pseudocholinesterase, and hemoglobinopathies are presented to show the relationship between genes and discernible phenotypes. The helical structure of DNA is examined; how DNA stores, transmits, and translates genetic information is explained.
- *The Gene Engineers.* Time-Life, 57 minutes, 16 mm, color. Discusses the scientific, moral, and legal questions raised by the new recombinant DNA technology.

15

THE RISE OF MOLECULAR GENETICS

Revision Highlights

This well-received chapter still emphasizes how concepts are developed in the scientific community. Better explanation of Avery's key experiments; simpler illustration of Hershey-Chase experiments. Much better illustrations of semiconservative replication (showing the base pairing). Better coverage of strand assembly at replication forks.

Chapter Outline

Objectives

1. Understand how experiments using bacteria and viruses demonstrated that instructions for producing heritable traits are encoded in DNA.
2. Know the parts of a nucleotide and know how they are linked together to make DNA.
3. Understand how DNA is replicated and what materials are needed for replication.

Key Terms

bacteriophages
cytosine
thymine
adenine
guanine
purines
semiconservative replication
DNA polymerases
continuous DNA assembly
discontinuous DNA assembly

Lecture Outline

I. Discovery of DNA Function
   A. Introduction

1. In 1868, Miescher first isolated deoxyribonucleic acid or DNA from pus and fish sperm.
2. The history of DNA demonstrates that an idea develops as a community effort.

B. A Puzzling Transformation
1. In 1928, Griffith studied *Streptococcus pneumoniae,* which causes a form of bacterial pneumonia. **(Fig. 15.1, p. 200; TM 71)**
2. Strain S produced smooth colonies and was pathogenic.
3. Strain R produced rough colonies and was not pathogenic.
4. Heat-killed colonies of S were not pathogenic.
5. Live R + heat-killed S was pathogenic.
6. Genetic material from the dead S transformed the living R bacteria into a pathogenic form.
7. In 1944, Avery and his coworkers used enzymes to demonstrate that the transforming factor was DNA. But at this time most biologists believed that genes were made of protein, and few took notice of Avery's work.

C. Bacteriophage Studies
1. Delbruck, Hershey, and Luria studied viruses called bacteriophages that infect bacteria.
2. Bacteriophages adsorb to a host cell, inject their contents into the cell, and the host soon begins to produce viral proteins and nucleic acids. **(Fig. 15.2a, p. 201; TM 72)**
3. In 1952, Hershey and Chase conducted experiments that confirmed Avery's conclusion. **(Fig. 15.3, p. 202; TA 50)**
4. By first growing bacteriophages in ^{35}S or ^{32}P, it was demonstrated that DNA—not protein—enters the host cell.
5. These experiments convinced biologists that heritable traits are encoded in DNA.

II. DNA Structure

A. Components of DNA
1. Long before 1952, it was known that DNA is a polymer or nucleic acid built from four kinds of nucleotides.
2. A nucleotide consists of a five-carbon sugar, a phosphate group, and a nitrogen-containing base. **(Fig. 15.4, p. 203; TM 73a)**
3. Each nitrogen base is either adenine (A), guanine (G), thymine (T), or cytosine (C). **(Fig. 15.5, p. 203; TM 73b)**
4. Pyrimidines consist of one ring, while the purines have a double-ring structure.
5. By 1949, Chargaff demonstrated that
   a. The four bases differ in relative amounts from one species to another.
   b. In any DNA molecule, the amount of A = T, and the amount of G = C.
   c. Could the arrangements of the four bases represent the hereditary instructions?
6. Wilkins, Franklin, and others studied DNA with x-ray diffraction.
   a. This technique can reveal the position of groups of atoms in a molecule.
   b. It revealed DNA to be long and thin, with a uniform diameter of 2 nanometers.
   c. DNA also showed repeats of 0.34 nanometer and 3.4 nanometers.
   d. DNA could be helical.

B. Patterns of Base Pairing
1. In the early 1950s, Watson and Crick deduced the structure of DNA.
2. If DNA has a uniform diameter, then it could be a double helix with A pairing with T, and C with G.
3. Each pair of nucleotides on opposite strands could be held together by hydrogen bonds, like the rungs of a ladder.
4. The DNA of different species show variation in the sequence of base pairs.

III. DNA Replication

A. How DNA Is Duplicated **(Fig. 15.8, p. 205; TA 51)**
1. First, the two strands of DNA unwind and expose their bases.
2. Then, free (complementary) nucleotides pair with exposed bases.
3. Thus, replication results in DNA molecules that consist of one "old" strand and one "new" strand.

4. That is, each old strand of DNA serves as a template to form the new strand.
5. This method is called semiconservative replication.

B. A Closer Look at Replication
1. Origin and Direction of Replication **(In-text art, p. 206; TM 74)**
   a. Replication begins at an origin on the DNA molecule as the double helix unwinds.
   b. Viral or bacterial DNA has a single origin while eukaryotes may have many origins.
   c. Unwinding occurs in both directions away from the point of origin.
   d. Strand assembly occurs behind each fork as the double helix continues to unwind.
2. Energy and Enzymes for Replication **(Fig. 15.9, p. 206; TA 52)**
   a. Unwinding requires many kinds of enzymes.
   b. DNA polymerases assemble the nucleotides into nucleic acids and "proofread" the new bases for mismatched pairs, which are replaced with correct bases.
   c. The energy to drive replication is derived from splitting away phosphates from triphosphates.

Suggestions for Presenting the Material

- This chapter amplifies the information on nucleic acids presented at the close of Chapter 4. Depending on the amount of information you presented in your lectures at that time, some of this chapter could be repetitious.
- For best success in presenting this chapter, use diagrams, models, and overhead transparencies when discussing the structure and replication of DNA.
- Students find it hard to understand and identify the components of DNA, so begin this section of the text by making sure that they understand what deoxyribose (the five-carbon sugar in DNA), phosphate groups, and the four nitrogen-containing bases each look like. Briefly show overheads of the molecular structures of each of these three major players, and then introduce the term *nucleotide.*
- Ask students to think about the benefits and drawbacks of DNA as a genetic material.

Classroom and Laboratory Enrichment

- Use large three-dimensional models to show DNA structure.
- Ask students to work in teams of two. Give each student a set of labeled paper shapes representing the sugars, phosphate groups, and each of the four bases present in DNA. Ask each student to construct a short segment of a DNA strand while their partner builds the complementary strand of the DNA double helix. Then ask students to demonstrate the semiconservative replication of DNA.
- Use a film or a film loop to demonstrate the semiconservative nature of DNA replication.
- DNA is described as a "double helix" or "twisted ladder." An inexpensive device that can show this structure very well is a plastic parakeet ladder that is flexible enough to be twisted from "ladder" configuration to "helix."
- Prepare a chronological listing of the dates, people, and significant contributions to the discovery of the structure of DNA.
- In the "Term Paper Topics, Library Activities, and Special Projects" section for Chapter 10 of this manual, mention is made of the popular high school text *Modern Biology* by Moon, Mann, and Otto. Another distinction in that 1956 edition was the complete absence of any mention of DNA. Use this as a dramatic example of how much we know now compared to then.
- Redraw Figure 15.6 ("exploded" DNA structure) so that each component is on separate acetate sheets bound at the edge. To use, simply lay down each sheet as you "assemble" DNA. (You can probably make this visual aid by some inventive use of a copy machine.)

Ideas for Classroom Discussion

- What are some reasons why DNA might be double-stranded instead of single-stranded?
- When during the cell cycle does DNA replication occur?
- What are some advantages of semiconservative replication?
- Why don't the different species of single-celled and multicellular organisms have different nucleic acids for coding hereditary information? Why do they all use DNA as the hereditary material?
- What experiments done before the structure of DNA was known showed that nucleic acid was the carrier of heredity?
- How did the Hershey and Chase experiment settle the question of which molecule—DNA or protein—carries heredity?
- Why should the term *DNA relative* replace the more popular term *blood relative* when referring to human kinship?
- Which of the following is a more likely source of altered DNA sequences?
  a. New copy has error made during replication from correct original.
  b. New copy has error faithfully copied from incorrect original.
- A *casual* reading of any one of an number of biology texts would imply that Fred Griffith was a pioneer in DNA research. Is this an accurate assessment?
- What would the shape of a DNA molecule be like if purines paired with purines and pyrimidines paired with pyrimidines?

Term Paper Topics, Library Activities, and Special Projects

- What happens if DNA is damaged? How does a cell "recognize" an error in base pairing? Does an organism have ways of repairing such damage?
- Describe the research tools (such as radioactive labeling) that have been used in the past and are being used today to learn about DNA structure and function.
- Describe how the semiconservative nature of DNA replication was discovered.
- Learn more about the collaborative nature of scientific discovery using the discovery of DNA as an example. Could such a discovery have been made at the time by only one individual working alone?
- The shape and structure of bacteriophages is reminiscent of a piece of hardware designed for outer space exploration. To make it more visually appealing, construct a model from inexpensive materials.
- Prepare a synopsis of James Watson's account of the discovery of DNA as recorded in the book *The Double Helix* published by Atheneum.
- Rosalind Franklin collected data critical to the elucidation of DNA structure. However, she is hardly mentioned in textbook accounts. Locate a biography of her, and speculate on why she is lesser known than her collaborators.
- Using the diagram in Figure 15.8 as a starting point, show how copies of DNA from your great-grandparents are present in you.

Films, Filmstrips, and Videos

- *DNA: Blueprint of Life.* Wiley, 1968, 17 minutes, 16 mm, color. The DNA molecule, protein structure, and protein synthesis are explained using animated models. The film begins by indicating how DNA and chromosomes determine the phenotypes of organisms. The biochemical mechanisms of transcription, translation, and mutation are related to evolution by natural selection.

- *DNA: Laboratory of Life.* NGS, 1985, 21 minutes, 16 mm, color. Explore biochemical genetics with Dr. Linus Pauling. Winner, 1985 Blue Ribbon, American Film Festival; Cine Golden Eagle.
- *The Double Helix.* MGHT, 12 minutes, 16 mm, color.
- *The Gene Engineers.* Time-Life, 57 minutes, 16 mm, color. Discusses the scientific, moral, and legal questions raised by the new recombinant DNA technology.

16

PROTEIN SYNTHESIS

Revision Highlights

Internal reorganization and major rewriting. There is now a clear overview of the three stages of protein synthesis, with emphasis on the central roles of RNAs. Next, details of transcription, including transcript processing. Details of translation begin with an explanation of the genetic code, then codon-anticodon interactions and ribosome structure, followed by descriptions of initiation, chain elongation, and chain termination. Summary illustration of protein synthesis moved up to precede section on mutation. Illustration of frameshift mutation now in this last section.

Chapter Outline

Objectives

1. Understand how earlier experimentation led to our current understanding of biochemical genetics.
2. Know how the structure and behavior of DNA determine the structure and behavior of the three forms of RNA during transcription.
3. Know how the structure and behavior of the three forms of RNA determine the primary structure of polypeptide chains during translation.
4. Know that there are exceptions to the universality of the genetic code and know how geneticists try to account for their existence.

Key Terms

electrophoresis
template
ribonucleic acid
uracil
transcription
RNA transcripts
translation
ribosomal RNA
ribosome
messenger RNA
transfer RNA
complementary to
intervening DNA
introns
exons
genetic code
codon
ribosome
polysome
initiation
chain elongation
chain termination
gene mutation
mutagens
mutation rate
sequence
crossing over
recombination
chromosomal aberration

Lecture Outline

I. Introduction
  A. DNA is like a book of instructions written in the alphabet: A, T, G, and C.
  B. But merely knowing the letters does not tell us how the genes work.

II. One Gene, One Polypeptide
  A. In early 1900s, Garrod studied a metabolic disorder and concluded that genes function through the synthesis of a specific enzyme. Intermediate molecules collected when an enzyme was defective.
  B. Three decades later, Beadle and Tatum studied bread mold (*Neurospora crassa*) and came to the same conclusion.
    1. The mold would normally grow on a minimal medium of sucrose, salts, and biotin.
    2. A nutritional mutant was found that also required B_6; another mutant required B_1.
    3. Each inherited mutant corresponded to a defective gene . . . hence, one gene, one enzyme.
  C. Studies of sickle-cell hemoglobin showed that abnormal hemoglobin has a valine instead of a glutamate. **(Fig. 16.3, p. 209; TM 76)**
    1. Because two genes code for hemoglobin, the theory was modified to one gene, one polypeptide.
    2. Thus, one gene codes for the amino acid sequence in a polypeptide.

III. The Path from Genes to Proteins
  A. Enter RNA
    1. Recall that Watson and Crick discovered that DNA was a double helix, and that during replication one strand serves as the template for the new, complementary strand.
    2. Because protein synthesis occurs in the cytoplasm of eukaryotes, another molecule must carry the instructions of the DNA to the site of protein synthesis—RNA or ribonucleic acid.
    3. RNA consists of a single strand of nucleotides.
    4. Each nucleotide consists of a ribose sugar, a phosphate group, and one of four bases: A, C, G, and uracil (U) instead of T, where A pairs with U. **(Fig. 16.1, p. 208; TM 75a)**
    5. Thus, RNA could form using a portion of DNA as a template.
  B. Overview of Protein Synthesis **(Fig. 16.4, p. 210; TA 53)**
    1. The retrovirus responsible for AIDS is an exception to direction of information flow.
    2. During transcription, a portion of DNA unwinds and serves as a template to produce an RNA transcript.
    3. Each region of DNA can be transcribed thousands of times in the life of a cell.
    4. During translation, three types of RNA convert the message of DNA into the sequence of amino acids in a polypeptide.
      a. Ribosomal RNA (rRNA) combines with proteins to form ribosomes.
      b. Messenger RNA (mRNA) carries the "blueprint" to the ribosome.
      c. Transfer RNA (tRNA) brings the correct amino acid to the ribosomal position required for protein synthesis.

IV. Transcription **(In-text art, p. 211; TM 75b)**

A. Synthesis of RNA

1. An RNA transcript is complementary to its DNA template. **(Fig. 16.5, p. 211; TA 54)**

B. RNA Transcripts **(Fig. 16.6, p. 212; TA 55)**

1. Exons are the parts of a gene that are translated.
2. Introns, or intervening DNA, are parts of a gene that do not get translated into a polypeptide.
3. Because exons and introns are both transcribed into RNA, the primary mRNA transcript contains more than the code for the primary structure of the polypeptide.
4. After the mRNA leaves the gene, a "cap" is added to the front, and a "poly-A tail" is added to the end.
5. The cap may function as a start signal for translation.
6. The function of the tail is unknown.
7. The introns are removed and the exons are spliced together, forming the mature mRNA transcript.

V. Translation

A. The Genetic Code

1. Nucleic acids must construct "words" from four kinds of nucleotides to designate each of the twenty amino acids found in polypeptide chains.
2. A sequence of three nucleotides (a triplet) provides sixty-four choices (4^3)—more than enough to specify twenty amino acids.
3. Crick, Brenner, and others deduced that the nucleotide bases are read three at a time and that a start signal establishes the correct "reading frame."
4. The genetic code consists of sixty-one triplets that specify amino acids and three that serve to stop protein synthesis.
5. Each triplet that codes for an amino acid is a codon.
6. The code is universal for all life forms with few exceptions.

B. Codon-Anticodon Interaction

1. Each kind of tRNA has an anticodon that is complementary to a mRNA codon. **(Fig. 16.9, p. 214; TM 77)**
2. After the mRNA arrives in the cytoplasm, an anticodon on a tRNA bonds to the codon on the mRNA, and thus a correct amino acid is brought into place. **(In-text art, p. 215; TM 79)**
3. Often, only the first two bases of an anticodon must be precisely complementary; the third base may vary—the "wobble" effect.

C. Ribosome Structure **(Fig. 16.10, p. 215; TM 78)**

1. A eukaryotic cell may have tens of thousands of ribosomes, and each one has two parts.
2. The large and small subunit are both made of rRNA + proteins.
3. A cluster of ribosomes on the same mRNA is a polysome.

D. Stages of Translation **(Fig. 16.12, p. 216; TA 56)**

1. In initiation, an "initiation complex" forms before chain elongation can occur.
   a. The complex consists of mRNA, both ribosomal subunits, and tRNAs.
   b. The start codon is at the P site, and it is bound to an initiator tRNA.
2. In chain elongation, tRNAs bring appropriate amino acids to the A site.
   a. An enzyme forms a peptide bond between the new amino acid and the growing polypeptide chain.
   b. The bond between the old amino acid and its tRNA is broken, and the old tRNA leaves the complex.
   c. The tRNA holding the growing polypeptide chain moves into the P site, and the A site can now receive another amino acid bound to a tRNA.
3. In chain termination, a stop codon moves into the A site and triggers the detachment of the polypeptide chain.
   a. Enzymes called release factors are needed.

b. The polypeptide can join the cytoplasmic pool of proteins or be further processed by the cytomembrane system.

4. See Figure 16.13 for a summary of these developments.

VI. Mutation and Protein Synthesis

A. Mutation at the Molecular Level

1. A gene mutation is a change in one to several bases in the nucleotide sequence of DNA.
2. Bases can be added, deleted, or replaced.
3. Mutagens include viruses, chemicals, and ultraviolet radiation.
4. Spontaneous mutation can arise from replication errors, or can be "frameshift mutations" due to insertions or deletions. **(Fig. 16.14, p. 219; TM 80)**
5. Genetic instructions read incorrectly due to frameshift results in abnormal proteins.

B. Mutation Rates

1. The average rate is one mutation per gene per million replications.
2. Because it is unlikely for two genes to mutate at the same time, two antibiotics may be used together.

C. Mutation and Evolution

1. While mutations are rare and most are harmful, beneficial mutations have been selected for by evolution.
2. Some mutations have resulted in genetic regions with no known function.

D. Perspective

1. All life on earth shares the same chemical heritage—DNA is the source of the unity of life.
2. Mutations, genetic recombination, and other mechanisms are the source of life's diversity.
3. The changing environment is the testing ground for the novel combinations of DNA that appear.

Suggestions for Presenting the Material

- The subject of protein synthesis is a difficult one even when presented on a introductory level. Begin by very briefly summarizing the process (you can use the same linear summary drawn by Francis Crick in 1956, shown in the text as "Overview of Protein Synthesis"). You may also want to review protein structure (Chapter 4).
- Students should be able to achieve a good understanding of protein synthesis if they begin by visualizing it as two major steps, transcription and translation, rather than getting lost in complex details. The events of protein synthesis can be effectively presented with visual aids such as overhead transparencies, films, and models. Students need to have some kind of mental picture in order to understand what happens during the making of a protein.
- To emphasize the link between genes and polypeptide production, briefly present a review of what students learned about heredity in earlier chapters before you cover the work of Beadle and Tatum and the discovery of the structure of hemoglobin.

Classroom and Laboratory Enrichment

- Use models, transparencies, and films to show protein synthesis.
- Give students a "dictionary" of the genetic code (Figure 16.8) either as a handout or as an overhead transparency, and ask them to decipher a nucleotide sequence within a piece of mRNA. This can also be part of a quiz.
- Using the overhead transparency of Figure 16.13, cover several of the labels and ask students to review the steps of protein synthesis aloud in class.
- Use a model of DNA to demonstrate the changes in hereditary material described in Table 16.1.

- The following items may help your students remember the difference between "transcription" and "translation":
  a. *Transcription* involves the transfer of information from one form to another *in the same* language, for example, an office memo in shorthand transcribed into typed copy but both in English; likewise a section of genetic code in DNA is copied to RNA (both nucleic acids).
  b. *Translation* is the transfer of information in *one language* to *another language,* for example, a story in French to English; likewise genetic code in RNA is transferred to amino acids (nucleic acid to protein).
- As an aid to the understanding of protein synthesis, the following analogy in which the process is compared to the construction of a building may be useful.

a. DNA "sealed" in the nucleus	a. Master blueprints that never leave the architect's office
b. mRNA that leaves nucleus to go to ribosome	b. Blueprint copies that are taken to the job site
c. Ribosomes and rRNA	c. The construction site
d. Enzymes	d. Construction workers
e. tRNA carrying amino acids	e. Trucks carrying materials
f. Amino acids	f. Building materials

Ideas for Classroom Discussion

- Ask students to compare and contrast: transcription and translation; codons and anticodons; rRNA, mRNA, and tRNA.
- Why are ribosomes essential to protein synthesis?
- Why is transcription necessary? Why don't cells use their DNA as a direct model for protein synthesis?
- Describe the three stages of translation.
- In what ways are the instructions encoded in DNA sometimes altered?
- How might alterations in DNA structure be harmful to a species? How might such alterations be beneficial? What type of genetic change is most important for evolution?
- In most species, mutation is usually not considered an important evolutionary force. Why?
- Using the blackboard or overhead transparency, demonstrate how gene mutations such as a base substitution, a frameshift, or a transposition will produce abnormal proteins.
- In what ways does RNA differ from DNA?
- Which of the RNA's is "reusable"?
- Why do you think DNA has *introns,* which are transcribed but removed before translation begins?
- How does a *chromosomal aberration* differ from a *gene mutation*?
- If all DNA is made of the same basic building units (sugar, phosphate, and nitrogenous bases), then how can DNA differ in, say, a human and a bacterium?
- This chapter refers to the participants and process involved in protein synthesis as if they have been seen doing their work; have they? How then do we know all of this information is accurate?
- How can you explain the occurrence of birth defects (caused by altered genes) in children and grandchildren of the victims of the atomic bombs that destroyed Hiroshima and Nagasaki, Japan, when the victims themselves were only mildly affected?

Term Paper Topics, Library Activities, and Special Projects

- Describe experiments performed by Khorana, Nirenberg, Ochoa, Holley, and others to decipher the genetic code.

- Discover why repeated applications of a single drug or pesticide can result in resistance among bacterial strains and species of insects. Why does this pose a problem? What steps can be taken to avoid resistant strains of pathogenic bacteria and disease-carrying insects?
- Learn more about the discovery and treatment of sickle-cell anemia.
- What kinds of substances act as chemical mutagens? What are some of the effects mutagenic substances can have? What kinds of mutagenic agents might be found in industrial waste?
- Prepare a visual aid chart that graphically depicts the series of errors (in DNA, mRNA, tRNA, amino acids) that lead to the production of the abnormal hemoglobin in sickle-cell anemia.
- The progress in molecular biology has proceeded from deciphering genetic codes to the construction of man-made genes by machine. Report on the construction and use applications of such devices.
- Investigate reports of "gene replacement" as a preventative of possible genetic abnormalities. In what organism has it been tried? Was it successful? What are the difficulties of this procedure?

Films, Filmstrips, and Videos

- *Enzyme Defects and DNA.* MIFE, 1976, 15 minutes, 16 mm, color. The one-gene/one-polypeptide concept is emphasized. This film is presented on a high verbal level. The genetic basis and clinical descriptions of PKU, galactosemia, pseudocholinesterase, and hemoglobinopathies are presented to show the relationship between genes and discernible phenotypes. The helical structure of DNA is examined; how DNA stores, transmits, and translates genetic information is explained.
- *Genetics: Functions of DNA and RNA.* CORF, 1968, 13 minutes, 16 mm, color.
- *The Living Cell: DNA.* EBEC, 1976, 20 minutes, 16 mm, color. Simulated photomicroscopy presents the nature of the DNA code.

17

CONTROL OF GENE EXPRESSION

Revision Highlights

Major rewrite and internal reorganization. Nonthreatening introduction to concept of selective gene expression. New overview of mechanisms of regulation (defining promoter, repressor protein, activator protein, operator). Clearer picture of negative control (of lactose operon and tryptophan operon) and positive control in prokaryotes. New introduction to gene expression in eukaryotes, with examples of roles of hormones and other signaling molecules. Levels of control listed and diagrammed. Rest of chapter deals with best-understood mechanisms (transcriptional controls): those inherent in the chromosome's structural organization, as evidenced by lampbrush chromosomes, chromosome puffing, and X chromosome inactivation. New illustration of human female affected by anhidrotic ectodermal dysplasia. Brief section on alternative processing of transcripts, which is emerging as a major control mechanism. (A gene coding for a contractile protein is the example; alternative processing leads to variations that may account for variations in different types of muscles in the body.) Updated Commentary on regulatory genes and cancer.

Chapter Outline

GENE REGULATION IN PROKARYOTES
- **Mechanisms of Regulation**
- **Negative Controls**
- **Positive Controls**

GENE REGULATION IN EUKARYOTES
- **Selective Gene Expression**
- **Levels of Gene Control**
- **Chromosome Organization and Gene Activity**
- **Transcript-Processing Controls**
- **Gene Control of Cell Division**
- ***Commentary:* Altered Regulatory Genes and Cancer**

SUMMARY

Objectives

1. Know the various ways that gene activity (replication and transcription) are turned on (activated) and off (inactivated).
2. Understand how operon controls regulate gene expression in prokaryotes.
3. Be able to visualize the organization of DNA in eukaryote chromosomes and understand how it affects gene expression.
4. Understand how differentiation proceeds by selective gene expression during development.
5. Understand how cancer may result when regulatory genes are altered.

Key Terms

Escherichia coli
promoter
repressor protein
activator protein
operator
negative control
positive control
regulator gene
lactose operon
operon
tryptophan operon
differentiate
selective gene expression
signaling molecule
Dictyostelium discoideum
fruiting body
transcriptional controls
transcript processing
transport controls
translational controls
posttranslational controls
histones
nucleosome
looped domains
lampbrush chromosomes
chromosome puffs
Barr body
anhydrotic ectodermal dysplasia
tumor
metastasis
oncogenes
proto-oncogenes

Lecture Outline

I. Introduction
   A. Because all cells in your body have the same genetic instructions, only a relatively small number of genes are active at any given time in any given tissue (e.g., only red blood cells activate hemoglobin genes).
   B. Gene control occurs through molecules that interact with DNA, mRNA, or polypeptide chains. The molecules include enzymes and hormones.

II. Gene Regulation in Prokaryotes
   A. Mechanisms of Regulation
      1. Most controls affect the role of transcription.
      2. Some promoters may bind RNA polymerase more strongly than others.
      3. Some regulatory proteins may bind to the promoter or operator (between the promoter and structural genes).
         a. A repressor protein may exert negative control over genes and inhibit transcription.
         b. An activator protein may exert positive control and promote transcription.
   B. Negative Controls
      1. Lactose Operon **(Fig. 17.2, p. 223; TM 81)**
         a. Lactose, found in the milk of mammals, is also food for *E. coli.*
         b. Jacob and Monod found that *E. coli* only produces lactose-metabolizing enzymes when lactose is present.
         c. *E. coli* DNA includes a promoter, an operator, and three genes associated with lactose metabolism.
         d. A gene (or genes) together with its promoter and operator = operon.
         e. A regulator gene codes for a repressor protein.
         f. When lactose concentrations are low, the repressor binds to the operator, blocks RNA polymerase, and inhibits transcription.
         g. When present, lactose binds to the repressor and prevents it from binding to the operator, and transcription proceeds.
      2. Trytophan Operon
         a. An amino acid, tryptophan, can serve as a repressor to block its own synthesis.
   C. Positive Controls
      1. Activator proteins bind to operators and enhance the binding of RNA polymerase.
      2. Transcription slows down when the activator is removed.

III. Gene Regulation in Eukaryotes
   A. Selective Gene Expression
      1. All cells grow, they differentiate; that is, they become specialized.
      2. Differentiation arises through selective gene expression.
      3. In vertebrates, hormones and other signaling molecules alter gene expression.

4. A signaling molecule is produced in one cell and alters gene expression in target cells.
5. Target cells have receptors for the appropriate signaling molecules.
6. Somatotropin secreted from the pituitary gland increases protein synthesis in most body cells.
7. While all cells in your body have the same genes, some cells activate and suppress a fraction of the genes to produce the unique properties of different tissue and cell types.

B. Levels of Gene Control **(Fig. 17.4, p. 226; TM 82)**
   1. Transcriptional controls affect when and how a gene is transcribed.
   2. Transcript processing controls modify the initial mRNA transcripts.
   3. Transcript controls determine which mRNAs will be sent to the cytoplasm for translation.
   4. Translational controls determine which mRNAs in the cytoplasm will be translated.
   5. Post-translational controls modify the structure and function of polypeptides.
   6. Because 90+% of mammalian DNA is never transcribed, most regulatory proteins are likely to be activators.
   7. In mammals, the structural organization of the DNA affords some measure of control.

C. Chromosome Organization and Gene Activity **(Fig. 17.5, p. 227; TA 57)**
   1. Each eukaryotic chromosome contains one long DNA molecule.
   2. Proteins, including histones, organize the DNA during the cell cycle.
   3. Some histones act as spools to wind the DNA into units called nucleosomes.
   4. Folding of the beaded chain results in more organized states.
   5. Lampbrush Chromosomes
      a. Lampbrush chromosomes are found in amphibian eggs during prophase I of meiosis.
      b. They have decondensed chromosomes that resemble bristle brushes.
      c. DNA is selectively loosened by enzymes; thus genes become available for transcription.
   6. Chromosome Puffs
      a. Chromosome puffs occur in the salivary glands of some fly larvae.
      b. DNA replication produces polytene chromosomes.
      c. Puffs are caused by the hormone ecdysone.
      d. The amount of transcription occurring is correlated with puff size.
   7. X Chromosome Inactivation
      a. In mammalian females, one X chromosome is inactivated in every cell in the embryo.
      b. The inactivated X chromosome is condensed and seen as a Barr body.
      c. Because in some cells the paternal X chromosome is inactivated, while in other cells the maternal X chromosome is inactivated, each adult female is a mosaic of X-linked traits—Lyonization.
      d. The mosaic effect is seen in the black and yellow patches of a calico cat, which is heterozygous for coat color.
      e. In humans, anhidrotic ectodermal dysplasia influences the pattern of sweat glands.

D. Transcript-Processing Controls **(Fig. 17.12, p. 230; TM 83)**
   1. Transcript-processing controls are not well understood.
   2. Artificially produced mRNA transcripts that lack introns are not sent to the cytoplasm for translation.
   3. The same primary RNA transcript can be processed in different ways to produce distinct proteins.

E. Gene Control of Cell Division
   1. While not well understood, the rate of cell division for different tissues under different conditions varies.
   2. When controls are lost, cancer can occur.

Suggestions for Presenting the Material

- This chapter builds upon information that students learned in previous chapters about gene structure and function. Terms such as *DNA, mRNA, transcription,* and *translation* must be familiar before beginning this chapter.

- Emphasize to the students that the control of gene expression is an extremely complex subject area, one which is best approached by first studying some fairly simple and well-understood examples in prokaryotes.
- Give students opportunities to learn and use new words such as *promoter, operator,* and *operon.* Gene control among eukaryotes will be easier to understand if students view it as a series of levels (Figure 17.4).

Classroom and Laboratory Enrichment

- Use visual aids such as overhead transparencies to illustrate the lactose operon, chromosome organization, and levels of gene control (Figure 17.4).
- Use models to show induction and repression of gene expression in the operon.
- Modify the overhead transparency of Figure 17.2 by obscuring the labels. Ask students to identify each item on the figure.
- Prepare a summary table that lists the following:

Type of Control	Specific Example	Found In		
		Prokaryote	*Eukaryote*	*Both*
a. transcriptional				
b. transcript processing				
c. translational				
d. posttranslation				

- If you can arrange a demonstration of chromosome puffs in fruit flies that may be reared by the resident geneticist, do so. If the timing is inconvenient for actual student viewing, make a visual record of the procedure for presentation later.

Ideas for Classroom Discussion

- Scientists know much more about controls over gene expression among prokaryotes than among eukaryotes. What are some reasons why research in this area is more difficult among eukaryotic species than it is among prokaryotic species?
- Distinguish between negative gene control and positive gene control.
- What is the role of gene control in causing cancer? How are some viruses known to be linked to cancer?
- Do you think cancer-causing genes could someday be repaired?
- What would be the hypothetical effect on the lactose operon of a modified lactose molecule? Do you think it would still bind to the repressor?
- What is the "economic" advantage to a prokaryotic cell of possessing inducible enzymes?
- There is far more DNA in eukaryotic cells than scientists can label as "necessary." Why do *you* think it is there? Do you think its function is eluding us for now?
- Why do eukaryotic cells "need" *histones, nucleosomes,* and *looped domains*?
- Can you think of a more modern term to substitute for "lampbrush" in describing the looped domains that extend from chromosomes that are in transcription?
- Is there a significant distinction between "lampbrush chromosomes" and "chromosome puffs"?

Term Paper Topics, Library Activities, and Special Projects

- Learn more about current research efforts attempting to uncover the mysteries of differentiation.
- Describe the operon and its function.
- Learn more about oncogenes and cancer.
- Discover more about the discovery and diagnostic uses of the Barr body in female mammalian cells.
- What are some common substances that act as mutagens? How are some of these mutagens known to cause cancer? Are there substances that will block the effects of mutagens?
- Learn more about tumor cell lines used to study cancerous cell growth in vitro in the laboratory.
- As a student project, encourage the construction of a model of the chromosomes as illustrated in Figure 17.5. Perhaps some wire and plastic spools could be the building materials.
- Locate the original article by Francois Jacob and Jacques Monod proposing the lac operon. How have the details changed?
- Likewise see if you can locate the original research publications of Murray Barr and Mary Lyon. Notice the dates of these publications. Were they before or after the publication of DNA structure by Watson and Crick in 1953?

18

RECOMBINANT DNA AND GENETIC ENGINEERING

Revision Highlights

Entirely rewritten chapter. Starts with nonthreatening, historical perspective on genetic recombination before getting into the new technologies. Highlights critical distinction between homologous recombination and site-specific recombination, the latter illustrated by (1) transpositions, (2) plasmid insertion into bacterial chromosome during conjugation, and (3) insertion of viral DNA into bacterial chromosome. Examples of recombinant DNA technology show how restriction fragments are produced, how a DNA library is prepared and cloned, then how cloned DNA of interest is identified by nucleic acid hybridization techniques and cDNA probes. Gene amplification through polymerase chain reaction is described; problems associated with expressing cloned human genes are addressed. New section on gene sequencing methods and gene sequencer machines. Section on genetic engineering include a brief history of the "ice-minus bacteria" controversy. Chapter concludes with a balanced look at social implications of human gene therapy.

Chapter Outline

NATURAL RECOMBINATION MECHANISMS
- **Transposable Elements**
- **Plasmids**
- **Viruses as Transposable Elements**

RECOMBINANT DNA TECHNOLOGY
- **Producing Restriction Fragments**
- **Preparing and Cloning a DNA Library**
- **Identifying the Cloned DNA of Interest**
- **Selected Gene Amplification**
- **Expressing the Cloned Gene**
- **Gene Sequencing**

GENETIC ENGINEERING: RISKS AND PROSPECTS
- **Genetically Engineered Bacteria**
- **Genetically Engineered Plants**
- **Genetically Engineered Animals**
- **Human Gene Therapy**

SUMMARY

Objectives

1. Know how genetic recombination occurs naturally.
2. Understand what plasmids are, how they are instrumental in conferring resistance to drugs, and how they may be used to insert new genes into recombinant DNA molecules.
3. Know how DNA can be cleaved, spliced, cloned, used as a probe, and extracted.
4. Be aware of several limits and possibilities for future research in genetic engineering.

Key Terms

recombinant DNA technology
homologous recombination
reciprocal
transposable elements
transposons
insertion sequence
plasmids
F plasmids
R plasmids
site-specific recombination
antibiotic
genome
restriction enzymes
DNA ligase
cloning vector
DNA library
cloned DNA
reverse transcriptase
complementary DNA, cDNA
nucleic acid hybridization techniques
cDNA probe
autoradiography
polymerase chain reaction
primers
promoter
ice-minus bacteria
halophytes
glycophytes
gene therapy
eugenic engineering

Lecture Outline

I. Introduction
  A. For at least 3-1/2 billion years, mutation, crossing over, random gene mixing at fertilization, and hybridizations between species have contributed to the diversity of life on earth.
  B. By domesticating animals and plants, humans have been manipulating the genetic character of many species for at least ten thousand years.
  C. Modern medicine is a form of genetic manipulation, for it preserves genotypes that would be lost.
  D. Today, recombinant DNA technology can splice a piece of DNA from one species into the DNA of another species.
    1. Synthetic genes can make silk without silkworms.
    2. Bacteria can be turned into factories that produce human insulin.
  E. Genetic engineering has great promise for agriculture, medicine, and industry.
  F. It has also raised ecological, social, and ethical questions.

II. Natural Recombination Mechanisms
  A. Introduction
    1. Crossing over produces recombinant DNA molecules by homologous recombination.
    2. In this case, there is a reciprocal exchange.
  B. Transposable Elements **(In-text art, p. 235; TM 84)**
    1. In the 1940s, McClintock demonstrated that parts of DNA "jump" to new locations in the same DNA or a different DNA molecule.
      a. Kernels of corn with the same genotype were colored, colorless, or spotted.
      b. Random insertion of transposable elements caused mutations in some cells.
    2. Twenty years later, similar elements were found in *E. coli* and called transposons.
    3. Some complex transposons consist of a pair of insertion sequences and a gene that confers antibiotic resistance. **(In-text art, p. 236; TM 84)**
  C. Plasmids
    1. Plasmids are small pieces of circular DNA found in bacteria.
    2. Each plasmid has a single origin and insertion sequences.
    3. A plasmid can replicate independently or become integrated into the chromosome.
    4. Plasmids contain nonessential genes, and can even be spread between bacterial species.
    5. F plasmids code for proteins that promote conjugation—a transfer of DNA. **(Fig. 18.5, p. 236; TM 85)**
      a. F^+ cells can be donors.
      b. F^- cells can be recipients.
      c. Some F plasmid genes code for the proteins needed to make a pilus—a long structure that can attach to a recipient.
      d. Attachment triggers the events needed to make both cells F^+.

6. Plasmids can also become integrated into the chromosome of the recipient cell (e.g., strain Hfr of *E. coli*).
   a. Plasmid insertion into a bacterial chromosome is an example of site-specific recombination. **(Fig. 18.6, p. 237; TM 86)**
   b. Many Hfr *E. coli* strains have integrated F plasmids and serve as donors in conjugation. **(Fig. 18.7, p. 237; TM 87)**
7. R plasmids carry conjugation-promoting genes plus at least one transposon with a gene that confers resistance to an antibiotic. This is a clinical problem for many diseases, such as gonorrhea, typhoid, and meningitis.

D. Viruses as Transposable Elements
   1. Site-specific recombination also occurs between viral DNA and bacterial chromosomes.
   2. Lambda bacteriophage can kill *E. coli* by a lytic pathway.
   3. Occasionally, the lambda bacteriophage enters a lysogenic pathway and transfers the viral DNA in a latent state to progeny cells.
      a. A viral enzyme cuts the bacterial chromosome, the viral DNA inserts, and the cuts are sealed.
      b. The viral DNA can remain incorporated in the bacterial DNA for many generations or leave and start an infectious cycle.
      c. This is a common form of recombination in bacteria, yeasts, fruit flies, and mammals.

III. Recombinant DNA Technology

A. Introduction
   1. By the early 1970s, studies with *E. coli* and bacteriophages taught scientists how to use enzymes to cut and splice pieces of DNA.
   2. It became possible to isolate pieces of interest, and to map the genome of many species.

B. Producing Restriction Fragments
   1. Bacteria have nucleases called restriction enzymes.
   2. Nucleases "protect" bacteria from invading DNA sequences.
   3. The cuts can be blunt or off-center.
   4. The fragments can base-pair with any DNA cut with the same enzyme.

C. Preparing and Cloning a DNA Library **(Fig. 18.9, p. 239; TA 58)**
   1. Use a restriction enzyme to cut the DNA of *E. coli* (perhaps 1,000 fragments result) and plasmids.
   2. Mix all fragments and add DNA ligase.
   3. Thus, we have made plasmids spliced with different fragments of DNA.
      a. The plasmid is a cloning vector.
      b. The collection of DNA fragments produced by the restriction enzyme and incorporated into the cloning vector is the DNA library.
   4. Treatment of *E. coli* cells with calcium salts will permit them to take up the cloning vectors.
   5. Cells that take up the cloning vectors are transformed.
   6. Each transformed cell can produce cloned DNA.
   7. Cutting genomic DNA into a large number of fragments is "shotgun cloning."
   8. Different enzymes can be used to try to ensure that your library will contain the gene(s) of interest.
   9. Alternatively, one can begin with mRNA and use reverse transcriptase to make the desired gene.
   10. DNA polymerase will make the single strand of DNA double-stranded.
   11. Any DNA copied from an mRNA transcript is called cDNA. **(Fig. 18.10, p. 240; TM 88)**
   12. A cDNA library can also be inserted into plasmid vectors and cloned.

D. Identifying the Cloned DNA of Interest
   1. It is necessary to identify the bacteria that have taken up the recombinant plasmids.
   2. One can use plasmids with genes for antibiotic resistance, and use the antibiotic to kill all cells without a cloning vector.
   3. One can also use a radioactively labeled cDNA probe with a nucleotide sequence complementary to the gene of interest.

a. Nitrocellulose filter paper is used to make a replica.
b. The paper is treated with NaOH, which lyses the cells and denatures the DNA into single-stranded forms.
c. The radioactive cDNA probe will only pair with the complementary sequence.
d. Autoradiography will pinpoint the colony of interest.

E. Selected Gene Amplification
1. If nucleotide sequences on both sides of a gene are known, a cloned gene can be amplified by the polymerase chain reaction. (Fig. 18.12, p. 241; TM 89)
2. Short nucleotide sequences, complementary to the flanking regions, are synthesized to serve as primers.
3. The DNA of interest is heated to denature it into a single-stranded form.
4. Primers and DNA polymerase are added to produce double-stranded DNA.
5. The preparation is heated again, more primers are added, and so on.
6. The DNA polymerase is not denatured because it is derived from bacteria that live in hot springs.

F. Expressing the Cloned Gene
1. Expressing a mammalian gene in a bacterium is not easy but is possible.
2. Bacteria lack the enzymes to remove the introns.
3. If one uses cDNA, it must lie between a promoter and a stop signal.

G. Gene Sequencing
1. One can synthesize artificial genes, if one knows the nucleotide sequence of the gene in question.
2. One way to learn the nucleotide sequence of a gene is to label cloned genes with ^{32}P and divide the genes into four samples.
   a. Each sample is treated with a chemical that randomly removes one of the four nucleotides (e.g., A).
   b. Mild treatment results in fragments of different lengths.
   c. Similar fragments are produced for the three other samples.
   d. Each sample is subjected to gel electrophoresis, which separates fragments by size.
   e. Autoradiography identifies each fragment, and the sequence can be deduced.
3. Gene sequencer machines can sequence 10,000 nucleotides of DNA per day. For the human genome, at that rate it will take 100,000 years.
4. Many laboratories are collaborating to map the human genome.

IV. Genetic Engineering Risks and Prospects

A. Genetically Engineered Bacteria
1. The public is fearful that they will be dangerous.
2. Even after a harmful gene was removed from a bacterium that lives on the leaves of strawberries, the public objected to experiments—years of litigation were needed before the experiments proceeded.

B. Genetically Engineered Plants
1. *Agrobacterium tumefaciens* is a bacterium that causes crown gall tumors.
2. One can insert DNA fragments into its plasmid, after removing the tumor-causing genes.
3. This provides a way to introduce foreign genes into plant tissue that can be used to increase food production or quality.
4. Perhaps the genes of halophytes and glycophytes can be combined to produce salt-resistant, high-yield strains?

C. Genetically Engineered Animals
1. In 1982, the rat gene for somatotropin production was introduced into mouse eggs, and the rat gene was expressed.
2. In large mammals, a variety of disorders—such as arthritis—developed after genetic engineering experiments.
3. Gene incorporation could activate oncogenes, cause mutations, or alter gene expression.

D. Human Gene Therapy
The technology is still at the research stage.

Suggestions for Presenting the Material

- Help students to see the relevance of this subject by telling them about some of the products (such as insulin) that are produced as a result of genetic engineering.
- Begin by reminding students that genetic recombination occurs naturally in all organisms during meiosis. Emphasize that even though examples of genetic research in bacteria may seem obscure and of little relationship to more complex eukaryotic genomes, such experimentation yields results of great value to humans.
- Ask questions to ensure that students are indeed comfortable with new terms such as *transposition, transposons*, and *plasmids*. Use visual aids whenever possible when discussing the different types of recombination covered in this chapter.

Classroom and Laboratory Enrichment

- Illustrate the cleavage and splicing steps of recombinant DNA formation using overhead transparencies or models.
- Use overhead transparencies to show the exchange of F plasmids or R plasmids among bacteria.
- Demonstrate transfer of a plasmid for antibiotic resistance from one strain of bacteria to another using selected strains of *E. coli*. (Laboratory kits containing all necessary materials are available.)
- Fabricate workable models of recombination between phage and bacterial DNA as depicted in Figure 18.6.
- Prepare a summary table of the recombination methods listed in this chapter. Include the following information:
  a. Natural versus man-made
  b. Examples of organisms
  c. Usefulness
- Ask two groups of students to prepare brief arguments *for* and *against* the continuation of genetic engineering research and development.

Ideas for Classroom Discussion

- Where (and when) does genetic recombination naturally occur?
- What are some characteristics of an organism ideally suited for research in genetic engineering?
- What are "jumping genes"?
- Discuss the benefits of genetic engineering versus potential risks.
- Why are restriction enzymes useful tools for genetic engineering? How is DNA ligase used in genetic engineering?
- Do you think that new genomes resulting from genetic engineering should be patented? Who should receive monetary benefits from such discoveries—the research scientists performing the work or their academic institutions?
- What do you think may happen in areas of the world where different antibiotics are being used in ever-increasing amounts?
- Distinguish between the *lysogenic* and *lytic* pathways in a bacteriophage life cycle.
- What advantages would insulin produced by genetic engineering have over preparations from animal sources in the treatment of human diabetes mellitus?
- Is genetic engineering a new concept in nature or just the human application of a natural mechanism already in operation? Explain with examples.

- How is the movement of a portion of DNA in the process called *transposition* different from *translocation*?

Term Paper Topics, Library Activities, and Special Projects

- Compile a history of research efforts in genetic engineering.
- Describe problems that have resulted from the standard prophylactic use of antibiotics among farm animals such as poultry, pigs, and cattle.
- Learn more about the research of Barbara McClintock.
- Describe the safeguards currently followed in labs doing work in genetic engineering.
- List and describe companies currently doing research in genetic engineering. Investigate the current financial worth of some of these companies.
- Discuss the growing problem of antibiotic resistance among the different species of bacteria responsible for causing diseases such as gonorrhea, typhoid, and meningitis.
- Trace the history of the development and production of interferon, insulin, or any other substance produced using techniques of genetic engineering.
- In the early days of genetic recombination research, fears of creating "monster" bacteria that could run amok in the human population were quite real. However, after careful evaluation of all laboratory procedures used in thousands of experiments, these fears seem exaggerated. Why do *you* feel this thinking has changed?
- It is known that the HIV virus that causes AIDS can delay its deadly effects for some time. Search the literature to find out if this is an instance of the virus entering a latent state before resuming its attack.

Films, Filmstrips, and Videos

- *Changing Genetic Messages.* NGS, 1988, 16 minutes, sound filmstrip. DNA, gene splicing, plasmids, restriction enzymes, insulin-producing bacteria, treatment of genetic diseases, and the revolution in agriculture are included.
- *The Gene Engineers.* Time-Life, 57 minutes, 16 mm, color. Discusses the scientific, moral, and legal questions raised by the new recombinant DNA technology.
- *Issues in Genetic Engineering.* NGS, 1988, 16 minutes, sound filmstrip. Creating new life-forms and releasing altered bacteria are discussed, along with the benefits and risks.

19

PLANT CELLS, TISSUES, AND SYSTEMS

Revision Highlights

Reorganization and simplification. Starts with general characteristics of monocots and dicots, then overview of flowering plant body, with simple definitions of primary and secondary growth. Simpler coverage of plant tissues, with related illustrations grouped on the same page. Types of meristems and their location are defined and illustrated in a brief section that precedes text on tissue organization of stems, leaves, and roots. Clear coverage of primary shoot system, with new side-by-side comparisons of monocot and dicot stems (Figures 19.9 and 19.10). Better coverage of secondary growth and structure of woody stems.

Chapter Outline

THE PLANT BODY: AN OVERVIEW
- **Monocots and Dicots**
- **Shoot and Root Systems**

PLANT TISSUES
- **Ground Tissues**
- **Vascular Tissues**
- **Dermal Tissues**
- **How Plant Tissues Arise: The Meristems**

THE PRIMARY SHOOT SYSTEM
- **Stem Primary Structure**
- **Formation of Leaves and Buds**
- **Leaf Structure**

THE PRIMARY ROOT SYSTEM
- **Taproot and Fibrous Root Systems**
- **Root Primary Structure**

SECONDARY GROWTH
- **Seasonal Growth Cycles**
- **Vascular Cambium Activity**
- **Cork Cambium Activity**
- **Early and Late Wood**

SUMMARY

Objectives

1. Describe the generalized body plan of a flowering plant.
2. Define and distinguish among the various types of ground tissues, vascular tissues, and dermal tissues.
3. Explain how plant tissues develop from meristems.
4. Know the functions of stems, leaves, and roots.
5. Explain what is meant by *secondary growth* and describe how it occurs in woody dicot roots and stems.

Key Terms

vascular plants
nonvascular plants
gymnosperms
angiosperms
monocots
dicots
shoot system
root system
tissue
primary growth
secondary growth
ground tissues
vascular tissues
dermal tissues
parenchyma
collenchyma
sclerenchyma
sclereids
fibers
xylem
phloem
tracheids
vessel members
vessel
sieve-tube members
sieve cells
companion cells
epidermis
cuticle
root
hair cells
guard cells
periderm
apical meristem
primary meristematic tissues
protoderm
ground meristem
procambium
lateral meristems
vascular cambium
primary xylem and phloem
secondary xylem and phloem
vascular bundle
cortex
pith
leaf primordium, -dia
node
internode
axillary bud primordia
leaf, leaves
blade
petiole
veins
palisade mesophyll
spongy mesophyll
stoma, stomata
primary root
lateral roots
taproot system
adventitious root
fibrous root system
root cap
root epidermis
pericycle
root apical meristem
vascular column
endodermis
Casparian strip
herbaceous plants
woody plants
annuals
biennials
perennials
vertical system
ray system
cork
bark
early wood
late wood

Lecture Outline

I. The Plant Body: An Overview
  A. Introduction
    1. There are more than 275,000 species of plants—most are vascular, and fewer than 30,000 are nonvascular (algae and bryophytes).
    2. Plants live in fresh water and sea water, on land, and high above the forest floor attached to other plants.
    3. Their size varies from microscopic algae to giant redwoods.
    4. Angiosperms are vascular plants that are flowering plants, and produce seeds enclosed in tissue layers.
    5. Gymnosperms are vascular plants that are conifers, and produce "naked" seeds.
  B. Monocots and Dicots (two classes of flowering plants) **(Fig. 19.2, p. 249; TA 59)**
    1. Monocots include grasses, lilies, irises, cattails, and palms.
    2. Dicots include common trees and shrubs.
    3. Monocot seeds have one cotyledon and dicot seeds have two.
  C. Shoot and Root Systems
    1. Shoot systems are stems and leaves with a vascular system.
      a. Stems are frameworks for upright growth and to display flowers.
      b. Parts of the system store food.
    2. Root System
      a. The root system absorbs water and minerals and conducts them to aerial plant parts.
      b. It stores food and anchors the plant.
    3. Tissues form in the embryo, within the seed.
    4. During germination, cells divide and germinate at the root and shoot tips; this is primary growth.
    5. Many dicots and a few monocots also undergo secondary growth, which results in a woody increase in girth.

II. Plant Tissues

A. Ground Tissues

1. Parenchyma consists of living thin-walled cells found in most plant parts.
   a. It functions in the healing of wounds and often in regenerating plant parts.
   b. It is involved in photosynthesis, storage, and secretion.
2. Collenchyma cells become thickened at their corners and help to strengthen the plant (examples include celery strings).
3. Sclerenchyma cells have thick secondary walls impregnated with lignin.
   a. Fibers are long and tapered (e.g., hemp and flax used to make papers, textiles, thread, and rope).
   b. Sclereids are found in seed coats, nut shells, and pears.

B. Vascular Tissues

1. Both xylem and phloem contain conducting cells, fibers, and parenchyma.
2. Xylem conducts water and minerals, and mechanically supports the plant.
   a. Tracheids and vessel members (mostly in angiosperms) conduct water.
   b. Both cell types have thick, multilayered walls impregnated with lignin and other substances.
   c. Both are dead at maturity; their walls have pits and pit-pairs.
   d. Vessel members are shorter cells and have perforation plates at their ends.
3. Phloem transports sugars and other solutes.
   a. Phloem contains live conducting cells.
   b. Angiosperms contain sieve tube members with sieve plates.
   c. Gymnosperms and ferns have sieve cells.
   d. Companion cells, adjacent to the sieve tube members, help to move sugars and direct the activities of mature sieve tube members that lack nuclei.

C. Dermal Tissues

1. The epidermis covers the primary plant body.
2. A cuticle, composed of waxes and cutin, forms a noncellular surface coat on aboveground parts.
3. Root hair cells are long, thin-walled cells and enhance absorption of water and nutrients.
4. Guard cells govern water loss and CO_2 uptake.
5. The periderm replaces the epidermis when roots and stems undergo secondary growth.
   a. The periderm consists of an outer cork tissue, a cork cambium, and an inner layer of parenchyma cells.
   b. Its cell walls are impregnated with suberin, a waxy secretion that waterproofs.

D. How Plant Tissues Arise: The Meristems **(Fig. 19.8, p. 253; TA 60)**

1. Cells in root and shoot apical meristems are perpetually young and retain the capacity to divide again and again.
2. Some daughter cells in root and shoot meristems become committed to producing three kinds of primary meristematic tissues—they divide and differentiate into the primary tissues of the plant.
   a. Protoderm gives rise to the epidermis.
   b. Ground meristem gives rise to the ground tissue.
   c. Procambium gives rise to the primary vascular tissues.
3. Secondary growth originates at two types of lateral meristems.
   a. Vascular cambium forms new vascular tissue.
   b. Cork cambium forms the periderm.

III. The Primary Shoot System

A. Stem Primary Structure

1. A vascular bundle contains primary xylem and phloem.
2. In most monocots, the vascular bundles are scattered throughout the ground tissue. **(Fig. 19.9, p. 254; TA 61)**
3. In conifers and most dicots, the vascular bundles form a ring that divides the ground tissue into the cortex and pith. **(Fig. 19.10, p. 254; TA 62)**

B. Formation of Leaves and Buds

1. Leaves develop from leaf primordia as lateral outgrowths of stem apical meristems.
2. A node is the point where a leaf or leaves attach to the stem.
3. An internode is the region on the stem between two nodes.
4. Meristematic cells form axillary bud primordia—each bud develops into a branch or a flower.
5. Buds form at nodes, in the upper axils of leaves.

C. Leaf Structure
   1. Monocot and Dicot Leaves
      a. Dicot leaves have a blade and a petiole.
      b. Monocot leaves form a sheath around the stem and lack a petiole.
      c. Leaves may be simple or compound with leaflets.
   2. Leaf Internal Structure
      a. Under the upper epidermis, the palisade mesophyll—with loosely packed, photosynthetic cells—is found.
      b. Then, one finds the spongy mesophyll containing many air spaces.
      c. The lower epidermis contains stoma.

IV. The Primary Root System
  A. Introduction
     1. A four-month-old rye plant has a root system with a surface area of 639 square meters (764 square yards).
     2. The primary root system functions in the absorption and storage of photosynthetic products (carrots provide an example).
  B. Taproot and Fibrous Root Systems
     1. The primary root emerges upon germination.
     2. In most dicots and gymnosperms, the primary root decreases in diameter, and lateral roots emerge to form a taproot system (e.g., the carrot and the dandelion).
     3. In short-lived monocots, adventitious roots from the stem replace the primary root (e.g, grasses).
  C. Root Primary Structure **(Fig. 19.19, p. 259; TA 63)**
     1. Cells in the apical meristem divide and then differentiate behind the meristematic region.
  D. Root Cap **(Fig. 19.19, p. 259; TA 63)**
     1. The root cap protects the apical meristem and pushes through the soil.
  E. Root Epidermis **(Fig. 19.19, p. 259; TA 63)**
     1. Behind the elongating portion of the root, some epidermal cells develop protuberances called root hairs.
  F. Vascular Column
     1. The vascular column is arranged as a central cylinder or ring.
     2. Air spaces in ground tissues permit gas exchange.
     3. Water moves into the vascular column by way of plasmodesmata.
     4. The endodermis—the innermost layer of the cortex—helps control water movement.
     5. The suberized Casparian strip forces water to move through the cytoplasm of the endodermal cells, and all solutes are subject to transport mechanisms.
     6. Within the endodermis is the pericycle—it is meristematic and can give rise to lateral roots and the vascular and cork cambia.

V. Secondary Growth
  A. Seasonal Growth Cycles
     1. Seasonal growth cycles proceed from seed germination, to seed formation, to death.
     2. Herbaceous plants show no secondary growth.
     3. Woody dicots and gymnosperms show secondary growth.
     4. Annuals such as corn live one season and usually show no secondary growth.
     5. Biennials such as carrots live two growing seasons—first they flower, then they form seeds and die.
     6. Perennials live many years and may have secondary growth.
  B. Vascular Cambium Activity

1. Vascular cambium forms secondary xylem and phloem.
2. The xylem cells rapidly divide, force the vascular cambium outward, and can crush the older phloem cells.

C. Cork Cambium Activity
1. Increases in girth and can kill cells in the cortex and outer phloem.
2. In response, pericycle cells form cork cambium, which in turn forms periderm, the corky covering that replaces epidermis.
3. Living phloem is a thin zone under the periderm. **(Fig. 19.26, p. 263; TA 64)**

D. Early and Late Wood
1. In regions with cool winters or dry spells, the vascular cambium is inactive part of the year.
2. Early wood contains xylem with large diameters and thin walls.
3. Late wood contains xylem with small diameters and thick walls.
4. The tree rings we see appear as alternating light bands of early wood and dark bands of late wood.

E. Summary of root primary and secondary growth. **(Fig. 19.28, p. 265; TM 90)**
F. Summary of stem primary and secondary growth. **(Fig. 19.29, p. 265; TM 91)**

Suggestions for Presenting the Material

- Help students to place the kinds of plants they will be hearing about in this chapter into perspective by briefly introducing the terms *gymnosperm* and *angiosperm*.
- The information in this chapter is very visual in nature, so use overhead transparencies, Kodachrome slides, models, and diagrams whenever possible. Even if students are seeing many of these structures in lab while they are covering this chapter in lecture, they will gain much reinforcement by seeing diagrams and photos of plant cells, tissues, and systems in lecture.
- Emphasize the link between structure and function. Stress the differences between plants and animals, particularly in regard to growth. The concept of growth only at plant meristems, so different from the way in which animals grow, is initially puzzling to many students. Distinguish between the terms *herbaceous* and *woody, apical* and *lateral, primary* and *secondary.* Remind students that the structures discussed in this chapter are vegetative structures (define this term); structures used for sexual reproduction will be covered in Chapter 21.
- Introduce useful and familiar plant examples as you go through this chapter. Students are aware, of course, of the importance of plants in our day-to-day existence, but many forget the role plants play in our lives. Use "grocery store" examples; students will be intrigued to hear more about the plants and plant parts that they have taken for granted. Bring in plants and fruits and vegetables whenever possible to demonstrate plant structures.
- Another area that is familiar to students, particularly those who have done woodworking, is wood. You can use many examples in this area to introduce points on plant structure and growth.

Classroom and Laboratory Enrichment

- Distribute unusual types of plant tissues around the lecture classroom or lab room. Some possible examples include a fresh celery stalk cut in two (notice strands of collenchyma running along the "ribs" of the stalk), hemp used in making rope (fibers), cotton bolls (fibers), nutshells (sclereids).
- Obtain information about wood from lumber and paper companies. Several companies can supply posters and other materials about trees used for lumber and paper.
- In the lecture classroom, use an overhead transparency of a woody twig and ask students to point out structures such as nodes, internodes, apical meristem, terminal bud, axillary bud primordia, and terminal bud scale scars.

- Demonstrate dendrochronology. Using an overhead transparency (or, in small groups of students, an actual portion of a tree cross-section), discuss how interpretation of growth rings can reveal drought, fire, loss of trees due to disease, windfall, or harvesting, and periods of normal growth. Show students a tree-boring device; let them interpret trunk samples obtained with it.
- Pass varied samples of leaves around the room. Ask students how leaf morphology varies with climate.
- If the number of students in the class permits it and the physical environment is suitable, take a short field trip to examine the plants and trees on your campus.
- Make ample use of plant structure models that will emphasize the 3-D quality lacking in flat photos. This is especially true of conducting tissue.

Ideas for Classroom Discussion

- How did companion cells acquire their name?
- How does growth in plants differ from growth in animals? What would humans look like if they grew like plants?
- Compare the life styles of plants and animals. In what ways does the nonmobility of plants influence their structure?
- What is the difference between primary growth and secondary growth? What kind(s) of growth occur(s) in a maple tree? Where would you look in the maple tree to find primary growth? Secondary growth? What is the difference between an apical meristem and a lateral meristem?
- What causes "growth rings" in wood? Does one ring always represent one growing season? Does some wood lack growth rings? Why or why not?
- What is girdling? Ask students to think of some examples of girdling that they might have seen (damage done to the bark of small trees by electric or gas-powered weed trimmers is one example). Ask how girdling might occur in nature (nibbling of bark by deer or porcupines is one possibility).
- What is the purpose of the root cap? Is there such a thing as a "shoot cap"? Why or why not? Is there any structure on the shoot tip that is analogous to the root cap?
- What plants have root systems and shoot systems that can be used as food? Many of these plants have modified structures such as roots (for example, carrot), stems (for example, ginger), or leaves (for example, celery, an enlarged petiole) that are eaten.
- Present some familiar examples of common and/or economically important plants such as corn, tomatoes, lettuce, rice, and wheat, and ask students to classify them as monocots or dicots.
- How useful are plant parts such as leaves, stems, and roots when one is attempting to identify a plant? Are such plant parts a reliable indicator of species?
- Ask your students if pines, firs, and spruces are monocots or dicots. (The answer, of course, is that conifers are gymnosperms and hence are not classified as either monocots or dicots.) The ensuing discussion will help students to learn the differences, not yet covered in the text, between gymnosperms and flowering plants.
- When you eat an apple, what ground tissue are you eating?
- What is the biggest monocot you can think of? (Some possible answers might be bamboo or palm.)
- What is cork? Where does the cork used in wine bottles come from?
- Plants are the source of many useful products including some insecticides. Do you know the source of pyrethrum? (answer: mum flowers) Of nicotine? (answer: tobacco leaves) Of rotenone? (answer: *Derris* root)
- Why are plants "taken for granted" in our culture?
- In what ways are animals dependent on plants for survival?

Term Paper Topics, Library Activities, and Special Projects

- What are some of the woody plant species used for lumber? Find the names of as many as you can; you may be surprised at the number! Describe some of these species and discuss their uses.
- Many plant parts provide dyes that can be used to color fabrics. Prepare a demonstration or an exhibit of plant species used as sources for dyes.
- Learn more about the history of paper making, and discuss how modern papers are made.
- Describe some of the morphological adaptations found among plants of extreme climates (for example, tundra, deserts, bogs).
- Describe some of the plant species whose leaves, stems, and/or roots can be used as sources for drugs. Just a few of the many examples to include in your research are coca plant, marijuana, *Ephedra,* foxglove, and opium poppy.
- Describe the morphology, habitat, and geographic range of the insectivorous plant species found in the United States.
- Can plant parts such as leaves, roots, and stems be used to grow new plants? Discuss techniques of vegetative propagation; describe its role in commercial greenhouses and nurseries.
- Investigate the effects on plants that the gypsy moth has had in the northeastern United States.
- What factors make the vast majority of the United States suitable for intensive agriculture?

Films, Filmstrips, and Videos

- *Plant Cells and Tissues.* EI, videocassette.
- *Plant Tissue Culture: The Basic Concepts.* EI, videocassette.
- *Root Structure and Function.* EI, filmstrip or slides.

20

WATER, SOLUTES, AND PLANT FUNCTIONING

Revision Highlights

Mostly minor editing on this well-received chapter.

Chapter Outline

ESSENTIAL ELEMENTS AND THEIR FUNCTIONS

- **Oxygen, Carbon, and Hydrogen**
- **Mineral Elements**

WATER UPTAKE, TRANSPORT, AND LOSS

- **Water Absorption by Roots**
- **Transpiration and Water Conduction**
- **Cohesion Theory of Water Transport**
- **The Dilemma in Water and Carbon Dioxide Movements**

UPTAKE AND ACCUMULATION OF MINERALS

- **Active Transport of Mineral Ions**
- **Controls Over Ion Absorption**

TRANSPORT OF ORGANIC SUBSTANCES IN PHLOEM

- **Translocation**
- **Pressure Flow Theory**

SUMMARY

Objectives

1. Know which elements are essential to plant health.
2. Explain how water is absorbed, transported, used, and lost by a plant.
3. Describe how the intake of CO_2 is connected with water loss.
4. Explain how essential mineral ions are taken up by a plant.
5. Know how translocation of organic substances occurs, according to the pressure flow theory.

Key Terms

plant physiology
mineral ions
macronutrients
micronutrients
nodules
turgor pressure
mycorrhiza
transpiration
cohesion theory of water transport
cuticle
stoma, -ata
abscisic acid
translocation
source regions
sink regions
pressure flow theory

Lecture Outline

I. Introduction
  A. The minerals, water, and gases that plants need may often be in short supply.
  B. Many structures and functions of plants are in response to scarce resources.
    1. The central vacuoles increase cell volume and hence increase the surface area for absorbing materials.
    2. Thin flat leaves are adapted for absorbing sunlight and carbon dioxide.
    3. Root systems grow so as to increase the absorptive surface.
  C. Plant physiology considers how plant systems respond to the environment.

II. Essential Elements and Their Functions
  A. Oxygen, Carbon, and Hydrogen
    1. These elements make up 96% of a plant's dry weight.
    2. Oxygen comes from water (also the source of H), O_2, and CO_2, which is also the source of carbon.
  B. Mineral Elements
    1. Thirteen essential inorganic substances are available to plants in ionized form ("mineral ions").
    2. Macronutrients are each at least 0.1% of dry weight.
    3. Micronutrients are a few parts per million of a plant's dry weight.
    4. Nitrogen is available to plants by the actions of microbes that convert nitrogen in the air to NH_4^+. Many of these microbes live symbiotically in the nodules of the roots of legumes.
    5. Minerals often function to activate enzymes of protein synthesis, photosynthesis, and other pathways.

III. Water Uptake, Transport, and Loss
  A. Water Absorption by Roots
    1. Annual grasses have a branched fibrous root system near the soil surface.
    2. Dicots have a taproot system that penetrates more deeply.
    3. Billions of root hairs may develop in a single system.
    4. Roots grow toward regions of water and dissolved minerals.
    5. A mycorrhiza is an association between a fungus and a vascular plant that aids the higher plant in taking up water and mineral ions.
  B. Transpiration and Water Conduction **(Fig. 20.6, p. 271; TA 65)**
    1. Water moves from roots to stems, and then to leaves.
    2. Some water is used for growth and metabolism, but most evaporates into the air by transpiration.
    3. Water is pulled upward by continuous negative pressures that extend from leaf to root.
  C. Cohesion Theory of Water Transport **(Fig. 20.6, p. 271; TA 65)**
    1. Hydrogen bonds help to pull the continuous water column up the plant.
    2. Transpiration causes the state of tension.
  D. The Dilemma in Water and Carbon Dioxide Movements **(Fig. 20.7, p. 273; TA 66)**
    1. If water loss by transpiration exceeds water uptake by the plant, dehydration and death can result.
    2. The cuticle, a waxy covering, conserves water but limits diffusion of CO_2 into the leaf.
    3. Transpiration and CO_2 uptake occur at leaf openings called stomata.
    4. When the two guard cells of each stoma swell with water, a gap is produced.
    5. When the two guard cells of each stoma lose water, the stoma closes.
    6. A stoma opens and closes according to how much water and CO_2 are present in the guard cells.
    7. Stomata open at sunrise because in response to photosynthesis using up CO_2, K^+ is pumped into the cells and water follows.
    8. At night, CO_2 levels increase, K^+ leaks out, and water follows—the stomata close.
    9. Abscisic acid is a hormone produced under stress that causes guard cells to lose K^+.

IV. Uptake and Accumulation of Minerals
   A. Active Transport of Mineral Ions
      1. Water uptake requires an osmotic gradient.
      2. An osmotic gradient is produced if cells have a high solute concentration.
      3. Solutes, especially mineral ions, are actively pumped into cells by membrane pumps that use ATP.
      4. The ATP is produced from photosynthesis and aerobic respiration.
   B. Controls Over Ion Absorption **(Fig. 20.9, p. 274; TM 92)**
      1. Solute absorption and accumulation must be coordinated throughout the plant.
      2. For example, root cells receive sugar from leaves during the day.

V. Transport of Organic Substances in Phloem
   A. Transport of organic molecules requires that storage starch, fats, and proteins be converted to smaller subunits that are soluble and transportable.
   B. Translocation
      1. Organic molecules travel from photosynthetic sites to organs that need them by translocation.
      2. The term "translocation" is most often used to signify the transport of sucrose and other compounds through phloem.
         a. Sieve tube cells are alive at maturity, and are interconnected from leaf to root.
         b. The transport of water and organic molecules occurs at up to 100 centimeters per hour.
         c. Observations of aphids provided translocation information.
   C. Pressure Flow Theory **(Fig. 20.11, p. 276; TM 93)**
      1. Movement of molecules through phloem follows a "source-to-sink" pattern.
      2. Leaves are sources and sinks may be fruits, seeds, or roots.
      3. Translocation through the phloem depends on pressure gradients between source and sink regions.
      4. Active transport moves solutes into sieve tubes and water follows.
      5. Active transport moves solutes into sink tissues and water follows.

Suggestions for Presenting the Material

- Students should be able to make an easy transition to this material if the role of function in determining plant morphology was stressed when the previous chapter was discussed. Emphasize that plants are supported by a "skeleton" formed by a continuous column of water. Review, if necessary, some of the terms learned earlier, such as osmosis and turgor pressure, that relate to water movement.
- To help students understand the large surface area of a plant's root system, provide data on the surface areas of some typical plant root systems. Ask them to guess the ratio of shoot surface area:root surface area of a typical plant.
- This chapter provides many good opportunities to discuss the selective role of the environment in shaping such features as stomata and root systems.

Classroom and Laboratory Enrichment

- Demonstrate the abundance and fragile nature of root hairs. Germinate radish seedlings in a petri dish lined with paper toweling. Pass the dish around the classroom, and allow each student to take a seedling and examine the root hairs. What happens to the root hairs minutes after the seedling is removed from the dish? Why?
- Demonstrate how species adapt to their surroundings by discussing the number, size, location, and distribution patterns of stomata in leaves of different species. Include some unusual examples, such

as aquatic plants with stomata on upper leaf epidermis, conifers with sunken stomata, and plants with pubescent leaves.

- Examine tomato seedlings suffering from a deficiency of one of the macronutrients. This can be prepared as a lab experiment of several weeks duration, or a demonstration (or 35 mm slides) of the results can be shown instead.
- Use a simple soil-testing kit in lab to test samples of several different local soils. What are some steps that could be taken to improve each of the soils tested, if necessary?
- Set up demonstrations of root pressure or transpiration in lecture or lab.
- Sketch a diagram of a sieve tube of the phloem on the board or on an overhead transparency (your sketch can be similar to Figure 20.11, except that solute molecules and water molecules should be omitted). Begin to add water molecules and solute molecules at the source. Then ask your students to tell you what will happen next step-by-step as sugars move from the source (for example, leaf cells) to the sink (for example, root cells). Ask your students to think of an easy-to-see example of the result of the pressure flow theory (a potato is one answer).
- Show 35 mm slides or transparencies of chloroplasts containing starch grains. Where did the starch come from? Would you be more likely to see such grains in the morning or in the afternoon? What will happen to the starch grains?
- In lab, provide prepared microscope slides of the undersides of plant species from different environments. In lecture, show 35 mm slides or diagrams of the lower leaf epidermis.
- Compare the rates of recovery after wilting among three tomato seedlings that have each been cut off at the base of the main stem as described below and then placed in a beaker of water: (1) seedling is cut; (2) seedling is cut while plant is briefly submerged underwater; (3) seedling is cut, then allowed to sit on desktop for fifteen minutes before being placed in water. Which seedling exhibits the least amount of wilting at the end of the lab period? The most? Why?
- Test slices of various fruits and vegetables for starch content by applying an iodine solution (should turn blue-black).
- Obtain a chart showing color photos of the symptoms of mineral deficiencies as listed in Table 20.2. Perhaps a plant nursery or fertilizer supplier can help you. If you cannot obtain a copy of some chart you would like, ask permission to photograph it for a slide.

Ideas for Classroom Discussion

- How do the tracheids and vessel members found in xylem conduct water even though they are dead at maturity?
- Desert plants must balance the need for carbon dioxide against the threat of desiccation. What are some adaptations of desert plants that allow them to open their stomata often enough to get the carbon dioxide sufficient for photosynthesis? Discuss how alternative photosynthetic pathways such as C4 and CAM photosynthesis have evolved in response to environmental pressures.
- What happens to transpiration rates on hot days? Dry days? Humid days? Breezy days?
- Ask students who have raised tomatoes or other garden plants if they have ever observed "midday wilt," a phenomenon in which even well-watered plants temporarily wilt during the late afternoon. Ask them why this happens. (Midday wilt occurs when transpiration exceeds the rate of water uptake.)
- What are some crop plants that are particularly adept at storing sugars or starches?
- Reports of topsoil erosion and mineral leaching are numerous. Are there finite amounts of soil and soil nutrients? Will we ever completely exhaust our sources of plant sustenance?
- How do plants combat insect pests in nature? How do we help plants resist insect attack?

Term Paper Topics, Library Activities, and Special Projects

- Discuss the role of each of the macronutrients in plant metabolism and growth.
- What role has the fibrous root system of grasses played in the establishment and maintenance of prairies in the United States? What happens to the species composition of prairies if such areas are interrupted by roads, farming, or railroads? Discuss the history of the American prairies.
- Describe the role of mycorrhizae in successful seedling growth among species of gymnosperms. How do commercial lumber companies ensure that the proper mycorrhizal fungi will be present on the roots of their tree seedlings?
- Learn more about soil testing. Describe how a typical soil-testing kit works.
- Visit a nursery or garden center where lawn and garden fertilizers are sold. List the N-P-K ratio for each of the different fertilizers. Explain differences among N-P-K ratios of fertilizers for lawns, vegetables, and flowers. Why do fertilizers for different purposes have such different N-P-K ratios? Summarize the roles of nitrogen, phosphorus, and potassium in plant functioning and development.
- What are the effects of acid rain on plant functioning?
- Discuss the effects of extremely cold climates, such as Arctic tundra, on the ratio of shoot systems:root systems. Why is so much of the plant underground in such climates?
- Locate a report (USDA documents?) showing the decline in soil fertility in the United States in the past 100 years.
- Insects such as aphids can be controlled by insecticides introduced into the plant via uptake by the roots. They are called "systemic" insecticides. How do these chemicals accomplish their control? Give an example.

Films, Filmstrips, and Videos

- *Carnivorous Plants.* NGS, 1974, 12 minutes, 16 mm. Shows the relationship between form and function. Demonstrates the physiological aspects of obtaining food.

21

PLANT REPRODUCTION AND EMBRYONIC DEVELOPMENT

Revision Highlights

Mostly minor editing on this chapter. Cleaner sporophyte-gametophyte distinction in introductory paragraphs on life cycles.

Chapter Outline

Objectives

1. Describe the typical patterns of life cycles in flowering plants.
2. Draw and label the parts of a perfect flower. Explain where gamete formation occurs in the male and female structures.
3. Define and distinguish between *pollination* and *fertilization.*
4. Trace embryonic development from zygote to seedling.
5. Describe the various styles presented by asexual reproduction in flowering plants.

Key Terms

flower	alternation of generations	receptacle	ovary
gametophyte		carpel	stigma

style	pollen tube	double fertilization	fruit
stamen	pollen grain	diploid	simple fruit
filament	nucellus	triploid	aggregate fruit
anther	integuments	vector	multiple fruit
petal	megaspores	coevolution	runners
sepals	embryo sac	nectary	parthenogenesis
microspores	endosperm	nectar	vegetative reproduction
pollen sac	female gametophyte	seed	clone
pollination	pollination		

Lecture Outline

I. Introduction
   A. Flowering plants, like humans, engage in sex and the female organs house the embryo during early development.
   B. Flowers attract pollinators that help bring sperm and egg together.
   C. Plants also reproduce asexually by mitosis.
   D. We will examine sexual reproduction and embryonic development in angiosperms.

II. Sexual Reproduction of Flowering Plants
   A. Life Cycles of Flowering Plants
      1. The sporophyte is diploid and consists of roots, stems, and leaves.
         a. The sporophyte produces flowers, within which some cells undergo meiosis to form haploid meiospores.
         b. Each meiospore divides by mitosis to form haploid gametophytes (male and female).
      2. Hence, there is an alternation of generations between the large sporophytes and the tiny gametophytes.
   B. Floral Structure
      1. Most floral parts are attached to the receptacle, the modified end of the floral shoot.
      2. A carpel is a female structure where eggs develop, fertilization occurs, and seeds mature.
      3. In the ovary of a cherry there is one ovule; peas have several ovules.
      4. The stigma is a "landing platform" for pollen.
      5. The style joins the stigma to the ovary.
      6. The stamen is a male structure and consists of a filament and an anther that contains pollen sacs.
      7. Collectively, petals form the corolla.
      8. Collectively, sepals form the calyx.

III. Gamete Formation
   A. Microspores to Pollen Grains
      1. In anthers, each diploid mother cells divides by meiosis to form four haploid microspores.
      2. Each microspore divides, and a pollen grain develops.
      3. One cell will form two sperm; the other will form the pollen tube.
      4. Each pollen grain is a sperm-bearing gametophyte.
   B. Megaspores to Eggs
      1. In the carpel, a mass of tissue forms an ovule that consists of an inner nucellus, integuments, and a micropyle.
      2. In the nucellus, a $2n$ mother cell divides by meiosis to form $1n$ megaspores.
      3. In most flowering plants, three of the four megaspores die.
      4. The remaining nucleus divides three times to form a single cell with eight nuclei.
      5. The nuclei migrate, and cytoplasmic division occurs to form the embryo sac—the female gametophyte.
      6. The endosperm mother cell contains two nuclei and will give rise to the endosperm.
      7. The egg cell is near the micropyle.

IV. Pollination and Fertilization
  A. Pollination and Pollen Tube Growth
    1. Pollination is the transfer of pollen to the surface of the stigma.
    2. After germination, a pollen tube forms, producing a path that the two sperm will follow to the ovule.
  B. Fertilization and Endosperm Formation
    1. The pollen tube enters through the micropyle and the sperm are released.
    2. Double fertilization occurs—one sperm fertilizes the egg, and a diploid zygote results.
    3. The other sperm fuses with the two nuclei of the endosperm mother cell, and a triploid primary endosperm cell results.
    4. What is the relationship between floral structure and pollinators?
  C. Case Study: Coevolution of Flowering Plants and Pollinators
    1. Origins of Pollination Vectors
      a. When plants first invaded the land some 400 million years ago, insects evolved to feed on this new niche.
      b. Pollen is a good source of protein and insects coevolved in response to the new food supply.
      c. Hence, in addition to passive agents like the wind, insects became vectors of pollen.
    2. Nectar as an Attractant
      a. A nectary secretes nectar—a fluid rich in sugar, proteins, and lipids.
      b. Between 60 and 40 million years ago, nectar began to attract insects, birds, and bats.
    3. Existing Flowers and Their Pollinators
      a. Red flowers attract birds.
      b. Many insects are attracted to odors.
      c. Wind disperses the flowers of many grasses.

V. Embryonic Development
  A. From Zygote to Plant Embryo
    1. The zygote differentiates such that the upper half contains most of the organelles; the upper half divides to give rise to the embryo.
    2. The lower half has a large vacuole and some daughter cells give rise to the suspensor, which transfers nutrients from the parent plant to the embryo.
  B. Seed and Fruit Formation
    1. After double fertilization, the ovule expands, the integuments harden and thicken, and the endosperm develops.
    2. The mature ovule is a seed; the integuments form the seed coat.
    3. In *Capsella,* the cotyledons absorb the endosperm.
    4. In corn, the endosperm remains filled with nutrients.
    5. A fruit = the seed + the mature ovary.
    6. Fruits function in seed dispersal and protection.

VI. Asexual Reproduction of Flowering Plants
  A. Strawberry plants reproduce by horizontal stems or runners.
  B. Oranges reproduce by parthenogenesis.
  C. Vegetative reproduction occurs in some wounded plants, such as the jade plant.
  D. Many food crops, such as MacIntosh apples and Bartlett pears, are from clones made from cuttings.
  E. Steward demonstrated that one can grow clones of carrots in tissue culture.

Suggestions for Presenting the Material

- Students frequently have trouble understanding the plant life cycle because it is so different from animal life cycles. Be sure to emphasize the importance of alternation of generations; spend plenty of time going over a diagram of the plant life cycle like the one shown in Figure 21.2. Compare it to a drawing or diagram of the human life cycle to give students a familiar reference point.

- Give examples of a sporophyte and a gametophyte in the life cycles of some common plants. Explain the difference between a spore and a gamete. It may be necessary to briefly review meiosis I and II and terms such as *haploid* and *diploid*, especially if it has been a long time since the students covered mitosis and meiosis.
- Students will find it easier to comprehend the development of male and female gametophytes if they have dissected a flower before this is discussed. If students wonder why the events of gamete formation and fertilization are so complicated, remind them that these are very advanced plants; plant evolution and diversity will be discussed in Unit Seven.

Classroom and Laboratory Enrichment

- Show line drawings or color photos of any angiosperm in flower, and ask students to point out the gametophyte and sporophyte portions of the plant.
- Use models, photos, and diagrams to present floral structure. Then dissect at least one type of flower, preferably one with large parts, in lab or lecture.
- Show time-lapse films of flower and fruit formation.
- Display some inflorescences of grasses (if available) in the lab or lecture classroom. Use diagrams or photos of corn inflorescences as examples of imperfect flowers; corn is a good choice because of the large "tassels" (staminate flowers) and because students are familiar with the ear of corn, the end product of many fertilization events.
- Learn the parts of a seed by dissecting bean and corn seeds after they have been softened in water for several hours.
- Use microscope slides and/or 35 mm slides of lily to discuss microspore and megaspore development.
- View germinating pollen grains under the microscope in lab or with a microprojector in lecture.
- Prepare a demonstration of different kinds of fruits, or ask students to bring fruits to class or lab. Students will be fascinated to see the relationship between the parts of the flower before fertilization and the subsequent fruit parts.
- Use techniques of vegetative propagation or tissue culture to grow new plants. This can be done by the students in lab or can be prepared ahead of time, ideally with different stages of growth represented, as a demonstration.
- Show scanning electron micrographs of the pollen grains of some familiar plants.
- Show close-up photos of bees and the pollen-carrying devices on their legs. Or if your group is small, place actual specimens of these insects under stereomicroscopes.
- Locate photos of flowers that are formed in various ways to lure insects to alight and feed.

Ideas for Classroom Discussion

- Can you tell by a flower's appearance if it is wind-pollinated or insect-pollinated? What are some clues suggesting that a flower is wind-pollinated? What are some floral features suggestive of insect pollination?
- Why do botanists use flowers rather than leaves or stems as indicators of species identity?
- Distinguish between megaspores and microspores.
- What is included in an ovule of a flowering plant? What events take place inside the ovule? What are some of the changes that take place inside an ovule after fertilization has occurred?
- Is a pollen grain analogous to a human sperm cell? Why or why not?
- Discuss floral diversity. Ask students to name different kinds of flowers, and then discuss the shapes, types of parts, and pollination of each.

- What is the difference between pollination and fertilization?
- What is double fertilization?
- Cut open apples, pears, green beans, strawberries, oranges, pineapples, and any other fruits or vegetables you have on hand in class or lab. Identify the floral origin of as many parts as you can. For example, which part of the fruit was the ovary wall? The carpel? The integument? The sepal? The stigma and style?
- Define the terms *fruit* and *vegetable.* What is the difference between these two terms? Why do we use the term *fruit* when we are talking about tomatoes and green beans? Ask students for examples of fruits, and use Table 21.2 in class to help classify each one.
- Why do many fruits change from green to red, yellow, or orange as they ripen? Of what adaptive value is the sweet smell produced by most fruits?
- Compare and contrast the advantages and disadvantages of sexual reproduction in flowering plants. What are some of the advantages and disadvantages that asexual reproduction offers to flowering plants? Why do so many species of flowering plants reproduce both sexually and asexually?
- Is there any truth to the disgusting allegation that honey is actually "bee vomit"?
- After studying this chapter, do you see flowers as providing beauty to enhance the human experience or do you perceive them as more functional that that? Explain.
- Why do you think there are poisonous seeds, berries, leaves, stems, and so on in the plant kingdom?

Term Paper Topics, Library Activities, and Special Projects

- Examine diversity among flowers. Describe the coevolution of flowering plants and insects.
- Identify as many of the flowering plants on campus as you can. If flowers are available, collect them, identify them, and bring them back to class or lab. One group of students could prepare a map of campus trees and plants and lead other groups in a campus plant walk.
- Select a well-known flowering plant species, and prepare a timetable showing when the reproductive events leading up to seed formation occur.
- Learn more about some of the reproductive isolating mechanisms that discourage self-pollination among many flowering plant species.
- Describe commercial uses of asexual reproduction in flowering plants.
- Describe the different uses of each part of the wheat grain. What is wheat germ? Wheat bran? How does white flour differ from whole wheat flour?
- Why are commercial bananas and many commercial varieties of grapes seedless? How are seedless varieties of normally "seedy" fruits such as watermelons created?
- Discuss the role of evolution in floral diversity. What is an example of a primitive flower? An advanced flower?
- Learn more about the role of pollen in human allergy. Collect daily and monthly information on the types and amounts of pollen found in your area. What species have the most pollen? What months of the year have the highest pollen counts? Should we regulate the planting of non-native ornamental vegetation in cities and towns in states such as Arizona where people have moved seeking low pollen counts?
- Discuss the use of pollen grains as indicators of the past vegetation history of an area.
- Describe different seed dispersal mechanisms used by flowering plants. Make a list of dispersal mechanisms found among local plants or plants on campus.
- Research the process by which bees make and store honey.
- Document the role of animals in seed dispersal and the extent to which this helps plants.

Films, Filmstrips, and Videos

- *Angiosperms—The Flowering Plants.* EBEC, 1962, 21 minutes, 16 mm, color. Describes the structural and reproductive characteristics that distinguish angiosperms from other plants. Animated and time-lapse photography are used to trace the processes of pollination, formation of seeds and fruits, seed dispersal, and plant growth in angiosperms. Illustrates a wide variety of angiosperms and explains their importance to man.
- *Development and Differentiation.* CRMP (MGHT), 1974, 20 minutes, 16 mm, color. Discusses the roles of DNA and the cytoplasm in development and follows a frog embryo through gastrulation.
- *The Growth of Plants.* EBEC, 1961, 21 minutes, 16 mm, color. Illustrates the growth process in plants and shows how cell division, elongation, and differentiation contribute to stem and root growth.
- *How Pine Trees Reproduce.* EBEC, 1963, 11 minutes, 16 mm, color.
- *Morphogenesis in a Marine Alga, Caulerpa.* HRW, 1972, 5 minutes, 16 mm, color. Growth and differentiation in a coenocytic marine alga are shown by time-lapse photography.

22

PLANT GROWTH AND DEVELOPMENT

Revision Highlights

Minor changes only. Apical dominance added to the chapter.

Chapter Outline

Objectives

1. Describe the general pattern of plant growth and list the factors that cause plants to germinate.
2. List the various chemical messengers that regulate growth and metabolism in plants. Explain how plants respond to changes in their environment.
3. Know the factors that cause a plant to flower, to age, and to enter dormancy. Describe each process.

Key Terms

seed germination
imbibition
radicle
primary root
elastic
plastic
true growth
hormone
target cell
auxins
gibberellins
gibberellic acid
cytokinins
zeatin
abscisic acid, ABA
abscission
ethylene
coleoptile
apical dominance
phototropism
flavoprotein
gravitropism
thigmotropism
biological clock
circadian rhythms
seasonal adjustments
photoperiodism
phytochrome
vernalization
long-day plants
short-day plants
day-neutral plants
florigen
senescence
dormancy

Lecture Outline

I. Seed Germination
   A. Seed germination is influenced by water, temperature, and light.
   B. Often the spring rains trigger germination—the imbibition of water due to hydrophilic groups of stored proteins.
   C. The seed splits and O_2 moves in and triggers aerobic metabolism.
   D. The embryo increases in size—the seedling results.

II. Patterns of Growth **(Fig. 22.2, p. 293; TM 94)**
   A. First, radicle cells grow and produce the primary root; the seed completes germination when the primary root protrudes.
   B. True growth is driven by water uptake; young cells have elastic walls that stretch and gain more wall material as water enters.

III. Plant Hormones
   A. Types of Plant Hormones **(Table 22.1, p. 294; TM 95)**
      1. A hormone is released from one cell and affects the growth and development of target cells.
      2. Target cells have receptors for the signaling molecules in question.
      3. Auxins, like indoleacetic acid, promote stem elongation; and synthetic auxins are used as herbicides to selectively kill dicots.
      4. Gibberellins, like auxins, also promote stem elongation.
      5. Cytokinins, like zeatin, stimulate cell division.
      6. Abscisic acid (ABA) promotes stomatal closure, seed and bud dormancy, and resistance to water stress.
      7. Ethylene (C_2H_4) stimulates the ripening of fruit.
   B. Examples of Hormonal Action
      1. In grasses, a coleoptile protects the tender leaves as they push through the ground. **(Fig. 22.3, p. 294; TA 67)**
      2. Sunlight stops the expansion of the coleoptile, and the leaves break through and begin photosynthesis.
      3. The auxin IAA stimulates coleoptile growth by making the cell walls more plastic.
      4. Gibberellin can overcome genetic dwarfism in corn.
      5. Hormones also promote growth of a soybean stem; an auxin is synthesized near the stem tip and in young leaves. **(Fig. 22.6, p. 296; TM 96)**
      6. Pinching off shoot tips results in increased branching below—caused by an unknown stem tip hormone that normally inhibits lateral bud growth. The natural effect is known as apical dominance.

IV. Plant Responses to the Environment
   A. Plants can interact with the environment and adjust their patterns of growth.
   B. For example, if a paper bag falls on top of a seedling, the shoot will bend and grow out from under the bag.
   C. The Many and Puzzling Tropisms
      1. Tropisms have unknown origins.
      2. Phototropism occurs when a plant grows or bends toward a source of light (this is related somehow to IAA).
      3. Gravitropism is a growth response to gravity—shoots grow up and roots with root caps grow down.
      4. Thigmotropism is unequal growth triggered by physical contact; it is found in climbing vines.
   D. Response to Mechanical Stress
      1. Rain, grazing animals, farm machinery, and winds can inhibit growth.
      2. Shaking plants can have the same response.

E. Biological Clocks in Plants
   1. Biological clocks are an internal time-measuring mechanism with a biochemical basis.
   2. For example, flowers open in the morning and close at night, independently of the light cycle.
   3. Circadian rhythms occur about every twenty-four hours.
   4. Somehow, the biological clocks must be reset as the seasons change.

F. Photoperiodism
   1. Photoperiodism is a biological response to a change in relative length of daylight and darkness in a twenty-four-hour cycle.
   2. Phytochrome, a blue-green pigment, is turned on by red light (Pr) and turned off by far-red light (Pfr). **(Fig. 22.12, p. 300; TM 97)**
   3. Photoperiodism controls stem elongation and branching, leaf expansion, and the formation of flowers, fruits, and seeds.
   4. Plants placed in the dark have greater stem elongation and less leaf expansion or stem branching.
   5. Phytochromes are embedded in cell membranes and may control which hormones can bind to membranes or move into cytoplasm.

V. The Flowering Process
   A. A mature flowering plant produces flowers, seeds, and fruits.
   B. Some plants need a cold winter to produce spring flowers; this is known as vernalization.
   C. Other plants use clues of daylength—phytochrome seems to play a role. **(Fig. 22.14, p. 301; TM 96)**
      1. Long-day plants, such as spinach, flower in the spring.
      2. Short-day plants flower in late summer or early fall (e.g., cocklebur).
      3. Day-neutral plants, such as the tomato, flower whenever they are mature; they are common near the equator.

VI. Senescence
   A. Plants really invest in reproduction; annuals and most perennials end up with dead leaves due to redistribution of nutrients to reproductive parts.
   B. Deciduous species transport the nutrients out of leaves that drop by abscission. Ethylene may be used to break down polysaccharides in the cell walls, resulting in dropping off of leaves, flowers, and fruits.
   C. Senescence is the total of processes leading to the death of a plant or plant part.
   D. It seems to be triggered by the growth of reproductive parts, since if flower buds are removed, stems and leaves stay healthy longer.
   E. More is involved; if one causes a cocklebur to flower, the leaves die regardless if young flowers remain on the plant.
   F. There seems to be a death signal that can even overcome cytokinins that delay senescence.

VII. Dormancy
   A. In autumn, daylength shortens and growth stops in many trees and nonwoody perennials.
   B. Dormancy occurs even though temperatures are moderate and water is available.
   C. In Douglas fir plants, a short period of red light during the dark period can overcome the effects of short daylength.
   D. The red light converts phytochrome into the active form.
   E. In spring, dormancy is broken by the number of hours of exposure to cold weather.
   F. Gibberellins can break dormancy in many species.
   G. Abscisic acid extends dormancy and can counteract the effects of gibberellins.
   H. Seeds may also require a period of cold or depend upon red wavelengths to break dormancy.
   I. Case Study: From Embryogenesis to the Mature Oak
      1. Ten million years ago, coast live oak evolved in the hills of central California.
      2. An acorn buried by a bird and forgotten grew into a new tree.

3. Red wavelengths of sunlight activated phytochromes that triggered hormonal events causing maturation.
4. Delicate relationships between the 300-year-old tree and the environment are eventually upset by human development, and the tree dies to become firewood.

Suggestions for Presenting the Material

- Discuss with your students once again how growth in plants differs from growth in animals.
- Introduce the term *hormone* and discuss the need for growth-regulating hormones. There are many lab experiments designed to investigate the role of hormones in plant growth and development; many of them are available from biological supply houses in the form of kits. These can be prepared ahead of time and used as demonstrations or performed by the students.
- Films and videotapes are also successful ways to present plant growth and development.

Classroom and Laboratory Enrichment

- For a demonstration of growth at apical meristems, mark the root tips or shoot apical meristems of sturdy just-sprouted seedlings (beans or peas are fine) at measured intervals with india ink. Keep the seeds moist in a petri dish lined with paper toweling. Measure the rates of growth after one week. Exactly where along the root or shoot is the increase in length occurring?
- Determine if the food stored in sprouting seeds is starch or sugar by treating the seeds with tetrazolium solution and Lugol's solution.
- Design and implement experiments involving seed germination. Vary conditions such as temperature or light, and examine the effects on germination rate.
- In species with hard seed coats, examine the effects of seed scarification (the removal or breakage of seed coats) on germination rate. Students can use a seed scarifier, if available, or gently rub seeds between blocks covered with a fine grade of sandpaper. Compare germination rates between seeds that have been scarified and those left unscarified.
- How does soaking seeds in water overnight before sowing affect germination rate? Compare germination rates between presoaked seeds and unsoaked seeds.
- Compare the cotyledons of monocots and dicots by examining recently sprouted corn and bean seedlings.
- How does complete darkness affect seedling growth? Compare the lengths and weights of bean or pea seedlings raised in light with those raised in complete darkness. What is the adaptive value of etiolation?
- Examine the effects of auxins, gibberellins, and cytokinins on plant growth and development.
- Investigate the role of the intact shoot tip in inhibiting lateral bud development in *Coleus*. What happens to lateral buds if the shoot tip is removed? Design an experiment in which applications of IAA dissolved in lanolin paste mimic the effects of the intact shoot tip.
- Discover the effects of different wavelengths of light on phototropism.
- Examine the effects of ethylene on fruit ripening or abscission.
- Locate and show a segment of a movie or videotape depicting growth and development of a seedling.

Ideas for Classroom Discussion

- How does growth in plants differ from growth in humans? Why can growth in plants be seen as a counterpart to movement in animals?

- How do hormone production and activity differ in plants and animals? What are some similarities shared by plants and animals? What are some differences?
- What is the adaptive value of seed dormancy?
- Why do grasses keep growing even after being repeatedly trimmed by lawn mowers or grazing animals? Where is the meristem located in shoots of grasses?
- What is "pinching back"? Why does it make plants bushier?
- What is "bolting"? Ask any of your students who have raised spinach or lettuce if they are familiar with this term. What can gardeners do to inhibit bolting?
- What is the biological reasoning behind the old saying "one bad apple spoils the whole bunch"? Discuss the role of ethylene in hastening fruit ripening.
- How could a sap-feeding insect actually stimulate plant growth? (answer: plant works harder to replace lost sap)
- How could a foliage-feeding insect increase growth of understory plants? (answer: increase light penetration)
- How could irrigation of cotton plants increase insect damage? (answer: more green leaves support more larvae)

Term Paper Topics, Library Activities, and Special Projects

- Describe how insect infestations or plant diseases can affect plant growth.
- How long can most seeds remain viable? Learn more about documented cases of extreme seed longevity. What steps can be taken to increase seed viability? What are some of the actions taken by seed companies to ensure maximum seed viability?
- Why are seedlings raised in complete darkness white or yellow instead of green? Describe the role played by sunlight in chlorophyll synthesis.
- What are some species whose seeds require stratification (cold, moist conditions for several weeks prior to sowing) for successful germination? What is the survival value of such a requirement?
- How are auxins, gibberellins, cytokinins, and abscisic acid used by commercial growers? How is ethylene used to hasten fruit ripening?
- Look up the geographic distributions of several long-day plants, short-day plants, and day-neutral plants. How is the effect of daylength on flowering related to the geographic range of a particular plant species?
- Describe synthetic auxins used as herbicides. How do these compounds actually work? Why does 2,4-D kill dicot weeds but not grasses? How are these compounds manufactured? What is their impact on the environment?
- Discuss the use of Agent Orange during the Vietnam War. What are its side effects? Was its use justified?
- Look up and report on what factors limit growth of plants above certain altitudes and how these factors specifically affect plant growth and development.
- What effect(s) would petroleum-solvent insecticides have on plants if they were used instead of insecticides formulated in water?

Films, Filmstrips, and Videos

- *Biological Rhythms: Studies in Chromobiology.* EBEC, 1977, 22 minutes, 16 mm, color.
- *The Evolution of Vascular Plants: The Ferns.* EBF (EBEC), 1962, 17 minutes, 16 mm, color. Describes the evolution of vascular systems in land plants. Explains the adaptive advantages of vascular systems in land plants and shows how biologists reconstruct the story of plants' evolution by studying fossils and living plants.

- *The Green Machine.* Time-Life, 49 minutes, 16 mm, color. Up-to-date film on plant physiology. Topics: photosynthesis, tropisms, plant hormones, cohesion theory, photoperiodism, phytochrome, and nastic movements.
- *The Growth of Plants.* EBEC, 1962, 21 minutes, color. Illustrates the growth process in plants and shows how cell division, elongation, and differentiation contribute to stem and root growth.
- *Plant Tropisms and Other Movements.* CORF, 1965, 11 minutes, 16 mm, color. The responses of various plants to internal or external stimuli are shown. Phototropism, geotropism, thigmotropism, nyctinasty, thermonasty, and nutational movements are shown and explored.

23

ANIMAL CELLS, TISSUES, AND ORGAN SYSTEMS

Revision Highlights

Rewritten and reorganized to include overview of levels of organization in animal body. Clear explanation of the critical relationship between organ systems and extracellular fluid. New "road map" (Figure 23.4) of human organ systems to be covered in subsequent chapters. Better introduction to cell junctions and extracellular matrix. New micrographs and diagrams of connective tissues and muscle tissues. Tighter introduction to concept of homeostasis.

Chapter Outline

ANIMAL STRUCTURE AND FUNCTION: AN OVERVIEW
- **Levels of Organization**
- **Organ Systems and the Internal Environment**
- **Major Organ Systems**
- **Some Anatomical Terms**

ANIMAL TISSUES
- **Tissue Formation**
- **Epithelial Tissue**
- **Connective Tissue**
- **Muscle Tissue**
- **Nervous Tissue**
- **Case Study: The Tissues of Skin**

HOMEOSTASIS AND SYSTEMS CONTROL

SUMMARY

Objectives

1. Understand the various levels of animal organization (cells, tissues, organs, and organ systems) and be familiar with the anatomical terms provided.
2. Know the characteristics of the various types of tissues. Know the types of cells that compose each tissue type and cite some examples of organs that contain significant amounts of each tissue type.
3. Describe how the four principal tissue types are organized into an organ such as the skin.
4. Explain how the human body maintains a rather constant internal environment despite changing external conditions.

Key Terms

tissue	organ system	vertebrates	interstitial fluid
organ	invertebrates	extracellular fluid	integumentary system

muscular system
skeletal system
nervous system
endocrine system
circulatory system
lymphatic system
respiratory system
digestive system
urinary system
reproductive system
germ cells
embryo
ectoderm
mesoderm
endoderm
somatic cells

cell junctions
epithelium,-lia
tight junctions
adhering junctions
communication junctions
simple epithelium
stratified epithelium
gland
endocrine gland
exocrine gland
connective tissue
connective tissue proper
dense connective tissue
loose connective tissue

adipose tissue
cartilage
bone
spongy bone tissue
red marrow
yellow marrow
compact bone tissue
blood
nervous tissue
neurons
muscle tissue
smooth muscle tissue
involuntary
skeletal muscle tissue
cardiac muscle tissue

epidermis
dermis
keratinization
homeostatic mechanisms
negative feedback mechanisms
receptors
stimulus
integrators
effectors
positive feedback mechanisms
feedforward mechanisms

Lecture Outline

I. Animal Structure and Function: An Overview
   A. Levels of Organization
      1. A tissue is an aggregation of cells and intracellular substances functioning for a specialized activity.
      2. Plants have dermal, vascular, and ground tissues.
      3. Animals have epithelial, connective, nerve, and muscle tissues.
      4. Different tissues result in a division of labor.
      5. All types of tissues can combine to form organs such as the heart.
      6. Organs may interact to form organ systems like the digestive system.
   B. Some Anatomical Terms
      1. The body contains five major cavities: cranial, spinal, thoracic, abdominal, and pelvic. **(Fig. 23.2, p. 309; TM 99)**
      2. Many animals show an anterior-posterior axis and a dorsal-ventral axis. **(Fig. 23.3, p. 309; TM 100)**
   C. Organ Systems and the Internal Environment
      1. The trillions of cells in our bodies must draw nutrients and dump waste into the same fifteen liters of fluid.
      2. The extracellular fluid consists of interstitial fluid (between cells and tissues) and plasma (blood fluid).
      3. The organ systems of an animal work together to maintain the extracellular environment required for life.
   D. Major Organ Systems **(Fig. 23.4, left and right, pp. 310, 311; TA 68, 69)**

II. Animal Tissues
   A. Tissue Formation
      1. Embryonic cells differentiate into three primordial tissue layers, which give rise to all tissue types.
         a. Ectoderm forms the skin and nervous system.
         b. Mesoderm forms muscles, connective tissues, skeleton, kidneys, and circulatory and reproductive organs.
         c. Endoderm forms the lining of the gut and major associated organs.
      2. Body tissues are composed of somatic and germ cells.
      3. Properties of individual cells promote recognition between similar cells and their adhesion to one another in tissues; for example, dissociated sponge cells of different types will recognize each other and aggregate into normal patterns.

4. Recognition proteins in the plasma membrane promote cell interactions like cell recognition and adhesion.
5. Cell junctions help cells to seal, adhere, and communicate.
6. Cells are embedded in an extracellular matrix that promotes adhesion.

B. Epithelial Tissue **(Fig. 23.6, p. 313; TA 70)**

1. In epithelial tissues, cells are linked by sealing junctions.
2. It contains one or many layers of cells.
3. One surface is free and the other adheres to a basement membrane.
4. Epithelial tissue covers external surfaces, internal cavities and tubes, and lines glands.
5. It functions for protection, absorption, and secretion.
6. Simple epithelium is one cell thick; for instance, it lines the air spaces of lungs.
7. Stratified epithelium has many layers—it forms skin.
8. Epithelial cells are flat (squamous), cuboidal, or columnar.
9. Endocrine glands secrete hormones directly into the blood.
10. Exocrine glands often secrete through ducts to free surfaces; they secrete mucus, saliva, wax, milk, and so on.

C. Connective Tissue **(Fig. 23.8, p. 315; TA 71)**

1. Connective tissue contains cells that produce ground substance and extracellular fibers.
2. Dense connective tissue connects different tissues, such as tendons and ligaments. It contains collagenous fibers in irregular or parallel arrays.
3. Loose connective tissue supports epithelia and organs and surrounds blood vessels and nerves. It contains collagenous and elastic fibers.
4. Adipose tissue cells contain a central vacuole for fat storage.
5. Cartilage contains collagenous and elastic fibers firmly positioned in a jelly-like ground matrix. It is a resilient tissue that can resist compression; it is found for example at the ends of bones and in parts of the nose.
6. Bone, or osseous tissue, forms bones that often attach at joints.
   a. Bone supports and protects soft parts of the body.
   b. It works with muscles to achieve movements.
   c. It consists of living cells, collagenous fibers, and a ground substance hardened by calcium salts.
   d. It contains canals and lacunae for osteocytes.
7. "Spongy" bone tissue may have red marrow that produces blood cells; adults have reserve yellow marrow.
8. "Compact" bone tissue has lamellae surrounding Haversian canals, which contain blood vessels and nerves.
9. Blood transports oxygen, wastes, hormones, and enzymes.
   a. Blood contains clotting factors to protect against bleeding.
   b. It contains components to protect against foreign invaders.

D. Muscle Tissue **(Fig. 23.10, p. 317; TA 72)**

1. Muscle tissue contracts in response to stimulation, then passively lengthens.
2. It functions for locomotion of body and food and for heat production.
3. Smooth muscle tissue contains spindle-shaped cells; it lines the gut, blood vessels, and glands, and is involuntary.
4. Skeletal muscle tissue attaches to bones for voluntary movement; it contains striated, multinucleated, long cells.
5. Cardiac muscle tissue is composed of short, striated cells.

E. Nervous Tissue

1. Different types of neurons detect stimuli, coordinate the body's responses, and send signals to muscles and glands to carry out responses.

F. Case Study: The Tissues of Skin

1. The skin has two layers, the epidermis and the dermis.
2. These are separated from the deeper tissues by the hypodermis.
3. The tissues of skin can differentiate to form feathers, hair, horns, beaks, and so on.
4. The epidermis of mammalian skin is stratified squamous epithelium and undergoes rapid cell division.

a. Keratinization of epidermal cells turns them into dead bags of keratin.
b. Keratin prevents dehydration and protects against microbial invasion.
c. Melanocytes, deep in the epidermis, produce melanin, which affords protection from ultraviolet light.
5. The dermis consists mostly of dense connective tissue.
a. The dermis is connected to the epidermis by a basement membrane.
b. During development, hair follicles and glands grow into the dermis from the epidermis.
c. Smooth muscle attached to the hair follicles enables hairs to stand on end.
d. Sensory endings detect temperature, touch, pressure, and pain.
6. The subcutaneous layer under the skin is loose connective tissue.
7. Blood vessels extend into the dermis (not epidermis) and can help regulate body temperature.

III. Homeostasis and System Control
A. How do tissues and organs interact to maintain a stable internal environment?
B. Homeostatic mechanisms operate to maintain chemical and physical environments within tolerable limits.
C. A common homeostatic mechanism is the negative feedback mechanism. **(Fig. 23.13, p. 320; TM 101)**
1. It is similar to the functioning of a thermostat in a heating system.
2. It requires receptors, integrators, and effectors.
a. Sensory cells are the receptors.
b. The brain and spinal cord are the integrators.
c. The effectors are muscles and glands.
D. Positive feedback mechanisms may intensify the original signal; sexual arousal is an example.
E. Feedforward mechanisms may anticipate a change—for instance, if a drop in temperature is detected.

Suggestions for Presenting the Material

- Although this chapter includes brief discussions of cell functions, embryonic tissues, and human organ systems, the two main topics are *tissues* and *homeostasis.*
- The presentation of tissues will be more meaningful if 2 x 2 transparency slides are shown during your lecture. If slides are not available, they can be readily made by photographing the color photos in the textbook using a 35 mm camera with a macro lens.
- Even though the structure of bone and the types of muscles are introduced here, the skeletal system and muscle physiology are topics discussed more fully in Chapter 28.
- It is difficult for students to think of bone as a living tissue. What most of us have seen in the dog's mouth is the nonliving portion of a tissue one author has called "living reinforced concrete."
- Skin structure and function is best presented by using an overhead transparency. This will allow the instructor to relate the layers of the integumentary system to the function of the structures in each layer.
- There is really no need to read the litany of body systems listed in this chapter because each will be discussed in later chapters.
- *Homeostasis* is a concept that should be thoroughly introduced here because it can be the common thread interwoven throughout the remaining lectures on body systems.
- Although it is overworked, the furnace/air conditioner thermostat is still the most familiar example of a homeostatic device. Because many students may never have seen the innards of such a device, a simple sketch on the overhead would be helpful.
- During your lecture, give some specific examples of the metabolic chaos that would result if homeostatic mechanisms were inoperative.

Classroom and Laboratory Enrichment

- Select several 2 x 2 transparencies of various types of tissues for projection onto a large screen. Ask the students to identify the type of tissue and where it is found in the body.
- Invite an athlete who has suffered a knee injury and has had corrective surgery to describe the damage and reconstructive process.
- During the discussion of bone, pass a cleaned bone (from a meat market) around the classroom. Comment on the nonliving and living composition of the bone.
- To show that bone is hard but not indestructible, treat bone fragments with a variety of common laboratory acids and alkalis.
- Have prepared microscope slides of each tissue type available for student viewing.
- Exhibit a model of the skin as a representative organ.
- Use a model of bone to illustrate its structure.
- Exhibit a vertebrate embryonic or fetal skeleton that is specially stained to show the cartilage.
- Demonstrate that bone has both organic and mineral components by soaking one chicken bone in acetic acid (to remove minerals) and by heating another in an oven at a high temperature (to remove organic material).

Ideas for Classroom Discussion

- How would you answer someone who says, "People in the tropics have such dark skin because they are out in the sun a lot making their skin tan fast and heavily."
- Survey the class concerning attitudes on cosmetic tanning. Do students realize the dangers? Does society expect one to sport a "tanned and healthy look"?
- Why is blood—a liquid—considered a *connective* tissue?
- Is there any validity to the cynic's observation that "Beauty queens are just exposing a lot of well-placed dead cells"?
- Rising blood sugar levels after a meal normally trigger insulin secretion, which in turn causes glucose to be converted to glycogen for storage. How is this similar to the response of an air conditioner thermostat to rising room temperature?
- What is liposuction? Does it permanently remove adipose tissue from the treated areas? Why is there a limit to the amount of adipose tissue that can be safely removed from the body at one time?
- Astronauts who orbited the earth early in the space program experienced considerable loss of bone mass under gravity-free conditions. How was this remedied in subsequent flights?

Term Paper Topics, Library Activities, and Special Projects

- Although it is not recognized as such, the skin constitutes the largest organ in the body. Prepare an analysis of the functions of the skin.
- Research the effects of tanning beds and natural sunlight on the skin. What factors contribute to the development of skin cancer?
- Osteoporosis is a topic of current interest especially to women. Can calcium supplements prevent, or do they simply delay, the stooped posture so prevalent in elderly women?
- Describe the mechanism of a modern furnace/air conditioner thermostat, and relate its function to the homeostatic mechanism that regulates the osmolarity of the blood in response to a meal high in salt.
- What roles, if any, do estrogen replacement therapy and exercise play in the prevention of osteoporosis?

- We are warned to protect our skin from the sun by using sun-block preparations. What chemical substances are effective in these preparations and how do they work?

Films, Filmstrips, and Videos

- *The Blood.* EBEC, 1971, 16 minutes, 16 mm, black and white. Live action and microphotography help tell the story of the highly complex tissue that is the lifeline of the body. The composition of the blood, its circulation, and the work that it does to support cell processes are clearly illustrated. Blood typing and the test for the Rh factor are also shown.
- *How Blood Clots.* BFA, 1969, 13 minutes, 16 mm, color. Animation and cinemicrography are used to illustrate the human circulatory system and its functions. The composition of blood and its separation into clots and serum are revealed through time-lapse photography. The role of platelets is shown in clot formation.
- *Incredible Voyage.* MGHF, 1975, 26 minutes, color. Describes the endoscope and shows pictures of different parts of the skeletal system, brain, heart, circulatory system, and digestive organs.
- *Man, the Incredible Machine.* NGS, 1975, 28 minutes, color. A fascinating photographic journey inside the human body.
- *The Skin: Its Structure and Function.* EBEC, 1983, 20 minutes, 16 mm, color. *Physiology Series.* Macrophotography and animation reveal skin structure and functions, emphasizing the role of skin in maintaining health and comfort.

24

INFORMATION FLOW AND THE NEURON

Revision Highlights

Revisions help make this inherently difficult subject (neural function) more accessible. Much better explanation of membrane excitability, resting membrane potential, graded potentials and action potentials. New, updated illustrations of action potential propagation; clearer picture of current flow and saltatory conduction. Tetanus used as example of interference with synaptic integration.

Chapter Outline

NEURONS: FUNCTIONAL UNITS OF NERVOUS SYSTEMS
- **Classes of Neurons**
- **Structure of Neurons**
- **Neuroglia**
- **Nerves and Ganglia**

ON MEMBRANE POTENTIALS
- **Membrane Excitability**
- **The Neuron "At Rest"**
- **Changes in Membrane Potential**

THE ACTION POTENTIAL
- **Mechanism of Excitation**
- **Duration of Action Potentials**
- **Propagation of Action Potentials**
- **Saltatory Conduction**

SYNAPTIC POTENTIALS
- **Chemical Synapses**
- **Excitatory and Inhibitory Postsynaptic Potentials**
- **Synaptic Potentials in Muscle Cells**

FROM SYNAPSE TO NEURAL CIRCUIT
- **Synaptic Integration**
- **Circuit Organization**
- **The Stretch Reflex**

SUMMARY

Objectives

1. Describe the visible structure of neurons, neuroglia, nerves, and ganglia, both separately and together as a system.
2. Describe the distribution of the invisible array of large proteins, ions, and other molecules in a neuron, both at rest and as a neuron experiences a change in potential.
3. Understand how a nerve impulse is received by a neuron, conducted along a neuron, and transmitted across a synapse to a neighboring neuron, muscle, or gland.
4. Outline some of the ways by which information flow is regulated and integrated in the human body.

Key Terms

neuron
system
sensory neurons
receptors
sensory stimuli
integrators
interneurons
motor neurons
effectors
innervate
dendrites
axon
input zone
trigger zone
conducting zone
output zone
neuroglia
nerves
pathways
tracts
nuclei
ganglion, -glia
polarity
action potentials
membrane excitability
open channels
gated channels
sodium-potassium pumps
voltage difference
resting membrane potential
receptor potential
synaptic potential
graded
polarized
depolarized
repolarized
threshold
positive feedback
all-or-nothing event
refractory period
Schwann cells
myelin sheath
node of Ranvier
saltatory conduction
chemical synapse
synaptic cleft
transmitter substance
presynaptic cell
postsynaptic cell
calcium
excitatory postsynaptic potential (EPSP)
inhibitory postsynaptic potential (IPSP)
neuromodulators
neuromuscular junctions
motor end plate
synaptic integration
tetanus
local circuit
reflex arc
reflex
stretch reflex
muscle spindles
monosynaptic pathways
withdrawal reflex
polysynaptic pathways

Lecture Outline

I. Neurons: Functional Units of Nervous Systems
   A. Introduction
      1. The neuron is the basic unit of communication.
      2. Each neuron acts with many others in a system to achieve a function like breathing.
   B. Classes of Neurons
      1. Complex animals have three classes of nerve cells (sensory neurons, interneurons, and motor neurons).
      2. Sensory neurons are receptors for specific sensory stimuli.
         a. Sensory stimuli are different forms of energy that can be detected by receptor cells.
         b. Each stimulus (such as light or pressure) causes an electrical disturbance at the receptor's surface; a message is then sent to the integrators.
      3. The brain and spinal cord are integrators and contain interneurons that influence the motor neurons.
      4. Motor neurons send information from integrator to muscle or gland cells (effectors).
   C. Structure of Neurons **(Fig. 24.1, p. 322; TM 102)**
      1. The cell body contains the nucleus and metabolic machinery for protein synthesis.
      2. Processes include dendrites, an axon, and axon terminals.
      3. In general, the dendrites are an input zone, and the axon is a conducting zone to the terminals.
   D. Neuroglia
      1. Neuroglia provide neurons with physical support, metabolic assistance, and protection.
      2. Some are macrophages and engulf debris and microbial invaders.
   E. Nerves and Ganglia
      1. Axons are bundled to form nerves. **(Fig. 24.2, p. 323; TA 73)**
      2. Nerves enter the brain and spinal cord as nerve tracts or pathways.
      3. In the brain and spinal cord, cell bodies are clustered into nuclei.
      4. In other body regions, they are called ganglia.

II. On Membrane Potentials
   A. Membrane Excitability **(Fig. 24.3, p. 325; TA 74a)**
      1. The resting membrane shows a polarity because there is more Na^+ outside than inside.

2. A stimulated neuron permits ions to flow across and reverse the polarity—an action potential.
3. This membrane excitability is influenced by
   a. Membrane lipids limit the flow of ions.
   b. Ions cross via ion channels that may be open all the time or gated and only open during action potentials.
   c. Membrane pumps, like the Na^+K^+ pump, actively transport ions and restore membrane potential.

B. The Neuron "At Rest"
1. At rest, there are more potassium ions inside the membrane than outside and far fewer sodium ions inside than outside.
2. At rest, channels for K^+ are open, and K^+ tends to leak out until there is no net movement of K^+.
3. At rest, Na^+ channels are shut and so do not contribute to the resting potential.

C. Changes in Membrane Potential
1. A stimulus, such as pressure, causes a brief change in membrane potential—a receptor potential.
2. The receptor potential is graded; it can vary in magnitude.
3. The receptor potential is local; the signal does not spread far unless it is intense or prolonged and causes an action potential to be initiated.

III. The Action Potential
A. An action potential is a pulse of electrical activity.
B. It is demonstrated in giant axons of squid by inserting electrodes and watching an oscilloscope.
1. At rest, the inside of the neuron membrane is more negative than the outside.
2. During an action potential, the membrane depolarizes.
3. After an action potential, the membrane is repolarized (resting condition).

C. Mechanism of Excitation **(Fig. 24.5, p. 326; TA 75a)**
1. If the graded signals reach a trigger zone and have at least a threshold level of energy, depolarization results from a brief flow of Na^+ into the neuron (caused by some gated Na^+ channels opening).
2. The action potential is "all or nothing."

D. Duration of Action Potentials
1. Most last a few milliseconds—several hundred per second.
2. Each action potential ends because depolarization
   a. Causes the Na^+ channels to close.
   b. Causes the K^+ channels to open, and the flow of K^+ out of the neuron restores the resting potential.
3. Gradual reduction in the gradients are restored by the Na^+K^+pumps.

E. Propagation of Action Potentials **(Fig. 24.6, p. 326; TA 76)**
1. An action potential propagates down the neuron without the intensity diminishing.
2. An action potential in one area causes the sodium channels in the adjacent region to open, and thus the current flows in a circuit between regions that differ in voltage.
3. The action potential flows away from the point of stimulation because of the refractory period—Na^+ gates are shut and the K^+ gates are open.
4. Later, K^+ channels close and the resting membrane state is restored. **(Fig. 24.7, p. 328; TA 74b)**

F. Saltatory Conduction
1. Peripheral nerves are covered by Schwann cells that form a myelin sheath.
2. Each sheath is separated by a node of Ranvier.
3. The action potentials jump from node to node by saltatory conduction, which is fast and efficient.

IV. Synaptic Potentials
A. Chemical Synapses **(Fig. 24.10, p. 329; TA 77)**

1. A chemical synapsis is a junction between two neurons separated by a synaptic cleft.
2. When an action potential arrives at the presynaptic cell, calcium channels open, and the Ca^{++} influx causes the release of transmitters into the synaptic cleft.
3. Transmitters diffuse across to the postsynaptic cells and bind with membrane receptors.
4. The binding of transmitter molecules to receptors causes channels to open, and a synaptic potential results.
5. The magnitude of the synaptic potential depends upon the amount of transmitter and the electrical state of the postsynaptic cell.

B. Excitatory and Inhibitory Postsynaptic Potentials

1. A synaptic potential can be excitatory or inhibitory.
2. In the brain and spinal cord, hundreds of excitatory potentials may be needed before a postsynaptic cell responds with an action potential.
3. Neuromodulators are signaling molecules that can regulate the responsiveness of target neurons to synaptic inputs.

C. Synaptic Potentials in Muscle Cells

1. At neuromuscular junctions, axon terminals are positioned in motor end plates. **(Fig. 24.11, p. 331; TA 78)**
2. Release of ACh by the motor neuron causes an action potential in the muscle cell and contraction results.

V. From Synapse to Neural Circuit

A. Synaptic Integration **(Fig. 24.12, p. 331; TA 75b)**

1. Synaptic integration is the combining of excitatory and inhibitory signals acting on adjacent membrane regions of a neuron.
2. Synapses closest to the trigger zone will have the greatest influence.
3. The summation is temporal and spatial.
4. *Clostridium* toxin can interfere with synaptic integration, and tetanus results.

B. Circuit Organization

1. Synapses link neurons into circuits of a few to millions of neurons.

C. The Stretch Reflex **(Fig. 24.13, p. 332; TA 79)**

1. A reflex arc is the basic unit of motor behavior.
2. The stretch reflex helps one to maintain an upright posture.
3. When a muscle is stretched, the reflex causes it to contract to its original length.
4. In leeches, sensory neurons synapse directly with motor neurons—a monosynaptic pathway.
5. In most animals, the withdrawal reflex is polysynaptic and involves a number of interneurons.

Suggestions for Presenting the Material

- One of the most effective comparisons of the nervous systems to anything man-made is to the worldwide telephone network. Although the analogy is not perfect, it does convey the truth that billions of individual phone sets can send impulses (communicate) with any other phone, or several phones at one time, via a connecting wire or microwave signal.
- Students ranging in scientific expertise from that of beginning freshman to third-year medical student all agree that the nervous system is one of the most difficult to comprehend at any level. Therefore, extra time and thorough explanations are especially needed in this chapter.
- Note also that the discussion of the nervous system is contained in two chapters—this one and the next. Most instructors prefer to begin with the structure and function of a generalized neuron, including action potentials, then proceed to synapse transmission.
- The organization of neurons into organs that function as integrated units is the subject of Chapter 25.
- The function of the neuron membrane in permitting passage of Na^+ and K^+ ions is at first confusing. Initially you may wish to focus on sodium only, then expand to the role of potassium.

- The changes in the membrane can be conveniently demonstrated using Figure 24.6 and the words *polarized, depolarized,* and *repolarized.*
- The concept of "all-or-nothing events" and "thresholds" can be illustrated by describing the use of a firearm. When the trigger is pulled and reaches the critical point (threshold) at which the hammer is released, the bullet leaves the barrel and travels the expected distance. Of course, the bullet either goes or stays (all-or-nothing), and the manner in which the trigger is activated (slowly or quickly) should not influence the speed of bullet travel.
- Emphasize the temporary nature of the acetylcholine bridge across the synapse by comparing it to a pontoon bridge used by the military to cross small streams and rivers.
- If the students can recite the sequence of structures through which an impulse passes during a *reflex arc* such as in Figure 24.13, they have a good grasp of the nerve conduction pathways. Add the ion flow across the membranes and the story is pretty well complete!
- Emphasize the difference between a neuron and a nerve. Students often have difficulty distinguishing between them.
- Prepare a transparency of an action potential recording (see Figure 24.5). While projecting it on a screen, describe how the different regions of the tracing are related to the flow of sodium and potassium ions.

Classroom and Laboratory Enrichment

- The concept of thresholds and all-or-nothing events can be demonstrated by using dominoes (or for large classes, several audio cassette cases). Line up about 20 dominoes placed on end and spaced about one inch apart. Ask a student to gently touch one end domino to begin the progressive fall. Emphasize that the student's touch (threshold stimulus) caused a standing row (polarized) to begin falling (depolarization) at a constant speed (all-or-nothing event). Pose the following question (and demonstrate the answer): Would a greater and faster stimulus cause more rapid falling? To demonstrate *repolarization,* a second student could begin resetting the dominoes even before the falling is complete.
- Arrange with a physics student or an instructor for a demonstration of an action potential as recorded on an oscilloscope screen.
- Permit students to demonstrate the knee jerk reflex arc by use of percussion hammers. It is important to ask the subjects to close their eyes to prevent "cheating."
- Show a film or video of an animation of nerve impulse transmission.
- Exhibit models of neurons and neuroglial cells.
- Provide microscope slides of longitudinal sections and cross-sections of nerves for student viewing.
- Have microscope slides of neurons available for laboratory demonstration.

Ideas for Classroom Discussion

- What would be the result of demyelination of axons such as occurs in multiple sclerosis?
- Upon hearing that salt was not good for him, a freshman college student began a fanatical program to eliminate all sodium chloride from his diet. By cooking his own meals, he was able to eliminate virtually all sodium. What complications could he expect as a result of his brash action?
- If neurons operate under the all-or-nothing principle, how are we able to distinguish soft sounds from loud sounds, or a gentle touch from a crushing blow?
- To most amateur musicians, the playing of 16th notes is a challenge, but to trumpet virtuoso Wynton Marsalis, 32nd and 64th notes are a breeze. Describe the action of nerves and tongue muscles that regulate the air flow through the mouthpiece.
- Why does a physician's tapping of the knee or elbow reveal the general status of the nervous system *in general,* not just the condition of those two joints?

- Why does saltatory conduction "afford the best possible conduction speed with the least metabolic effort by the cell"?
- Why does drinking large amounts of coffee or other caffeine-containing beverages tend to make a person "nervous" or "jittery"?

Term Paper Topics, Library Activities, and Special Projects

- The neurons of the human body can communicate one with the other much the same as telephones in your city can intercommunicate. In the telephone system, wires touch wires to pass the impulse, but neurons are not directly "wired." Investigate the effects on the body of the elimination of synapse function such as would be caused by organophosphate pesticides, which inhibit acetylcholinesterase.
- One of the most effective antidotes for the organophosphate poisoning referred to above is *atropine.* Investigate its mechanism of action. Based on what you find, could administration of atropine be harmful if OP poisoning *has not* occurred?
- Prepare a list of neurological disorders in which you focus on the specific cause of the difficulty in each case.
- Ask a resource person to explain the consequences of central nervous system damage as opposed to peripheral damage.
- Investigate some of the factors that determine the speed at which an impulse is conducted in a neuron.

Films, Filmstrips, and Videos

- *The Hidden Universe.* CRMP, 1978, 49 minutes, 16 mm, color. Explores the mysteries of the human brain; presented in a documentary format.
- *The Nerve Impulse.* EBEC, 1971, 22 minutes, 16 mm, color. Presents a physiological description of neuronal activity, all-or-none activity, and synaptic activity.

25

NERVOUS SYSTEMS

Revision Highlights

Well-received chapter; only notable change is a condensation of the section on invertebrate nervous systems (the topic is covered in detail in Chapter 41).

Chapter Outline

NEURAL PATTERNING: AN OVERVIEW

INVERTEBRATE NERVOUS SYSTEMS
- **Nerve Nets**
- **Cephalization and Bilateral Symmetry**
- **Segmentation**

THE VERTEBRATE PLAN
- **Evolution of Vertebrate Nervous Systems**
- **Functional Divisions of the Vertebrate Nervous System**

PERIPHERAL NERVOUS SYSTEM
- **Nerves Serving Autonomic Functions**
- **Cranial Nerves**

CENTRAL NERVOUS SYSTEM
- **The Spinal Cord**
- **Divisions of the Brain**
- **Hindbrain**
- **Midbrain**
- **Forebrain**

THE HUMAN BRAIN
- **The Cerebral Hemispheres**
- **Memory**
- **States of Consciousness**
- *Commentary:* **Drug Action on Integration and Control**

SUMMARY

Objectives

1. Contrast invertebrate and vertebrate nervous systems in terms of neural patterns.
2. Describe the organization of peripheral versus central nervous systems.
3. Identify the parts of primitive brains; then tell how the human brain is advanced beyond the primitive types.

Key Terms

theory of neural patterning
radial symmetry
nerve net
cephalization
bilateral symmetry
ganglion, ganglia
nerves
nerve cords
afferent
nerve cord
efferent
biofeedback
spinal cord
white matter
gray matter
hindbrain
medulla oblongata
pons
reticular formation
cerebellum
midbrain
tectum
olfactory bulbs
cerebrum
forebrain
thalamus
hypothalamus

cerebral cortex
limbic system
corpus callosum
motor centers
Broca's area
primary somatic sensory cortex
primary receiving centers
association centers
thinking
memory
memory trace
short-term storage
long-term storage
retrograde amnesia
electroencephalogram, EEG
alpha rhythm
slow-wave sleep
REM sleep
EEG arousal
reticular activating system, RAS
sleep centers
psychoactive drug
analgesic

Lecture Outline

I. Neural Patterning: An Overview
   A. The complex behaviors of animals—such as food gathering or locating a mate—require a complex nervous system.
   B. The theory of neural patterning states that
      1. Reflexes are stereotyped movements in response to sensory stimuli.
      2. The nervous system evolved through accretion.
      3. The oldest parts of the nervous system deal with reflexes.
      4. Newer layers form the basis for memory, learning, and reasoning.

II. Invertebrate Nervous Systems
   A. Nerve Nets
      1. Cnidarians are radially symmetrical and have the simplest nervous system.
      2. A nerve net creates reflex pathways between epithelial receptor cells and contractile cells.
   B. Cephalization and Bilateral Symmetry
      1. Complex nervous systems may have evolved from a cnidarian larval form (planula) that moved like a flatworm.
      2. Flatworms show cephalization and bilateral symmetry, which leads to paired structures.
   C. Segmentation
      1. The body is composed of repeating segments—annelids and arthropods.
      2. Each segment has a pair of nerves and a ganglion—a nerve cord extends through the segments and expands to form an anterior brain.

III. The Vertebrate Plan
   A. Evolution of Vertebrate Nervous Systems
      1. The vertebrate nervous system evolved by modification of bilateral symmetry, a notochord, and a hollow nerve cord.
      2. The notochord, a stiff rod, is replaced during development by vertebrae.
      3. The vertebral column permitted rapid movement and the evolution of strong jaws.
      4. In time, the anterior region of the nerve cord expanded into the brain. **(Fig. 25.3, p. 338; TM 103)**
   B. Functional Divisions of the Vertebrate Nervous System
      1. The central nervous system (CNS) = the brain + spinal cord.
      2. The peripheral nervous system (PNS) = nerves + ganglia.

IV. Peripheral Nervous System
   A. Afferent nerves carry sensory input.
   B. Efferent nerves carry motor output.
   C. The somatic system consists of efferent nerves going to the skeletal muscles.
   D. The autonomic system consists of efferent nerves going to the heart, smooth muscle, and glands.
   E. Nerves Serving Autonomic Functions
      1. When there is little stress, parasympathetic nerves tend to slow down overall body activity.
      2. During time of excitement or danger, sympathetic nerves prepare one for fight or flight.
      3. The actual rate is determined by the sum of opposing signals.

F. Cranial Nerves
   1. The PNS also includes twelve pairs of cranial nerves which directly connect to the brain.
   2. Some cranial nerves contain only sensory axons while others also contain motor axons.

V. Central Nervous System
   A. The Spinal Cord
      1. The spinal cord resides within stacked vertebrae and contains a central canal.
      2. Gray matter has cell bodies and dendrites concerned with reflexes.
      3. White matter contains axons of interneurons—often bundled into sensory nerve tracts that may end at the brain.
      4. Connections between sensory and motor neurons required for many reflexes occur in the spinal cord.
      5. Interneurons often lie between sensory and motor neurons.
   B. Divisions of the Brain
      1. The brain is an expanded extension of the spinal cord—both are covered by membranes.
      2. The brain contains fluid-filled ventricles that are continuous with the central canal of the cord.
   C. Hindbrain
      1. The hindbrain includes the medulla oblongata, cerebellum, and pons.
      2. It includes part of the reticular formation—a net of nerve cells involved with motor coordination, sleeping and dreaming, arousal, and complex reflexes.
   D. Midbrain
      1. The midbrain was originally for coordinating reflexes associated with visual input.
      2. The tectum integrates visual and auditory signals.
      3. In fish and amphibians, the tectum exerts major control over the body (if you remove the cerebrum from a frog, its behavior is hardly affected); in mammals, sensory information is rapidly sent to higher centers.
   E. Forebrain
      1. Early vertebrates evolved two olfactory lobes and a primitive cerebrum to integrate olfactory input and to select the proper motor responses—needed to acquire food.
      2. The thalamus relayed and coordinated sensory signals.
      3. The hypothalamus influenced thirst, hunger, and sex.
      4. In time, the cerebral cortex developed into an information-processing center.
      5. The limbic system, formed from parts of the cortex, thalamus, and hypothalamus, influences learning and emotions.

VI. The Human Brain
   A. The Cerebral Hemispheres
      1. A thin surface layer, the cerebral cortex, contains much of the gray matter.
      2. It is divided into right and left hemispheres, and the surface is highly folded.
   B. Nerve Tracts in the Hemispheres
      1. White matter consists of major nerve tracts; for example, the corpus callosum permits communication between hemispheres.
   C. Functional Regions of the Cortex **(Fig. 25.12, p. 345; TM 104)**
      1. Some regions are motor centers.
      2. Other regions are primary receiving centers for sensory input.
      3. In association centers, stored memory is added to primary sensory information.
   D. Memory exists as memory traces, but in what form is unclear.
      1. Short-term memory may be a fleeting stage of neural excitation.
      2. Long-term memory may depend on chemical or structural changes in the brain.
      3. Because one can remember information unused for decades, some memory traces must be immune to degradation—even though 50,000 neurons die each day.
      4. Disuse can cause a synapse to wither and sever the connection between the neurons.
   E. States of Consciousness
      1. States of consciousness are governed by the central nervous system and altered by psychoactive drugs. **(Commentary table, p. 349; TM 106)**

2. An electroencephalogram (EEG) is an electrical recording from the brain's surface.
3. EEG Patterns **(Fig. 25.14, p. 347; TM 105)**
   a. Alpha rhythm—relaxed, with eyes closed.
   b. Slow-wave sleep—80% of total sleeping time.
   c. REM sleep—rapid eye movements and vivid dreams.
4. The Reticular Activating System
   a. The reticular formation connects with the spinal cord, cerebrum, cerebellum, and itself.
   b. The reticular activating system (RAS) maintains wakefulness.
   c. One sleep center secretes serotonin that inhibits the RAS.
   d. Another sleep center secretes factors that counteract serotonin and promote REM sleep.

Suggestions for Presenting the Material

- If students are unfamiliar with the invertebrate animal representatives discussed early in this chapter, then transparencies, drawings, or specific references to the textbook pages would be warranted.
- Although Figure 25.2 does not present the body orientation sufficient to show it, you may wish to emphasize the difference between the *ventral* nerve cord of invertebrates and the *dorsal* one of vertebrates.
- Because both *notochord* and *nerve cord* are presented in this chapter, it is important to distinguish the former as a member of the support system of the body (notice spelling *CH*ord as in *CHORD*ata).
- The division of the human nervous system into component parts as presented in Figure 25.5 and accompanying text is of course an arbitrary one. You should emphasize the "oneness" of the system. The divisions are really ones made by us who need to study the interrelated functions.
- If you use Figure 25.5 as the basis for your lecture, you may wish to begin with the CNS then proceed to the PNS, which is the reverse of the sequence in the textbook.
- Some terms used to describe the divisions are not parallel. For example, *autonomic* (not a misspelling of "automatic" as some students think) is used to designate nerves NOT under voluntary control and *somatic* is used to designate nerves under voluntary control.
- If this is your first use of the word *antagonistic*, you may want to remove any negative connotations students have attached to the word during regular conversational use. Tell them there are several instances where body homeostasis is maintained by antagonistic nerves, hormones, and muscles.
- You might be surprised at the number of students who cannot distinguish the backbone (= vertebral, or spinal, column) from the spinal cord. Give them some assistance by referring to Figure 25.7.
- Emphasize the continuity of fiber tracts between brain and spinal cord. Stress the primary functions of the spinal cord as a reflex center versus the brain as a sense-interpretation and directed-response center.
- The extent to which each instructor requires the students to delve into brain regions and functions will vary. Some instructors may want to select the major brain regions (cerebrum, hypothalamus/pituitary, cerebellum, and medulla) for special emphasis.
- Material on the "memory," EEG, consciousness, and drugs will make an interesting capstone to your lectures but may be omitted, or assigned for reading only, if time is short.
- Students generally find it easier to distinguish between the sympathetic and parasympathetic divisions of the autonomic nervous system if they are told that the sympathetic division is involved in mobilizing "fight-or-flight" reaction while the parasympathetic division produces a general "slowing-down" and "business as usual" response.

Classroom and Laboratory Enrichment

- Use models of *Hydra*, earthworm, grasshopper, and starfish to demonstrate nervous system development.
- Many laboratories have preserved specimens of the human, or other vertebrate, brain, which are valuable aids to comprehending the size and arrangement of brain parts.
- Using live *Hydra* and *Dugesia* (planaria), test for nervous response to touch, vibration, light, mild acid or alkali, and heat. Are there differences between the two species with respect to degree and speed of response?
- After obtaining the proper permission if necessary, design a series of experiments to show the effects of beverage alcohol on mental function. Begin with a simple memory test using flash cards, or a before-and-after writing sample. Compare persons of different gender, body weight, and drinking habits.
- Use a dissectible model of the brain to illustrate the location of its parts.
- Exhibit a vertebral column/spinal cord/herniated disc model to demonstrate why so much pain and functional loss are associated with herniated discs.
- Show a film or video on brain function.
- Use a spinal cord/vertebral column cross-sectional model to illustrate the relation between the two structures.

Ideas for Classroom Discussion

- Do invertebrates, such as the cockroach, feel pain?
- How do invertebrate nervous systems differ from vertebrate ones *structurally*?
- The central nervous system (and closely associated ganglia) house the cell bodies of neurons. As opposed to the peripheral axons and dendrites, the cell bodies are not regenerated after traumatic injury. What advantages and disadvantages does this structural arrangement pose for humans?
- Why does a small speck of food stuck between your teeth feel like a large chunk when rubbed with your tongue?
- The exact mode of action of the famous, and now banned, insecticide DDT has never been elucidated (after nearly fifty years of research). However, textbooks describe it as a "central nervous system" poison. What does this imply?
- Explain why elderly people may be unable to remember what they ate for breakfast but can relate the details of a teenage romance.
- Discuss the characteristics of brain disorders such as Parkinson's disease and Alzheimer's disease.
- Distinguish between flaccid and spastic paralysis. What are their causes?

Term Paper Topics, Library Activities, and Special Projects

- Why are injuries to the central nervous system, such as gunshot wounds, more permanently debilitating than those to the peripheral system?
- What is the basis for "healing" accomplished by the practice of chiropractic? What are its strengths and weaknesses?
- Using the mode of action of organophosphate insecticides as a tool, delve into the similarities and differences between the physiology of insect and human nerve function.
- Prepare an argument for the suppression of a presently readily-available drug, say, alcohol; or prepare an argument for the legalization of marijuana.
- Explore the research relating dreams to actual events—past and future. What do dreams tell us about ourselves?

- Discuss the basis for the use of acupuncture for relieving pain or providing anesthesia.
- Research the "gate theory" of pain transmission.
- Investigate the use of biofeedback for controlling pain, heart rate, and other autonomic functions.

Films, Filmstrips, and Videos

- *Exploring the Human Brain.* EBEC, 1977, 19 minutes, 16 mm, color.
- *Fundamentals of the Nervous System.* EBEC, 1959, 17 minutes, 16 mm, color. The two major divisions of the nervous system and their functions are illustrated by means of live photography, photomicrography, animated drawings, and demonstrations. The film shows how information from the environment is conveyed to the brain and spinal cord and how the appropriate responses are conveyed to muscles and glands.
- *Human Body: The Nervous System.* CORF, 1980, 22 minutes, 16 mm, color. An animated film that describes how the structure of neurons allows them to transmit electrochemical impulses and function as part of the central, peripheral, and autonomic divisions of the human nervous system.
- *The Nervous System in Man.* IU, 1966, 18 minutes, 16 mm, color. Shows how the nervous and endocrine systems interact. Also shows types of neurons, divisions of the nervous system, talks about reflexes, and discusses the functions of various parts of the brain.
- *Nervous Systems in Animals.* IU, 1971, 17 minutes, 16 mm, color. The responses of several protistans, hydra, and planaria to stimuli are examined. Spinal cords, spinal nerves, and parts of the brain are pointed out in dissected fetal pigs, frogs, birds, cats, and humans. Basic elements of the neuron and pathway of the nerve impulse during a reflex are shown through still and animated drawings.

26

INTEGRATION AND CONTROL: ENDOCRINE SYSTEMS

Revision Highlights

Tighter writing. New table on releasing hormones; refined tables on hormone sources, targets, and effects. New photos of pituitary dwarfism, gigantism, acromegaly, and goiter. Updated text on adrenal gland and parathyroid glands. New section on pancreatic alpha, beta, and delta cell secretions and diabetes mellitus. New text and updated illustrations on molecular mechanisms of steroid and protein hormone action.

Chapter Outline

"THE ENDOCRINE SYSTEM"
HORMONES AND OTHER SIGNALING MOLECULES
NEUROENDOCRINE CONTROL CENTER
- **The Hypothalamus-Pituitary Connection**
- **Posterior Lobe Secretions**
- **Anterior Lobe Secretions**

ADRENAL GLANDS
- **Adrenal Cortex**
- **Adrenal Medulla**

THYROID GLAND
PARATHYROID GLANDS
GONADS
OTHER ENDOCRINE ELEMENTS
- **Pancreatic Islets**
- **Thymus Gland**
- **Pineal Gland**

LOCAL CHEMICAL MEDIATORS
- **Prostaglandins**
- **Growth Factors**

SIGNALING MECHANISMS
- **Steroid Hormone Action**
- **Nonsteroid Hormone Action**

SUMMARY

Objectives

1. Know the general mechanisms by which molecules integrate and control the various metabolic activities in organisms.
2. Understand how the neuroendocrine center controls secretion rates of other endocrine glands and responses in nerves and muscles.
3. Know how sugar and salt distribution is regulated in hormones.
4. Diagram the relationship between the various hormones that control reproduction in the human female and the human male.

Key Terms

signaling molecules
endocrine glands
hormones
neurosecretory cells
neurohormones
synapsing neurons
local mediator cells
exocrine glands
pheromones
neuroendocrine control center
posterior lobe
anterior lobe
intermediate lobe
antidiuretic hormone
oxytocin
releasing hormones
portal vessels
corticotropin-stimulating hormone, ACTH
thyrotropin-stimulating hormone, TSH
follicle-stimulating hormone, FSH
luteinizing hormone, LH
prolactin, PRL
somatotropin, STH (= growth hormone, GH)
pituitary dwarfism
gigantism
acromegaly
adrenal cortex
glucocorticoids
mineralocorticoids
parathyroid glands
homeostatic feedback loops
adrenal medulla
epinephrine
norepinephrine
thyroid gland
thyroxin
hypothyroidism
hyperthyroidism
goiter
calcitonin
parathyroid glands
parathyroid hormone, PTH
rickets
gonads
androgens
estrogens
pancreatic islets
alpha cells
glucagon
beta cells
insulin
delta cells
somatostatin
diabetes mellitus
thymus gland
pineal gland
nerve growth factor, NGF
steroid hormones
testicular feminization syndrome
nonsteroid hormones
second messengers
cyclic AMP, cAMP
amplify

Lecture Outline

I. "The Endocrine System"
  A. In the early 1900s, Bayliss and Starling first demonstrated that a hormone secreted into the blood triggers the secretion of pancreatic juices.
  B. Many vertebrate hormones are now known.
  C. Hormones are secreted from members of the endocrine system.
  D. Some neurons secrete hormones (the hypothalamus) or innervate endocrine glands.

II. Hormones and Other Signaling Molecules **(Table 26.1, p. 354; TM 107)**
  A. These respond to chemical changes by secreting signaling molecules that affect target cells.
  B. Endocrine glands and neurosecretory cells secrete hormones into the blood that reach distant targets.
  C. Synapsing neurons secrete transmitters that act immediately upon adjacent target cells.
  D. Many local mediator cells secrete substances that affect cells in the immediate vicinity.
  E. Exocrine glands secrete pheromones that have targets outside the body.

III. Neuroendocrine Control Center
  A. The Hypothalamus-Pituitary Connection
    1. The hypothalamus and pituitary act together as a neuroendocrine control center.
    2. The hypothalamus monitors internal organs and influences related behavior, such as hunger.
    3. The pituitary is connected to the hypothalamus by a stalk.
    4. The posterior lobe consists of nervous tissue and secretes two neurohormones.
    5. The anterior lobe consists of glandular tissue and secretes six hormones and controls the release of others.
    6. Nonhuman vertebrates have an intermediate lobe that secretes hormones that affect body coloration.
  B. Posterior Lobe Secretions
    1. Posterior lobe secretions include antidiuretic hormone (ADH), which helps to reduce fluid loss through the kidneys.
    2. Oxytocin stimulates uterine contraction during labor, and the secretion of milk.

C. Anterior Lobe Secretions

1. Releasing hormones from the hypothalamus stimulate or inhibit the secretions of hormones.
2. The first four hormones act on endocrine glands that produce other hormones.
3. Prolactin stimulates milk production.
4. Somatotropin influences overall growth; it induces protein and RNA synthesis, stimulates cell division, and causes the liver to release growth factors called somatomedins.
    a. Too little causes pituitary dwarfism.
    b. Too much causes gigantism.
    c. Too much as an adult causes acromegaly—thickened skin and bones.

IV. Adrenal Glands

A. Adrenal Cortex

1. Humans have one adrenal gland above each kidney.
2. Cells of the cortex produce glucocorticoids and sex hormones.
3. Glucocorticoids function in carbohydrate, protein, and lipid metabolism; they also play a role in inflammation.
4. Mineralocorticoids regulate the solute concentration of extracellular fluid.
5. Cortisol helps to control the concentration of glucose in the blood. **(Fig. 26.9, p. 360; TM 108)**
    a. Cortisol secretion is an example of a homeostatic feedback loop.
    b. Under stress, higher levels are secreted.

B. Adrenal Medulla

1. The adrenal medulla secretes epinephrine and norepinephrine.
2. It controls blood circulation and carbohydrate metabolism.
3. During stress, it enhances the fight or flight response of the sympathetic nervous system.
4. Its secretion is controlled by sympathetic nerves.

V. Thyroid Gland

A. The thyroid gland secretes thyroxine and triiodothyronine.
B. It influences metabolic rate, growth, and development.
C. Hypothyroidism in adults results in lethargy and weight gain; in infants, retardation and dwarfism can result.
D. Hyperthyroidism increases heart rate and blood pressure, and causes weight loss.
E. Goiter can result from a deficiency in iodine.
F. The thyroid gland secretes calcitonin, which causes calcium deposition in bones.

VI. Parathyroid Glands

A. The parathyroid glands secrete parathyroid hormone (PTH), which regulates blood levels of Ca^{++} for gene activation and muscle contraction.
B. Secretion from the parathyroid glands is regulated by the Ca^{++} level in the blood.
C. PTH removes calcium from bone, increases the kidney's reabsorption of Ca^{++}, and activates vitamin D.

VII. Gonads

A. The gonads secrete sex hormones, which affect reproduction and secondary sex characteristics.
B. They are influenced by FSH and LH.

VIII. Other Endocrine Elements

A. Pancreatic Islets

1. Alpha cells secrete glucagon, which raises glucose level in blood.
2. Beta cells secrete insulin, which stimulates glucose uptake and protein and fat synthesis and lowers the glucose level in the blood.
3. Delta cells secrete somatostatin, which can inhibit the secretion of glucagon and insulin.
4. Diabetes mellitus (type 1 diabetes) is caused by insulin deficiency.

a. Dehydration and degradation of proteins and fats result.
b. Coma and death can follow.
c. It is caused by genetic factors, viruses, and autoimmune responses.
d. It can be treated with injections of insulin.

5. Type 2 diabetics have normal or above-normal insulin but insufficient or abnormal receptors. This is treated by diet and drugs that stimulate insulin production.

B. Thymus Gland
1. This secretes thymosins that affect the function of some lymphocytes.

C. Pineal Gland
1. The pineal gland represents a third eye in lampreys.
2. In higher vertebrates, it secretes melatonin in the absence of light.
a. In winter, melatonin suppresses the sexual activity of hamsters.
b. In humans, decreased production of melatonin helps to trigger puberty.

IX. Local Chemical Mediators
A. Every mammalian tissue has mediator cells that detect local changes in the chemical environment and respond by counteracting or amplifying the signal.
B. Prostaglandins
1. Prostaglandins are always released by many tissues, but the rate of synthesis may vary.
2. Their production is stimulated by epinephrine and norepinephrine, which cause blood vessels or airways to constrict or dilate.
3. Prostaglandins cause cramping during the menstrual cycle and destruction of the corpus luteum if pregnancy does not follow ovulation.
C. Growth Factors
1. Growth factors include nerve growth factor needed for the growth of neurons in embryos.

X. Signaling Mechanisms
A. These only affect cells with receptors, and they affect different cells in different ways.
B. Steroid Hormone Action **(Fig. 26.12, p. 364; TA 80)**
1. A steroid hormone diffuses through the cell membrane and fuses with a specific receptor in the nucleus.
2. The hormone-receptor complex interacts with chromosomal proteins.
3. Gene activation results and particular mRNA transcripts are synthesized.
C. Nonsteroid Hormone Action **(Fig. 26.13, p. 364; TA 81)**
1. The hormone binds to a receptor in the plasma membrane.
2. Binding activates many molecules of adenyl cyclase.
3. Each adenyl cyclase converts many ATP molecules to cyclic AMP.
4. Each cyclic AMP activates many enzymes.
5. Each enzyme may then form many different enzymes.
6. Thus, second messengers greatly amplify the signaling molecule.

Suggestions for Presenting the Material

- The "core" of this chapter is the section describing the various endocrine glands, their secretions and functions.
- Two other important topics are appended to the opening (signaling molecules, Table 26.1) and closing (signaling mechanisms) of the chapter. Some instructors may want to move the signaling mechanisms to the introductory lecture.
- When discussing signaling mechanisms and Figures 26.12 and 26.13, emphasize the chemical nature of steroids (lipid-bilayer soluble) versus proteins (not lipid soluble) and the need for a second messenger, namely cyclic AMP.

- After suitable introductions have been made, there is no practical way to escape a presentation of the major glands and their secretions. Unfortunately, students soon recognize the "cataloging" approach and become restless. One solution is indicated below.
- Because the chapter contains excellent Figures on the pituitary (26.2–26.6) and Tables 26.3 and 26.4, these can be used to great advantage. One seasoned instructor asks students to bring their texts to class and invites them to follow along. He also prepares his own overhead transparencies of the tables to help everyone keep up the pace.
- This chapter presents a maddening array of new words—hormone names that are long and unfamiliar. As an aid to learning, subdivide the name and give the literal meaning of each portion, for example, adreno (adrenals)—cortico (cortex)—tropic (stimulate).
- "Antagonism" was mentioned in connection with the autonomic nervous system. Take this opportunity to point out antagonistic hormone pairs: calcitonin/parathyroid hormone; insulin/glucagon.
- Emphasize the necessity of learning both the hormone name and the abbreviation, which is often more commonly used than the name itself.
- Notice that even though the gonadal hormones are included in Table 26.4, they are not discussed until Chapter 35.
- Emphasize that the posterior pituitary gland does not synthesize the hormones it secretes.
- Point out that some organs function as both endocrine and exocrine glands.

Classroom and Laboratory Enrichment

- Human nature is such that students are very interested in the abnormalities that hyper- and hyposecretion of human hormones cause. You can stimulate interest in the total area of hormone control by showing 2 x 2 transparencies of the physical manifestations of such imbalances.
- Ask a local health scientist or practitioner to report on his/her experiences with hormone therapy.
- If a member of the class is willing to share his/her experiences as a diabetic, arrange for such a presentation before class begins and allow time for questions.
- Seek evidence of a class member who has experienced or witnessed an epinephrine-mediated "emergency response." Ask him/her to report.
- Survey local grocery stores to determine the relative stocks of iodized and noniodized salt. Are there implications for the unwary consumer?
- Use a dissectible mannequin or a dissected fetal pig to show the location of the various endocrine glands.

Ideas for Classroom Discussion

- Beverage alcohol inhibits the action of ADH. How is this unseen physiological event evidenced during a night of bar-hopping?
- Do hormones occur only in vertebrates? Have you ever heard of "ecdysone" in insects?
- Why does insulin have to be administered by injection rather than orally?
- Using knowledge gained in a freshman biology class, an athlete decided he might be able to raise his blood sugar quickly by injecting glucagon. This attempt is doomed for what reasons?
- What is the possible connection between the pineal gland and puberty?
- Some hormones seem to be doing another's duties, for example, sex hormones from the adrenals, blood sugar control by epinephrine, thyroxine regulation of growth. Why is this so?
- What are anabolic steroids? Why do some athletes use them? What are the dangers associated with their use?
- Oxytocin is commonly used to induce labor. How does it work?

- Why do certain hypoglycemics, who regularly ingest excessive amounts of sugar, frequently develop diabetes later in life?
- Untreated diabetes mellitus victims tend to be very thirsty and yet produce large volumes of urine. Why is this so?

Term Paper Topics, Library Activities, and Special Projects

- Discuss the ethical issues of administering somatotropin to persons of normal stature who wish to become "super athletes."
- Investigate and report on the fascinating discovery of the role of insulin by researchers Banting and Best.
- Select a major hormone and prepare an in-depth report on the abnormalities that may result from hypo- and hypersecretion.
- The role of hormones in insect development has been elucidated in only the past forty years. Check an insect physiology text, and prepare a chronology of this research.
- Report on current research designed to correct "type 1" diabetes, which is the result of autoimmune responses.
- Investigate which human hormones are now being produced using genetic engineering methods.
- Discuss the reason aspirin is effective as an analgesic, a fever reducer, and an anti-blood-clotting agent.
- Research the differences between "diabetes mellitus" and "diabetes insipidus."
- As children many of us were told that "sleeping makes you grow." Is there any scientific basis for this statement?

Films, Filmstrips, and Videos

- *Hormone Controls in Human Reproduction.* MGHT, 16 minutes, 16 mm, color.
- *Menstrual Cycle.* Lilly, 1971, 12 minutes, 16 mm, color. Explains the basic hormonal and histological changes that occur in the ovaries, uterus, and other organs and tissues during the normal menstrual cycle.
- *Principles of Endocrine Activity.* IU, 1960, 16 minutes, 16 mm, color.
- *The Endocrine System.* EBEC, 1982, 20 minutes, 16 mm, color. Using animation, this film describes the major endocrine glands and their hormones. Clarifies their contribution to growth, reproduction, metabolism, and homeostasis. Explains the regulatory mechanism of bio-feedback.

27

SENSORY SYSTEMS

Revision Highlights

Minor changes only. For example, details of organs of equilibrium in the human ear are moved to Figure 27.7 caption to keep text flowing.

Chapter Outline

SENSORY PATHWAYS
- **Receptors Defined**
- **Principles of Receptor Function**
- **Primary Sensory Cortex**

CHEMICAL SENSES

SOMATIC SENSES

THE SENSE OF BALANCE

THE SENSE OF HEARING
- **Which Animals Hear?**
- **Echolocation**

SENSE OF VISION
- **Invertebrate Photoreception**
- **Vertebrate Photoreception**
- **Rods and Cones**
- **Processing Visual Information**

SUMMARY

Objectives

1. Know what a receptor is and list the various types of receptors.
2. Contrast the mechanism by which the chemical senses work with that by which the somatic senses work.
3. Understand how the senses of balance and hearing function.
4. Describe how the sense of vision has evolved through time.
5. Draw a medial section of the human eyeball through the optic nerve, identify each structure, and tell the function of each.

Key Terms

neural programs	transducer	vestibular apparatus	scala tympani
stimulus	receptor potential	semicircular canals	round window
receptors	taste receptors	otolith organ	basilar membrane
sensory organs	taste buds	statocyst	organ of Corti
chemoreceptors	olfaction	hearing	tectorial membrane
mechanoreceptors	somatic senses	tympanic membrane	outer ear
photoreceptors	equilibrium	oval window	middle ear
thermoreceptors	hair cells	scala vestibuli	inner ear

cochlea	eyespots	compound eyes	accommodation
echolocation	eyes	ommatidium, -dia	rod cell
photoreception	cornea	mosaic theory	cone cell
phototaxis	retina	sclera	fovea
visual system	iris	choroid	rhodopsin
lens	pupil		

Lecture Outline

I. Introduction
   A. Some parts of our nervous system are silent until they receive a stimulus.
   B. Other neurons, like those concerned with breathing, never rest.
   C. During development, gridworks of neurons are laid down and interact to form neural programs, or patterns of activity that enable one to respond to situations that are likely to occur; for example, a snake can locate prey at night by sensing their body heat.

II. Sensory Pathways
   A. Receptors Defined
      1. A stimulus is any form of energy detected by a receptor.
      2. Receptors are the peripheral endings of sensory neurons.
      3. Sensory organs = receptors + epithelial + connective tissues; they amplify or focus the energy of a stimulus.
      4. Chemoreceptors detect ions or molecules; they include olfactory and taste receptors.
      5. Mechanoreceptors detect changes in pressure, position, or acceleration; they include receptors for touch, stretch, hearing, and equilibrium.
      6. Photoreceptors detect light.
      7. Thermoreceptors detect temperature; they include infrared receptors.
   B. Principles of Receptor Function
      1. A receptor is a transducer.
      2. The plasma membrane may be stimulated and a receptor potential results; the change is a graded response.
      3. While all action potentials never vary in magnitude, different sensations are experienced because the signals from different receptors end up in specific parts of the brain.
      4. Variations in stimulus intensity are encoded by the frequency of action potentials in an axon.
      5. Stronger stimuli may excite more receptors in a given area. **(Fig. 27.2, p. 369; TM 109)**
   C. Primary Sensory Cortex
      1. Sensory nerve pathways from different receptors lead to different parts of the cerebral cortex. **(Fig. 27.3, p. 370; TM 110)**

III. Chemical Senses
   A. Taste receptors in the mouths of vertebrates enable them to distinguish nutritious from noxious substances. **(Fig. 27.4, p. 370; TM 111)**
      1. Often part of taste buds are on the tongue.
      2. Taste receptors are often components of taste buds distributed mostly on the tongue.
   B. Olfactory receptors detect odors.
      1. Many animals use pheromones as social signals.
      2. Male silk moths can detect one molecule of bombykol (a sex attractant) in 10^{15} molecules of air.

IV. Somatic Senses
   A. "Somatic senses" refers to sensations of touch, pressure, temperature, and pain near the body surface.
   B. Free nerve endings in skin sense light pressure, temperature, and pain.
   C. Pacinian corpuscles sense deep, rapid pressure and vibration.

D. Pain is the perception of injury to some region of the body; even a leech has pain receptors.
   1. Feeling pain seems to require our limbic system.
   2. Much visceral pain is referred.

V. The Sense of Balance
   A. Most animals have a position known as equilibrium—a baseline against which displacement of a body part is measured.
   B. Hair cells in the fluid-filled vestibular apparatus contribute to our sense of balance.
   C. Jellyfish use a statocyst as an organ of equilibrium.

VI. The Sense of Hearing **(Fig. 27.9, p. 374; TA 82)**
   A. The sense of hearing also depends on the bending of hair cells under fluid pressure.
   B. Sound produces waves, and the peaks are perceived as loudness.
   C. The outer ear collects sound waves.
   D. The internal auditory canal channels the waves inward.
   E. The middle ear is a series of small bones that transfers vibrations from the eardrum to the inner ear.
   F. The inner ear contains a cochlea, where hair cells are stimulated.
   G. Echo location is used by bats, dolphins, and whales. High-frequency sound waves are emitted, and echoes return that can be used to detect predators, prey, and so on.

VII. Sense of Vision
   A. Introduction
      1. Photoreception uses pigments embedded in the membranes of receptor cells to absorb light energy and trigger action potentials.
      2. Even small invertebrates show phototaxis.
      3. A visual system includes structures that focus light onto photoreceptors and a neural grid in the brain that can interpret the patterns.
   B. Invertebrate Photoreception
      1. Ocelli or eyespots function in photoreception.
      2. Mollusks have eyes capable of forming clear images—some are fast moving and need to catch prey. **(Fig. 27.12, p. 377; TM 112)**
      3. Insects and crustaceans have compound eyes with units called ommatidia. **(Fig. 27.14, p. 378; TM 113)**
      4. The mosaic theory suggests that each ommatidium samples a small part of the visual field.
   C. Vertebrate Photoreception
      1. Almost all vertebrates have eyes capable of forming clear images.
      2. The eyeball has a lens, sclera, choroid, a receptor-packed retina, and a transparent cornea covering the front of the eye.
      3. Choroid tissue extends inward to form the iris; a clear fluid fills the space between cornea and iris.
      4. Adjustments of the lens produce accommodation. **(Fig. 27.17, p. 380; TM 114)**
   D. Rods and Cones
      1. Rods are sensitive to dim light and are abundant in the periphery of the retina.
      2. Cones respond to high-intensity light, contribute to sharp daytime vision, and are packed at the fovea. Three types respond to red, green, or blue light.
      3. Each rod has a stack of membranes with rhodopsin molecules.
         a. Light changes the form of rhodopsin, and a change in voltage across the membrane results.
         b. Rods signal bipolar cells, which relay the signal to ganglion cells.
         c. Axons from ganglion cells form the optic nerve that brings the signal to the thalamus, then to the cortex for further processing.
   E. Processing Visual Information
      1. This occurs in the retina as well as the brain.

Suggestions for Presenting the Material

- This chapter presents information that your students will find more familiar because of previous exposure to the material. Most junior and senior high school health and biology classes provide a fair introduction to the sensory organs.
- Assuming your students do possess basic knowledge of the senses, it remains for you to emphasize two areas. The first is to relate the sense receptor and its interpretation within the brain. This is the subject of the initial portion of the chapter. The second is to provide some depth to the students' understanding of sensory receptor mechanisms.
- Two of the more difficult questions beginning students pose are: "How do I distinguish, say, sight from sound?" and "How do I perceive varying intensities of a stimulus?" Emphasize the role of the brain as an interpreter of impulses directed to specific regions by specialized receptors. Also point out that the frequency of action potentials and the number of axons that "fire" provide the quality we call "intensity" of stimulus.
- Each of the senses provides unique input. Try to draw distinctions between those that operate rather independently (for example, sight) and cooperatively (for example, taste and smell).
- As you describe each sense, you should describe the structure of the sense organ, the mechanism of stimulus reception, and the interpretation of that stimulus. For example, the eye perceives light by reaction with chemicals on the retina to give the sensation of degrees of light and color.
- Be sure to point out to the students that incoming light must pass through several neuronal layers before it reaches the rods and cones. They generally get this backwards. Refer them to Figure 27.18c.
- Students frequently have difficulty understanding how contraction of the ciliary muscles causes the suspensory ligaments to relax and the lens of the eye to thicken. It is the opposite of what they expect. Explain how the contraction of these muscles essentially lessens the circumference of the circle of processes to which the suspensory ligaments are attached. As a result the ligaments relax.

Classroom and Laboratory Enrichment

- The simple detection of taste by a blindfolded person is still a student favorite. Ask volunteers to hold their nose and close their eyes while drops of various liquids (vinegar, onion, lemon, and so on) are placed on the tongue. Ask them to identify the substances, but don't respond until you have repeated the experiment but with the students using *both* smell and taste receptors.
- If your lecture room or laboratory can be sufficiently darkened, you can demonstrate the abilities of rods and cones by a simple demonstration. Pull the shades and turn out the lights, quickly pull a red cloth from your pocket, and ask students to identify the color. Substitute other colors, change the light intensity, and wait for iris accommodation as variations in the protocol.
- Attempt to find film footage or photo stills of the highly successful evasive maneuvers that moths are able to make by detecting bat echolocation signals. It is truly a remarkable performance!
- One of the usual practices in optometrists' offices is to take instant photos of the retina. Ask for permission to make 2 x 2 transparencies of several photos, perhaps some exhibiting defects, and show them to the class. If the doctor will speak to the class, even better!
- A model of a cross-section through the organ of Corti is an excellent aid to comprehending this rather complex structure.
- Use dissectible models of the ear and eye to illustrate their structure.
- Use a Snellen chart to demonstrate the visual acuity test.
- If available, have the students use the Ishihara color charts for color blindness. They enjoy this.
- Demonstrate the use of the ophthalmoscope for viewing the retina.
- Demonstrate the use of the otoscope for viewing the tympanic membrane.

Ideas for Classroom Discussion

- Why is it that the tastiest foods are bland and flat when eaten by a person with a bad head cold?
- It's trivia quiz time; name the sense that:
  a. is most easily fatigued (thank goodness!)
  b. can be dulled by smoking
  c. cannot be shut out easily
  d. has more receptors in more places than any other
  e. uses small bones
  f. operates like a camera
- Sometimes musicians are said to have a "trained ear." What does this expression really mean?
- What is the advantage to an insect of having a compound eye?
- Why do many persons in their mid-forties need to use bifocal lenses?
- Is there any scientific basis to that "carrots are good for your eyes" slogan?
- What is "motion sickness"? How can it be controlled?
- In our "civilized" world, many people experience hearing loss as a result of aging, a condition called presbycusis. A study revealed that this condition did not exist in a primitive Sudanese tribe, the Mebans. This study suggests that some environmental factor of the civilized world is responsible for this type of deafness. What is that factor?

Term Paper Topics, Library Activities, and Special Projects

- Scientists are gathering increasing amounts of evidence from laboratory studies and human testing that show gradual hearing loss caused by exposure to highly amplified music. Report on the dangers—are they real or imagined?
- One of the most intriguing subjects is the phenomenon of "phantom" pain. Explore its manifestations.
- Based on your library research, prepare an "awards" list for the animal group that exhibits the keenest of each of the five major senses (sight, hearing, smell, taste, touch).
- How do local anesthetics block the sensation of pain?
- Describe the various problems associated with vision that are correctable with lenses, surgery, drugs, or other means.
- How can animals such as fly larvae (maggots) respond to and move away from light when they have no eyes of any kind?
- How does a detector device that checks blood vessel patterns in the eye and compares them to known records provide a better security system for military installations than do fingerprints?
- Radial keratotomy, a surgical procedure for correcting myopia, is controversial. What does this procedure involve and what are the pros and cons of its use?
- Investigate the differences between sensorineural deafness and conductive deafness.

Films, Filmstrips, and Videos

- *The Ears and Hearing.* EBEC, 1969, 22 minutes, 16 mm, color. Visualizes the structure and functions of the human ear and how it transmits sound waves to the brain. Microphotography, animation, and models are used to explain the principles of sound perception. Various hearing malfunctions are explored, and the use of microsurgery helps explain and show how delicate surgery on the middle ear is carried out.
- *Eyes.* INFORM, 1975, 15 minutes, 16 mm, color. Tracks the evolution of organs that perceive light and form images from protists to mammals, including primitive eye-spots, compound eyes in

arthropods, fish, reptiles, birds, mammals and humans. Explains the role of eyes in survival, and provides images as they appear through the eyes of nonhuman animals.

- *Eyes and Seeing.* EBEC, 1968, 19 minutes, 16 mm, color. *Physiology Series.* Illustrations, models, and macrophotography present visual perception, image formation, and visual brain centers. Anatomy and function of parts of the eye are covered in detail. Experiments with animals in a visual perception laboratory distinguish clearly between the physical and electrochemical processes of seeing.
- *The Senses of Man.* IU, 1965, 18 minutes, color. Indicates the importance of internal and external receptors. Describes the general sense receptors of temperature, pressure, touch, and pain. Depicts the special senses of vision, hearing, taste, smell, and equilibrium.
- *The Sensory World.* IFB, 1976, 30 minutes, 16 mm, color. Clear demonstrations of how the eyes, touch receptors, and proprioceptors work, how multiple inputs to the brain can cause sensory confusion, how sound and vision function according to the wave theory, and what "masking" is. Colorblindness, image construction, and misleading illusions are also demonstrated and explained.
- *Survival and the Senses.* MCGH, 1973, 25 minutes, 16 mm, color. *Behavior and Survival Series.* Uses experiments, observations, and creative film techniques to explain how sense organs work, and how they relate to survival of the species.

28

MOTOR SYSTEMS

Revision Highlights

Largely rewritten. Simpler descriptions of human skeletal structure/function. More straightforward description of types of bones, bone development, and types of joints. New SEMs of osteoporosis. New painting and micrographs of fine structure of skeletal muscle (Figures 28.8 and 28.9). New section on skeletal-muscle interactions, including motor unit recruitment.

Chapter Outline

Objectives

1. Compare invertebrate and vertebrate motor systems in terms of skeletal and muscular components and their interactions.
2. Identify human bones by name and location.
3. Describe some types of injuries that are commonly encountered by people in exercise programs.
4. Explain in detail the structure of muscles, from the molecular level to the organ systems level. Then explain how biochemical events occur in muscle contractions and how antagonistic muscle action refines movements.

Key Terms

muscle fiber	endoskeleton	flat bones	sinovial joints
antagonistic system	axial skeleton	irregular bones	osteoarthritis
hydrostatic skeleton	appendicular skeleton	osteoblasts	rheumatoid arthritis
cuticle	long bones	osteocytes	osteoclasts
exoskeleton	short bones	fibrous joints	runner's knee

osteoporosis	sliding-filament model	sarcoplasmic reticulum	reciprocal innervation
myofibril	cross-bridges	transverse tubule system	motor unit
myofilaments	rigor	origin	twitch
actin filament	rigor mortis	insertion	tetanus
myosin filament	sarcolemma		muscle tone
sarcomere			

Lecture Outline

I. Invertebrate Motor Systems
   A. Sea anemones use circular and longitudinal muscles in an antagonistic system.
   B. In annelids, muscles contract against fluid-filled chambers and a hydrostatic skeleton results.
   C. Spiders use blood under high pressure to power their jumping legs.

II. Vertebrate Motor Systems
   A. Humans have an endoskeleton, and the interactions between bones and muscles form the basis of movements.

III. Skeletal Structure and Function
   A. The human skeleton is divided into the axial and appendicular skeletons.
   B. There are five types of vertebrae: cervical, thoracic, lumbar, sacral, and the coccyx. **(Fig. 28.5, p. 387; TA 83)**
   C. Types of Bones
      There are four types of bones: long (e.g., arms), short (e.g., wrist), flat (e.g., skull), and irregular (e.g., vertebrae).
   D. Development of Bones
      1. Some, like cranial bones, form directly from connective tissue in the embryo.
         a. Osteoblasts secrete fragments of bone tissue that gradually fuse together and remodel into bone.
         b. Mature osteoblasts trapped in their own secretion are called osteocytes.
      2. Most bones form a cartilage model made during embryonic life.
         a. Osteocytes form bone inside the cartilage.
         b. Calcium is deposited and the cartilage cells die.
         c. Intercellular material degenerates to form the marrow cavity.
         d. Bone also forms at the knobby ends until the only cartilage left is at the joints.
   E. Types of Joints
      1. Fibrous joints have no gap between the bones and hardly move; flat cranial bones are an example.
      2. Cartilaginous joints (such as intervertebral disks) also have no gap, but are held together by cartilage and can move a little.
      3. Synovial joints, the most common, move freely.
         a. They are held together by a capsule of dense connective tissue.
         b. The capsule produces synovial fluid that lubricates the joint.
      4. In osteoarthritis, the cartilage at the end of the bone has worn away.
      5. In rheumatoid arthritis, the synovial membrane becomes inflamed, the cartilage degenerates, and bone is deposited into the joint.
   F. Bone Tissue Turnover
      1. Bone is constantly being renewed for remodeling and to maintain Ca^{++} levels.
      2. Osteocytes resorb bone at surfaces during remodeling—under hormonal control to maintain Ca^{++} balance.
      3. Older American women commonly have osteoporosis.
         a. Osteoporosis is associated with decreases in osteoblast activity, sex hormone production, exercise, and Ca^{++} intake.
         b. Excessive protein uptake is also a factor.

IV. Muscle Structure and Function **(Fig. 28.8 left, p. 390; TA 84)**
   A. Each skeletal muscle may be composed of thousands of cells or muscle fibers.
   B. Each muscle contains myofibrils composed of actin and myosin filaments. **(Fig. 28.8 right, p. 391; TA 85)**
   C. The unit of muscle contraction is the sarcomere. **(Fig. 28.9, p. 392; TA 86)**
      1. Z lines serve to anchor actin filaments that form the I bands.
      2. An "A" band in the center of each sarcomere is the location of the myosin filaments.
   D. Mechanisms of Muscle Contraction
      1. A muscle can only achieve movement by contracting—sarcomeres decrease in length.
      2. The sliding-filament model states that the actin and myosin filaments slide over each other during contraction. **(Fig. 28.10, p. 393; TM 115)**
         a. Cross-bridges form between the heads of myosin molecules and actin filaments.
         b. The cross-bridges are then activated and tilt inward, and then the heads detach and reattach.
         c. The energy is derived from ATP.
         d. ATP binds to the heads and causes them to detach and be primed for a new cycle as the ATP is converted into ADP + inorganic phosphate.
         e. At death, there is no ATP to cause the heads to detach, and the body enters rigor mortis.
      3. Creatine phosphate, a stored form of muscle energy, helps to maintain constant levels of ATP.
   E. Control of Muscle Contraction
      1. Three types of membrane are needed: sarcolemma, sarcoplasmic reticulum, and transverse tubule system. **(Fig. 28.11, p. 394; TM 116)**
      2. The signal that triggers contraction occurs at the neuromuscular junction.
      3. An action potential is produced on the muscle cell that travels along the sarcolemma and transverse tubule system.
      4. In response, the sarcoplasmic reticulum releases Ca^{++}, which triggers muscle contraction. **(Fig. 28.12, p. 394; TM 117)**
      5. The pumps in the sarcoplasmic reticulum then sequester the Ca^{++} again, and the muscle relaxes.

V. Skeletal-Muscular Interactions
   A. The skeleton and muscles act as a system of levers in which rods (bones) move about fixed points (joints).
   B. During muscle contraction, the origin remains stationary and the insertion moves.
   C. A limb can be extended and rotated about a joint because of antagonistic pairs of muscles.
   D. Reciprocal innervation contributes to coordination.
   E. Because the axon terminals from one neuron may innervate several muscle cells, the muscle cells are part of the motor unit and contract at the same time.
   F. When one stimulus activates a motor unit, the muscle cells' contraction is called a twitch. **(Fig. 28.14, p. 395; TM 118)**
   G. A motor unit repeatedly stimulated maintains a state of tetanus.
   H. A muscle organ can contract with different strengths of contraction by motor unit recruitment.
   I. When awake, muscles are partly contracted to maintain posture; muscle tone is controlled by the cerebellum.

Suggestions for Presenting the Material

- Although this chapter's title may not immediately reveal it, the contents consist of the *skeletal* and *muscular* systems.
- The discussion of motor systems in invertebrates can be abbreviated here and expanded during your presentation of Chapter 41 (Animal Diversity).

- Students, especially nonscience majors, may question the value of learning the names of bones. However, with today's emphasis on the human body, the use of the anatomical names for body parts, including bones, is becoming more common.
- The formation, development, and replacement of bone is more meaningfully presented by referring briefly to the material in Chapter 23.
- It is nearly impossible to present the ultrastructure of muscle without a visual similar to Figure 28.9. Using the analogy of a rope (see the "Enrichment" section below) is very helpful.
- Don't miss the opportunity when discussing the sliding filament model to emphasize the molecular explanation for the fact that individual muscles can only pull, not push. Again the rope analogy is helpful.
- The dual role of calcium as the provider of bone hardness and as muscle facilitator should be made clear.
- The value of learning the names of the muscles may be questionable, but the popularity of body building has made the names of major muscles almost common knowledge.
- Emphasize that ATP is required for *both* muscle contraction and relaxation as well as for the active transport of calcium back into the sarcoplasmic reticulum.

Classroom and Laboratory Enrichment

- Ask a forensic pathologist to tell the class about the wealth of information that the skeletal remains can reveal about one's health and medical history at the time of death.
- Demonstrate the action of muscle by using a "muscle contraction kit" available from biological supply houses.
- If you have one available, the use of a real human skeleton rather than a transparency will stimulate better student interest and present a three-dimensional aspect to your lecture.
- Arrange for a body builder to appear before the class (more effective if unannounced). Describe the origin, insertion, and function of some major muscles as they are flexed.
- Students can visualize the ultrastructure of muscles more readily if the comparison to a large rope is made. The best is one used for boat anchorage because it is made of many subunits.
- If you can locate a ratchet mechanism (preferably not enclosed), your explanation of the actin/myosin crossbridges (Figure 28.10) will be more comprehensible.
- Show a video of arthroscopic surgery. Many orthopedic surgeons routinely produce such videos and give them to their patients.
- Use models of freely movable joints to illustrate their structure.
- Obtain a fresh beef knee joint to demonstrate the structure and tissues of that joint.
- If available, use a model of a sarcomere to facilitate the students' comprehension of that microscopic structure.

Ideas for Classroom Discussion

- Distinguish between hydrostatic skeleton, endoskeleton, and exoskeleton.
- The exoskeleton and muscle arrangement of ants allows them to accomplish extraordinary feats for their small size. Why are there not larger animals with similar motor system arrangements?
- Consider the present evolutionary state of the human knee. Is it sufficient for the punishment modern athletic activity places on it?
- Why is bone considered "nonliving" by those unfamiliar with its structure?
- What technological and research developments allowed the sliding filament *theory* of the 1960s to become the sliding filament *model* of the 1980s?

- Evaluate this statement: Muscles only pull. If this is true, how can you *push* a door open?
- Which muscles are collectively called the "hamstrings"? In which sports are they most likely to be injured and why?
- What is a "slipped" or herniated disc? What are its most common causes?

Term Paper Topics, Library Activities, and Special Projects

- A challenging task is the assembly of a disarticulated skeleton. If your department has one, attempt to lay the bones out on a lab table in their approximate position and articulation. This is a good group project.
- Investigate the difference(s) in the development of muscles for power (weight) lifting versus development for body sculpting and exhibition.
- Document the development of the sliding filament theory of muscle contraction and the research evidence that supports it.
- Report on the use (and possible misuse) of the terms *tetanus* and *tetany.*
- Report on the technique of arthroscopic surgery.
- Research the following spinal disorders: spina bifida, scoliosis, lordosis, kyphosis.
- Investigate the use of electrical stimulation to accelerate the healing of bone fractures.
- Discuss the muscle disease called myasthenia gravis, its suspected cause, and the type of treatment currently used.

Films, Filmstrips, and Videos

- *Bones and Joints.* BFA, 1971, 8 minutes, 16 mm, color. Describes the skeletal construction of the human body and shows how the skeletal systems of other animals are shaped according to the particular lifestyle.
- *Muscle: A Study of Integration.* CRMP, 1972, 30 minutes, 16 mm, color. Three types of muscle tissue found in humans and other vertebrates are described. Animated figures demonstrate the operations of the three types, and the microscopic structure and biochemical function involved in the sliding filament hypothesis of muscle contraction are explained.
- *Muscular and Skeletal Systems.* NGS, 1988, 20 minutes, 16 mm, color. Explores the architecture and function of striated, smooth, and cardiac muscle and examines the interaction of striated muscle with the human skeleton.
- *Muscular System.* CORT, 1980, 11 minutes, 16 mm, color. *Human Body Series* (2nd edition). X-ray and photomicrography, graphics, and live subjects present structure, functions, and chemical mechanisms of the three types of muscles in the human body.

29

CIRCULATION

Revision Highlights

Major revision. New micrographs of red blood cells and capillaries. New illustration of stem cell derivation of cellular components of blood (Figure 29.4). Refined text and illustrations on cardiovascular system. Simpler description of cardiac cycle. New illustration on blood pressure readings. New Commentary on cardiovascular disorders (including hypertension, coronary artery disease, heart attacks, strokes); describes risk factors (high blood pressure, cholesterol, obesity, smoking, etc.), distinguishes between thrombus and embolus, describes coronary bypass surgery and laser and balloon angioplasty, and covers arrhythmias (bradycardia, tachycardia, ventricular fibrillation), with examples of EKGs. New section on ABO and Rh blood typing. Revised text on lymphoid organs; new illustrations of lymph vessels.

Chapter Outline

CIRCULATION SYSTEMS: AN OVERVIEW

CHARACTERISTICS OF BLOOD

- **Functions of Blood**
- **Blood Volume and Composition**

CARDIOVASCULAR SYSTEM OF VERTEBRATES

- **Blood Circulation Routes**
- **The Human Heart**
- **Blood Pressure in the Vascular System**
- ***Commentary:* On Cardiovascular Disorders**
- **Regulation of Blood Flow**
- **Blood Typing**
- **Hemostasis**

LYMPHATIC SYSTEM

- **Lymph Vascular System**
- **Lymphoid Organs**

SUMMARY

Objectives

1. Describe the composition and functions of blood.
2. Explain how the cardiovascular systems of vertebrates differ from those of invertebrates.
3. Explain the factors that cause blood to exist under different pressures.
4. Describe the composition and function of the lymphatic system.

Key Terms

closed circulation system
open circulation system
lymph vascular system
plasma
plasma proteins
red blood cells
oxyhemoglobin
carbaminohemoglobin
cell count
leukocytes
stem cells

white blood cells
platelets
cardiovascular system
capillary bed
pulmonary circuit
systemic circuit
myocardium
pericardium
endocardium
endothelium
atrium, atria
ventricle
atrioventricular valve
semilunar valve
coronary circulation
aorta
cardiac cycle
systole
diastole
cardiac conduction system
sinoatrial node
cardiac pacemaker
atrioventricular node (= AV node)
intercalated disk
blood pressure
arteries
pulse pressure
arterioles
vasodilation
vasoconstriction
capillary
filtration
absorption
venules
veins
hypertension
coronary artery disease
heart attack
stroke
atherosclerosis
arteriosclerosis
atherosclerotic plaque
thrombus
embolus
angina pectoris
angiography
coronary bypass surgery
laser angioplasty
arrhythmia
electrocardiogram (EKG)
bradycardia
tachycardia
ventricular fibrillation
angiotensin
antibodies
agglutination (= clumping)
Rh blood typing
erythroblastosis fetalis
intrinsic clotting mechanism
extrinsic clotting mechanism
lymphatic system
lymph
lymph vascular
lymph capillaries
lymph vessels
right lymphatic duct
thoracic duct
lymphoid organs
lymph nodes
spleen
thymus

Lecture Outline

I. Circulatory Systems: An Overview
   A. Cell survival depends upon the diffusion of substances in and out of cells.
   B. In large animals, interior cells are too far from the body surface to exchange materials and require a circulation system.
   C. A circulation system consists of blood, a heart, and blood vessels.
   D. Most animals have a closed circulation system. **(Fig. 29.1, p. 397; TM 119a)**
   E. An open circulation system is found in arthropods and mollusks. **(Fig. 29.2, p. 398; TM 119b)**
   F. Even in a closed system, some fluid leaves capillaries and returns via the lymph vascular system.

II. Characteristics of Blood
   A. Functions of Blood
      1. Blood transports products and waste from cells.
      2. It transports nutrients and oxygen to cells.
      3. It contains phagocytic cells that fight infection.
      4. Blood helps stabilize internal pH.
      5. It equalizes body temperatures in birds and mammals.
   B. Blood Volume and Composition
      1. A 150-lb. man has a blood volume of about 5 quarts.
      2. Plasma contains water, and plasma proteins like albumin that influence how much water is in the bloodstream.
         a. Alpha and beta globulins transport lipids and fat-soluble vitamins.
         b. Gamma globulin contributes to the immune response.
         c. Fibrinogen functions in blood clotting.
         d. Plasma contains ions, simple sugars, amino acids, vitamins, hormones, gases, fats, phospholipids, and cholesterol.
      3. Red Blood Cells (Erythrocytes). **(Fig. 29.4, p. 400; TA 87)**
         a. In mammals, red blood cells are biconcave disks that transport oxygen.
         b. Red blood cells contain hemoglobin—an iron-containing protein that binds with oxygen to form oxyhemoglobin.
         c. They form in bone marrow, lose nuclei, and live 120 days.

d. Old cells are removed in the liver and spleen by phagocytic cells.
e. The red blood cell count is maintained by a feedback mechanism involving an enzyme secreted by the kidneys.

4. White Blood Cells **(Fig. 29.4, p. 400; TA 87)**
a. Leukocytes remove dead or worn-out cells and protect us against invading microbes and foreign agents.
b. Leukocytes are derived from stem cells in the bone marrow.
c. There are five kinds of leukocytes and two kinds of lymphocytes.

5. Platelets are bits of cytoplasm that function in blood clotting. **(Fig. 29.4, p. 400; TA 87)**

III. Cardiovascular System of Vertebrates

A. Introduction
1. Heart → arteries → arterioles → capillaries → venules → veins → heart.
2. Total cross-sectional area of a capillary bed is greater than that of the arterioles leading into it—thus, the rate of flow decreases, and the time for exchange of materials increases.

B. Blood Circulation Routes
1. Fish have two capillary beds before blood returns to the heart, hence, two sharp drops in blood pressure occur.
2. Because the mammalian heart is divided into two pumps (the pulmonary and systemic circuits), there is only one capillary bed before blood returns to the heart; thus, higher blood pressure results. **(Fig. 29.6c, p. 402; TA 88)**
3. An exception is the capillary bed in the digestive tract that enters the hepatic portal vein, which leads to another capillary bed in the liver.

C. The Human Heart
1. Heart Structure
a. The myocardium is cardiac muscle tissue.
b. The pericardium is a fibrous connective tissue cover.
c. The endocardium lines the lumen and consists of connective tissue and an endothelium.
d. Each half consists of an atrium and a ventricle separated by an atrioventricular valve.
e. Blood exits each ventricle through a semilunar valve.
f. Valves prevent backflow.
g. Heart muscle cells are serviced by coronary circulation = two coronary arteries + capillary beds.

2. Cardiac Cycle **(Fig. 29.8, p. 404; TM 120)**
a. The cardiac cycle consists of a sequence of contraction (systole) and relaxation (diastole).
b. As the atria relax and fill, the ventricles are relaxed.
c. Pressure forces the atrioventicular valves open, and the ventricles fill as the atria contract.
d. The ventricles contract, the atrioventricular valves close, and blood flows out the pulmonary artery and aorta.

3. Cardiac Conduction System
a. The nervous system controls rate and strength of heartbeat.
b. The cardiac conduction system initiates and conducts action potentials.
c. The sinoatrial (SA) node is the cardiac pacemaker; it causes the atria to contract at the same time.
d. Each wave also spreads to atrioventricular (AV) node; the bundle of His transmits impulse to ventricles.

4. Heart Muscle Contraction **(Commentary art, p. 411; TM 122)**
a. Cardiac muscle cells branch and the plasma membranes of adjacent cells are fused at intercalated disks.
b. Intercalated disks permit rapid conduction of the impulse such that all the cells seem to contract as a unit.

D. Blood Pressure in the Vascular System **(Fig. 29.11, p. 405; TM 121)**
1. Blood pressure drops along the way due to energy loss from resistance.

2. Arterial Blood Pressure **(Commentary art, p. 409; TA 89)**
   a. The arterial wall distends and then recoils in the response to the heart pumping blood.
   b. Thus, the pressure changes associated with the cardiac cycle are smoothed.
   c. Normal systolic pressure is 120 mm Hg, and normal diastolic pressure is 80 mm Mg; thus, the pulse pressure is 40 mm Hg.
   d. A sphygmomanometer is used to measure blood pressure.
3. Resistance of Arterioles
   a. Smooth muscle in arterioles controls the distribution of blood.
   b. Their diameter is controlled by neural and endocrine controls.
   c. Enlargement causes vasodilation.
   d. Contraction causes vasoconstriction.
4. Capillary Function
   a. A capillary is a 1-mm tube specialized for exchange of substances with interstitial fluid.
   b. The diameter of capillaries is so small that red blood cells travel single file.
   c. The density of capillaries is related to metabolic rate.
   d. Gases diffuse through the capillary wall.
   e. Proteins are exchanged by endocytosis and exocytosis.
   f. Bulk flow maintains blood pressure.
   g. Filtration is the outward movement of fluid due to pressure differences.
   h. Absorption is the inward movement of fluid due to differences in solute concentration.
   i. There is a net filtration of fluid, which returns by lymph vessels.
5. Venous Pressure
   a. Capillaries merge to form venule and venules merge into veins.
   b. Some diffusion occurs across the venule wall.
   c. Blood pressure in veins begins at 10–15 mm Hg less than arteries.
   d. In the right atrium, fluid pressure is close to 0 mm Hg.
   e. Valves help to prevent backflow. **(Fig. 29.15, p. 411; TM 123)**
   f. Veins are blood volume reservoirs because their walls can distend or contract.
   g. Body movements can effectively decrease vein diameter. **(Fig. 29.15, p. 411; TM 123)**

E. Regulation of Blood Flow
1. After you eat, blood is diverted to your digestive tract.
2. When you are exposed to the cold, blood is diverted to deeper tissue regions.
3. The medulla oblongata controls blood pressure by affecting the rate and strength of the heartbeat and the diameter of arterioles and veins.
4. Hormones can cause vasoconstriction or dilation of arterioles.

F. Blood Typing
1. Introduction
   a. All cells have surface proteins and other molecules that serve as markers.
   b. Antibodies recognize markers on foreign cells.
2. ABO Blood Typing
   a. ABO blood typing is based upon surface markers of red blood cells.
   b. Type A blood has A markers, type B has B markers, and type O has neither marker.
3. Rh Blood Typing **(Fig. 29.17, p. 413; TM 124)**
   a. An Rh^- person transfused with Rh^+ blood will produce antibodies to the Rh marker.
   b. Rh^- women have a risk in childbirth or pregnancy.
   c. Medical treatment can inactivate the Rh antibodies.

G. Hemostasis

Mechanisms to control bleeding:
1. Spasm of the smooth muscle in the damaged blood vessel stops blood flow for a few minutes.
2. Platelets clump to plug the rupture.
3. The blood coagulates and forms a clot; the clot then contracts.
4. Coagulation mainly involves an intrinsic clotting mechanism. **(Fig. 29.18, p. 414; TM 125)**

5. Extrinsic clotting mechanisms are triggered by substances outside of the blood and lead to thrombin formation.

IV. Lymphatic System
  A. Introduction
    1. The lymphatic system returns excess tissue fluid to the bloodstream via transport tubes.
    2. It contains lymphoid organs, which take part in defense responses.
  B. Lymph Vascular System
    1. The lymph vascular system includes lymph capillaries, lymph vessels, and ducts.
    2. It returns excess fluid and protein and transports fats, and foreign materials are brought to lymph nodes for disposal.
    3. Lymph capillaries end blindly; they lead to lymph vessels, which lead in turn to ducts that return the fluid to the bloodstream. **(Fig. 29.20, p. 415; TA 90)**
  C. Lymphoid Organs
    1. These include the lymph nodes, spleen, thymus, tonsils, and patches of lymphoid tissue in the digestive tract.
    2. They contain lymphocytes that help to fight infections.

Suggestions for Presenting the Material

- This chapter begins with a discussion of closed and open circulatory systems. Students will be familiar with a closed system because that is the type in the human body, but the open concept implies a great deal of "hemorrhaging," which sounds *terminal*, but in this context, is *normal.*
- Some instructors may wish to vary the sequence of topics presented in the text and begin with the "plumbing" prior to the blood composition. The text lends itself well to this arrangement. Nearly all lectures on circulation conclude with the *lymphatic system.*
- One of the strengths of the Starr and Taggart text is the inclusion of excellent summary tables such as Table 29.1, which reviews the components of the blood. These tables should be called to the students' attention, perhaps by using an overhead reproduction.
- When discussing the functions of blood, it may be helpful to refer to future lectures that will expand the topic. Such instances include plasma and nutrient transport (Chapter 32); red blood cells and respiration (Chapter 31); white blood cells and immunity (Chapter 30).
- Because of the complexity of the entire cardiovascular system, an introduction based on Figure 29.6 should survey the pathways throughout the body. The concept of oxygen-rich and oxygen-poor blood can be emphasized here. This can be followed by the heart pathways and contraction sequence.
- The heartbeat, pulse, and blood pressure are measurable quantities of interest to students; therefore, these deserve as much lecture time as possible.
- Capillary exchange is a topic that allows you to review previous lectures on diffusion and active transport.
- The mechanism of clotting, and the somewhat confusing terms used to describe it, may need special attention and/or simplification.
- Blood typing is confusing to students. This is especially true because most believe that the ABO markers and the Rh markers are intimately connected since blood type is nearly always expressed that way, O^+ for example. Stress the independent nature of these two sets of markers.
- Usually students feel they have a fair grasp of circulation until the lymphatic system is introduced. Then it is a "what's this for?" expression. Stress the necessity for such a system and the *one-way return* of fluid to the general blood supply.

Classroom and Laboratory Enrichment

- Ask a physician to demonstrate an artificial pacemaker; include comments on the limitations and usefulness of such an artificial device versus the natural SA node.
- Borrow a sphygmomanometer and ask a person skilled in its use to explain how blood pressure is determined using members of the class as volunteers.
- Determine the resting and active pulse rates for a variety of body types in the class. Can you make any generalizations from your data?
- Obtain a recording of heart sounds as they would be heard through a stethoscope. Relate sounds to events of the normal and abnormal cardiac cycle.

Ideas for Classroom Discussion

- How can a worker in a police crime lab determine if the blood found at a crime scene is human?
- How can you explain the fact that persons who die of heart attack (lack of oxygen to heart muscle) have perfectly adequate amounts of oxygenated blood in their heart chambers?
- The heart is really a "double pump." It is also, of course, divided into four chambers. Does this mean that one pump consists of atria, the other of ventricles, *or* does it mean the left and right sides are pumping to separated circulations? Explain your reasoning.
- Why is the lymph system such a "highway" for the spread of certain metastatic cancers?
- Why is hemophilia in *females* (although extremely rare) very often more fatal than in males?
- Evaluate the truthfulness of the statement "I'm a blood relative." Is there a more accurate expression that could replace this one?

Term Paper Topics, Library Activities, and Special Projects

- Evaluate the reported links between lipids and cardiovascular diseases.
- Survey the various corrective surgical procedures that are routinely performed on the heart and its vessels. Select one or two for an in-depth report.
- It is often said "the heart never rests." True, it does beat continuously from before birth until death, usually at old age. However, it does rest for 0.3 second after each beat. Assuming a steady pulse of 70 beats/minute, calculate the *total* amount of time the heart has rested in a person 75 years old.
- The Rh factor is named for the Rhesus monkey. Trace the historical discovery of this protein marker in primates and humans.
- Deaths of newborn babies from erythroblastosis fetalis (Rh factor related) have been virtually eliminated since the 1950s. How has this been accomplished?

Films, Filmstrips, and Videos

- *The Blood.* EBEC, 1971, 16 minutes, 16 mm, black and white. Live action and microphotography help tell the story of the highly complex tissue that is the lifeline of the body. The composition of the blood, its circulation, and the work that it does to support cell processes are clearly illustrated. Blood typing and the test for the Rh factor are also shown.
- *Blood: The Microscopic Miracle* (2nd ed.). EBEC, 1983, 22 minutes, 16 mm, color. Explores the structure and functions of blood in the human body. Describes blood as the major system of transport in the body and investigates briefly the role of blood transfusion in modern medicine.
- *Circulatory and Respiratory Systems.* NGS, 1988, 17 minutes, 16 mm, color. See, in chamber by chamber detail, the heart in action. Take a trip through arteries, veins, and capillaries. Travel the pulmonary circulation and the systemic circulation.

- *How Blood Clots.* BFA, 1969, 13 minutes, 16 mm, color. Animation and cinemicrography are used to illustrate the human circulatory system and its functions. The composition of blood and its separation into clots and serum are revealed through time-lapse photography. The role of platelets is shown in clot formation.
- *Transport Systems in Animals.* IU, 1971, 17 minutes, 16 mm, color. Examines the heart structures of the earthworm, clam, insect, axolotl, turtle, rat, and fetal pig. The open circulatory system of a damselfly nymph and the closed circulatory systems of the earthworm, chick embryo, perch, frog, and fetal pig are shown.
- *The Work of the Heart.* EBEC, 1968, 19 minutes, 16 mm, color. Pictures and drawings show how the heart and blood vessels circulate blood. Factors that affect the rate of the heartbeat are discussed.

30

IMMUNITY

Revision Highlights

Major revision, updating, reorganization. Updated, more accurate overview illustration and text on cell-mediated and antibody-mediated immune responses. Simpler text and illustration of molecular basis of self-nonself distinction. Includes information on natural killer (NK) cells and perforins. New SEMs of tumor cell being destroyed by killer T cell. Much better text and illustrations of antibody structure and synthesis; straightforward picture of clonal selection theory and antibody diversity. New Commentary on AIDS, with replication cycle, micrographs of HIV. New summary table of white blood cells and their characteristics.

Chapter Outline

NONSPECIFIC DEFENSE RESPONSES
- **Barriers to Invasion**
- **Phagocytes**
- **Complement System**
- **Inflammation**

SPECIFIC DEFENSE RESPONSES: THE IMMUNE SYSTEM
- **The Defenders: An Overview**
- **Recognition of Self and Nonself**
- **Primary Immune Responses**
- ***Commentary:* Cancer and the Immune System**
- **Control of Immune Responses**
- **Antibody Diversity and the Clonal Selection Theory**
- **Secondary Immune Response**

IMMUNIZATION

ABNORMAL OR DEFICIENT IMMUNE RESPONSES
- **Allergies**
- **Autoimmune Disorders**
- **Deficient Immune Responses**
- **Case Study: The Silent, Unseen Struggles**
- ***Commentary:* Acquired Immune Deficiency Syndrome (AIDS)**

SUMMARY

Objectives

1. Describe typical external barriers that organisms present to invading organisms.
2. Understand the process involved in the nonspecific inflammatory response.
3. Understand how vertebrates (especially mammals) recognize and discriminate between self and nonself tissues.
4. Distinguish between antibody-mediated and cell-mediated patterns of warfare.
5. Describe some examples of immune failures and identify as specifically as you can which weapons in the immunity arsenal failed in each case.

Key Terms

complement system
inflammatory response
mast cells
histamine
vertebrate immune system
specificity
memory
macrophage
helper T cells
B cells
killer T cells
natural killer (NK) cells
suppressor T cells
memory cells
antibodies
lymphokines
interleukins
perforin
nonself cells
self cells
major histocompatibility complex, MHC markers
antigens
primary immune response
plasma cells
immunoglobulins
clonal selection theory
clone
immune therapy
interferons
monoclonal antibodies
secondary immune response
memory lymphocytes
immunization
vaccine
passive immunity
allergy
Ig E
asthma
hay fever
autoimmune response
rheumatoid arthritis
acquired immune deficiency syndrome
human immunodeficiency virus

Lecture Outline

I. Introduction
   A. Smallpox once was a serious worldwide problem and killed about half of those infected.
   B. In 1796, Jenner demonstrated that inoculation with cowpox could protect against smallpox.
   C. Later, Pasteur developed similar vaccinations—they mobilize the immune response.

II. Nonspecific Defense Responses
   A. Barriers to Invasion
      1. Intact skin is an important barrier.
      2. Ciliated, mucous membranes sweep out bacteria and particles.
      3. Exocrine glands secrete lysozymes that degrade the cell wall of bacteria.
      4. Gastric fluid in the stomach kills many pathogens.
      5. The normal microbial flora of the gut and vagina keep the growth of pathogens in check.
   B. Phagocytes
      When the skin is cut, phagocytes destroy foreign cells.
   C. Complement System **(Fig. 30.2, p. 419; TM 126)**
      1. The complement system is a set of plasma proteins that enhance nonspecific and specific defense responses.
      2. It becomes activated when it encounters foreign cells.
      3. It attracts phagocytes and can promote lysis of foreign cells.
   D. Inflammation
      1. While complement proteins are being activated, basophils and mast cells secrete histamine.
      2. Histamine dilates blood vessels, and fluid seeps into the area.
      3. Clotting mechanisms keep blood vessels intact and wall off the infected or damaged area.
      4. Warmth, redness, and swelling result.

III. Specific Defense Responses: The Immune System **(Fig. 30.3, p. 421; TA 91)**
   A. Introduction
      1. Phagocytes summoned to an inflammation site are indiscriminate engulfers and may not be enough to check the spread of an invader.
      2. Macrophages and T and B lymphocytes of the vertebrate immune system may be needed.
      3. Interactions among these cells are the basis of the vertebrate immune system.
      4. This system shows specificity and memory.
   B. The Defenders: An Overview

1. Macrophages are phagocytic and alert helper T cells to the presence of specific foreign agents.
2. Helper T cells stimulate division of B and killer T cells.
3. B cells produce antibodies.
4. Killer T cells and natural killer cells destroy body cells infected with viruses or fungi and may kill cancer cells.
5. Suppressor T cells slow down or prevent immune responses.
6. Memory cells (derived from T or B cells) quickly respond to subsequent invasions.
7. Antibodies bind with foreign targets and tag them for destruction by phagocytes or complement.
8. Lymphokines and interleukins are secretions used by white blood cells to communicate with each other.
9. Thus, there are cell-mediated and antibody-mediated responses. **(Fig. 30.3, p. 421; TA 91)**

C. Recognition of Self and Nonself
   1. Lymphocytes use two kinds of receptors to recognize the following: **(Fig. 30.4, p. 422; TM 127)**
      a. Proteins encoded by the major histocompatibility genes (used to recognize "self"—MHC markers).
      b. Millions of antigens of "nonself" origin can be recognized—but how can so many proteins be encoded?
   2. Antibodies consist of four polypeptide chains.
   3. Variable regions of antibodies are coded by gene segments that undergo recombination in each lymphocyte.

D. Primary Immune Responses
   1. The first encounter with a bacterium may result in an inflammatory response.
   2. A macrophage may engulf a bacterium, and lysosomes fuse with and kill the invader.
   3. The antigens move to the plasma membrane and bind with MHC markers; this triggers helper T cells.
   4. Upon coupling to a helper T cell, the macrophage secretes interleukin-1.
   5. In response, helper T cells secrete lymphokines; this stimulates helper T cells to divide and calls B and T cells into action.
   6. Antibody-Mediated Immune Response **(Fig. 30.5, p. 423; TA 92)**
      a. A virgin B cell has thousands of antibody molecules at its surface.
      b. If these receptors join with antigens, we have a "sensitized" B cell with antigen-MHC complexes at its surface.
      c. In response, a helper T cell secretes lymphokines and the B cell divides and the progeny develop into plasma cells.
      d. Plasma cells secrete antibodies—2,000 molecules per second—and die in less than a week.
      e. Antibodies bind with antigens and mark them for disposal by other agents.
      f. Different immunoglobulins (Ig) enlist different cells and substances.
      g. The IgM and IgG antibodies activate macrophages and the complement system.
      h. The IgE antibodies stimulate mast cells to secrete histamine.
      i. Targets of antibodies include bacteria and extracellular phases of viruses, fungi, and protozoa.
   7. Cell-Mediated Immune Response
      a. Cells infected with an intercellular parasite display antigen-MHC complexes at their surface.
      b. Killer T cells will recognize these complexes, punch a hole in the cell, and kill it before the parasite can reproduce.
      c. After some killer T cells are activated to a specific antigen, cell division will produce an army of them.
      d. Killer T cells also attack organ transplants and mutant and cancerous cells.

E. Control of Immune Responses
   1. Cell-mediated and antibody-mediated immune responses are regulated events.
      a. An example is feedback inhibition in antibody production.

F. Antibody Diversity and the Clonal Selection Theory **(Fig. 30.6, p. 426; TA 93)**
A lymphocyte activated by a specific antigen will divide and give rise to a clone of cells that are specific only to that antigen. **(Fig. 30.7a, p 427; TA 94)**

G. Secondary Immune Response
1. A secondary immune response is a greater and longer response than a primary response. **(Fig. 30.8, p. 428; TM 128b)**
2. It is caused by a large population of memory cells. **(Fig. 30.7b, p. 427; TM 128a)**

IV. Immunization
A. Immunization involves a deliberate production of memory cells by a vaccine—a primary immune response.
B. It may use killed or weakened bacteria or viruses.
C. One can also incorporate antigen-encoding genes from one pathogen into a different organism. For example, the vaccinia virus has been used to immunize animals against hepatitis B, influenza, and rabies.

V. Abnormal or Deficient Immune Responses
A. Allergies
1. An allergy is a secondary immune response to a harmless substance.
2. Exposure triggers production of IgE antibodies, which cause the release of histamines and prostaglandins.
3. A local inflammatory response results; death can even occur.

B. Autoimmune Disorders
In autoimmune disorders lymphocytes turn against the body's own cells; rheumatoid arthritis is an example.

C. Deficient Immune Responses
1. When cell-mediated immunity is weakened, infections that would normally not be serious become life-threatening.
2. In acquired immune deficiency syndrome (AIDS), the cause is the HIV virus. **(Commentary art, p. 431; TA 95)**

D. Case Study: The Silent, Unseen Struggles
1. Suppose that you step on a tack and several thousand bacteria are brought into your body.
2. Blood pools and clots around the wound.
3. Basophils and mast cells secrete histamine; vasodilation and greater permeability increase the supply of the complement system.
4. Phagocytes engulf some foreign objects and bacteria, but the bacterial divisions outpace the phagocytes.
5. Some memory lymphocytes join with antigens and become activated—they divide rapidly in response.
6. By the third day, a peak of antibody production turns the tide, and continues for two weeks until all invaders are killed.
7. Memory lymphocytes continue to circulate and wait for a future battle.

Suggestions for Presenting the Material

- A quick glance at Figures 30.3 and 30.5 will convince anyone that the subject of this chapter is a difficult one both to teach and to learn.
- The understanding of defense against foreign organisms by the human body is complicated by the fact that so many mechanisms and factors are operating at the same time.
- You may wish to organize your lectures around "lines of defense." For example, the "nonspecific defense responses" are sometimes referred to as the first and second lines of defense. Furthermore, the "barriers to invasion" would be a first line and "inflammation" (including "phagocytes" and "complement") would be a second line.

- The third line of defense—the immune system itself—is more specific and more intricate. Because of this complexity, either or both of the figures referred to above should be used as a guide throughout your lecture(s).
- Students sometimes have difficulty distinguishing "antibody" from "antigen." This may help: antigen is short for *anti*body *gen*erator.
- The function of the *thymus gland* was mentioned briefly in Chapter 26. Here, details of its role in immunity are given.
- The topics of *immunization* and *immune diseases* are always of interest to students and should be given sufficient time to allow for student discussion.

Classroom and Laboratory Enrichment

- The dramatization of the suffering and death of a victim of AIDS (or a similar disease) will serve as an attention-getter for this topic—even more so if the victim was a personal acquaintance of the instructor.
- Tracking down the cause of an annoying allergy can involve some real detective work. Survey the class for such an experience, and ask for a brief oral report if the person is willing to share his/her experience.
- With the assistance of a microbiology student, prepare petri dishes onto which smears from the human mouth, nose, hands, head, as well as commonly touched surfaces, are made. Identify the microorganisms present in each location.
- Are insect bodies filth carriers? Attempt to answer this by letting different insects including a cockroach, house fly, and cricket crawl over the surface of an agar-filled petri dish.

Ideas for Classroom Discussion

- Where does our term *vaccination* derive its meaning? Was it first used as a medical term as it now is?
- Every year a small number of children die from diseases that develop as a result of vaccines received to protect them. It seems to be an inherent hazard associated with mass preventative inoculation. Is it worth the risk? Can you debate both sides of the issue?
- Thomas Malthus (Chapter 2) proposed three "grim reapers" that would restrain human population growth. One of these was "pestilence," or disease. How effective is disease as a population limiting factor in the developed countries versus the underdeveloped countries?
- If there are so many infectious people as patients in hospitals, why aren't doctors and nurses continuously ill?

Term Paper Topics, Library Activities, and Special Projects

- Controversy still surrounds the polio vaccines—Salk and Sabin. Explore the details of how each of these vaccines is made and used. Include the advantages and disadvantages of each.
- Research on the cause and treatment of AIDS has been rapid and continues to progress. Report on the latest strategies. Use *Scientific American,* October 1988 issue, for background material.
- Although vaccines are available throughout the world for the prevention of measles, diphtheria, and polio, there are about twenty other infectious diseases for which vaccines could be developed but there seems to be little incentive to do so. Report on the reasons why this is so.
- Polio and smallpox have been conquered by effective vaccination programs. Prepare a report on the development of the vaccine for either of these diseases; be sure to include the chronology of events leading up to the marketing of the vaccine.

Films, Filmstrips, and Videos

- *AIDS: What Everyone Needs to Know* (Revised). IU, 1987, 19 minutes, 1/2 inch VHS, color. Describes the symptoms of AIDS and ARC; discusses modes of transmission and ways to avoid being infected.
- *Microorganisms That Cause Disease.* MLA-Wards, 1977, 11 minutes, 16 mm, color. A clear and comprehensive showing of the five principal groups of pathogenic microorganisms: bacteria, viruses, fungi, rickettsiae, and protozoa. Stresses that microorganisms that are pathogenic cause infectious diseases through destruction of cells. Electron micrographs, photomicrographs, and animated diagrams show the structure of various microorganisms.
- *Our Immune System.* NGS, 1988, 25 minutes, 16 mm, color. Meet the phagocytes, lymphocytes, and T cells of our immune system and learn how they recognize and combat invaders. From allergies to rheumatoid arthritis to AIDS, this film examines challenges to the immune system, and how science is helping. Excellent microphotography.

31

RESPIRATION

Revision Highlights

Only minor changes in this well-received chapter; for example, new illustration of ventilation (Figure 31.9).

Chapter Outline

SOME PROPERTIES OF GASES

RESPIRATORY SURFACES

- Integumentary Exchange
- Gills
- Tracheas
- Lungs

OVERVIEW OF VERTEBRATE RESPIRATION

HUMAN RESPIRATORY SYSTEM

- Air-Conducting Portion
- Gas Exchange Portion
- Lungs and the Pleural Sac

VENTILATION

- Inhalation and Exhalation
- Lung Volumes

GAS EXCHANGE AND TRANSPORT

- Gas Exchange in Alveoli
- Gas Transport Between Lungs and Tissues

MATCHING AIR FLOW AND BLOOD FLOW DURING VENTILATION

- Neural Control Mechanisms
- Local Control Mechanisms
- Hypoxia

HOUSEKEEPING AND DEFENSE IN THE RESPIRATORY TRACT

- When Defenses Break Down
- The Heimlich Maneuver

SUMMARY

Objectives

1. Understand the behavior of gases and the types of respiratory surfaces that participate in gas exchange.
2. Understand how the human respiratory system is related to the circulatory system, to cellular respiration, and to the nervous system.
3. List some of the things that go awry with the respiratory system and describe the characteristics of the breakdown.

Key Terms

respiration
rate of diffusion
partial pressure
integumentary exchange
gill
external gills

internal gills
countercurrent flow
tracheas
lung
airways
blood vessels
pulmonary ventilation
nasal cavities
pharynx
larynx
epiglottis
true vocal chords
glottis
trachea
bronchus, -chi
respiratory bronchioles
alveolus, -oli
alveolar ducts
alveolar sac
diaphragm
pleural sac
intrapleural space
pleurisy
oxyhemoglobin, HbO_2
carbaminohemoglobin, $HbCO_2$
carotid bodies
aortic bodies
hypoxia
carbon monoxide poisoning
emphysema
lung cancer
Heimlich maneuver

Lecture Outline

I. Introduction
   A. Aerobic metabolism requires oxygen.
   B. Respiration is acquiring oxygen and eliminating carbon dioxide.
   C. All systems use diffusion, which may be enhanced by bulk flow.

II. Some Properties of Gases
   A. Diffusion depends upon differences in concentrations between two regions.
   B. Partial pressure is the pressure exerted by one gas in a mixture.
   C. Total atmospheric pressure at sea level = 760 mm Hg.
   D. If oxygen is 21% of pressure, 760 x 0.21 = 160 mm Hg (= partial pressure of oxygen).
   E. A gas will diffuse from a region of higher partial pressure to a region of lower partial pressure.

III. Respiratory Surfaces
   A. Integumentary Exchange
      1. In small animals with low metabolic rates, the body surface is used for integumentary exchange.
      2. Gases diffuse through a thin, moist epidermis.
      3. Integumentary exchange is found in most annelids, some arthropods, some mollusks, and amphibians.
   B. Gills
      1. A gill has a moist, thin, epidermis.
      2. Adult fishes have internal gills arranged such that water flow is opposite from blood flow (this is called countercurrent flow).
   C. Tracheas
      Tracheas are air-conducting tubes found in insects, centipedes, and some mites and spiders.
   D. Lungs
      1. The lungs contain internal respiratory surfaces shaped as a cavity or sac.
      2. In fishes, 450 million years ago, primitive lungs began a path of evolution into swim bladders or complex respiratory structures. **(Fig. 31.4, p. 438; TM 129)**
      3. Paired lungs are the respiratory surfaces in all reptiles, birds, and mammals. **(Fig. 31.5, p. 439; TM 130)**

IV. Overview of Vertebrate Respiration
   A. Airways carry gases to and from one side of the respiratory surfaces.
   B. Blood vessels carry gases to and from the other side of internal environments.
   C. Pulmonary ventilation moves air by bulk flow into and out of lungs.
   D. Pulmonary circulation enhances diffusion of dissolved gases in lung capillaries.

V. Human Respiratory System
   A. Air-Conducting Portion
      1. Through nasal cavities, air enters or leaves the respiratory system; hair and cilia filter dust and particles. Blood vessels warm air and mucus moistens air.

2. Air moves to the pharynx → epiglottis → larynx with vocal cords—open space is the glottis → trachea → bronchi → bronchioles → lungs.

B. Gas Exchange Portion

1. Terminal bronchioles divide into respiratory bronchioles that have outpouchings called alveoli.

C. Lungs and the Pleural Sac **(Fig. 31.8, p. 442; TM 131)**

1. Human lungs are a pair of organs in the rib cage above the diaphragm.
2. They contain more than 300 million alveoli (50–100 square meters surface area) + blood capillaries.
3. Each lung lies in a pleural sac consisting of epithelium + loose connective tissue.
4. Inflamed, swollen pleural membranes with painful breathing is called pleurisy.

VI. Ventilation

A. Inhalation and Exhalation **(Fig. 31.9, p. 442; TA 96)**

1. To inhale, the diaphragm contracts and flattens, and the volume of the chest cavity increases.
2. Muscles move the rib cage and contribute to an increase in chest cavity volume.
3. To exhale, the muscles relax and elastic lung tissue recoils, as the chest cavity volume decreases.

B. Lung Volumes

1. At rest, 500 mm of air enter or leave as tidal volume.
2. The vital capacity is the maximum capacity in a deep breath.
3. At rest, about 150 mm remain in air-conducting tubes; hence, 350 mm of fresh air reach the alveoli per breath.

VII. Gas Exchange and Transport

A. Gas Exchange in Alveoli **(Fig. 31.10, p. 443; TM 132)**

1. Gas in alveoli diffuses across the alveolar wall, a thin film of interstitial fluid, and the capillary wall. **(Fig. 31.11, p. 443; TM 133)**

B. Gas Transport Between Lungs and Tissues

1. While diffusion can move enough oxygen into the lungs, not enough oxygen can dissolve in the blood to meet the needs of the body; hemoglobin increases the oxygen-carry capacity by seventy times.

C. Oxygen Transport

1. Each hemoglobin molecule combines with four oxygen molecules to form oxyhemoglobin.
2. The amount of oxygen that combines with hemoglobin depends upon the partial pressure of the gas.
3. Oxyhemoglobin gives up oxygen when the partial pressure is low, CO_2 production and temperatures are high, and the pH declines—conditions characteristic of active tissues.

D. Carbon Dioxide Transport

1. CO_2 diffuses into the lungs as dissolved CO_2, $HbCO_2$, and HCO_3^-.
2. Most is transported as bicarbonate ions because . . .
   $CO_2 + H_2O \rightleftarrows H_2CO_3 \rightleftarrows HCO_3^- + H^+$
3. In red blood cells, carbonic anhydrase speeds up this reaction 250 times.
4. HCO_3^- tends to diffuse out of the red blood cells into the plasma.
5. In the lungs, because of the low partial pressure of CO_2 in the alveoli, the reactions proceed in reverse.

VIII. Matching Air Flow and Blood Flow During Ventilation

A. Gas exchange in the alveoli is most efficient when air flow equals the rate of blood flow.

B. Neural Control Mechanisms

1. Control oxygen and carbon dioxide levels in arterial blood.
2. Involve respiratory centers in the brain, chemoreceptors in the brain and arteries, and respiratory muscles.
3. The medulla oblongata coordinates contractions needed for inhalation or exhalation.

4. The respiratory centers in other parts of brain fine-tune contraction.
5. Chemoreceptors in the medulla are stimulated by a shift in pH.
6. Other chemoreceptors (carotid bodies and aortic bodies) are stimulated by decreased oxygen in arterial blood.

C. Local Control Mechanisms
1. The diameter of bronchioles and blood vessels increases or decreases in response to needs.

D. Hypoxia
1. Hypoxia occurs when tissues do not receive enough oxygen.
2. At high altitudes the partial pressure of oxygen is lower than at sea level, so that hyperventilation results.
3. CO combines with hemoglobin 200 times faster than oxygen; CO poisoning can result.

IX. Housekeeping and Defense in the Respiratory Tract

A. Introduction
1. Large particles are filtered out by the nose.
2. Small particles are filtered out by cilia in air-conducting tubes.

B. When Defenses Break Down
1. Bronchitis
   a. Smoking and other irritants increase mucus secretion and diminish cilia function and numbers.
   b. Coughing attempts to clear mucus—scar tissue results that can obstruct the respiratory tract.
2. Emphysema
   a. Emphysema results when extensive scar tissue builds up and gas exchange is compromised.
   b. Emphysema is related to environmental conditions, diet, infections, and genetics.
3. Effects of Cigarette Smoke
   a. These effects prevent the cilia from beating and stimulate mucus production.
   b. Coughing can contribute to bronchitis and emphysema.
   c. Cigarette smoke kills phagocytic cells in respiratory epithelium.
   d. It contains compounds, like methylcholanthrene, that are modified in the body to form carcinogens.
   e. Smoking causes 80% of lung cancer deaths.

C. The Heimlich Maneuver **(Fig. 31.15, p. 448; TM 134)**
1. Strangulation can result when food enters the trachea instead of the esophagus.
2. The Heimlich maneuver can elevate the diaphragm and use air to dislodge the obstruction.

Suggestions for Presenting the Material

- The presentation of respiration can be effectively accomplished by following the sequence of the text. The outline begins with a survey of respiratory surfaces, then focuses on human respiratory organs and function, and concludes with mention of respiratory problems.
- When describing various respiratory surfaces, emphasize the one common feature they all share—moisture.
- Distinguish the arthropod *tracheae* from other respiratory surfaces by its independence from the circulatory system.
- The most logical way to present the human respiratory system is to display Figure 31.6 and follow the pathway of an inhaled breath.
- The names of most of the structures in Figure 31.6 are familiar to students except the most critical structures—the *alveoli.* These deserve the special attention afforded them in the section "Gas Exchange and Transport."

- As you describe the *glottis* and *epiglottis*, you may wish to include a note on the *Heimlich maneuver* described at the end of the chapter.
- Emphasize the correct pronunciation of *larynx* (lair—inks) and *pharynx* (fair—inks).
- Be sure to stress the passivity of the lungs (compare to balloon) during respiration, and emphasize the role of the diaphragm and rib muscles.
- Highlight the differences between the ways oxygen and carbon dioxide are transported in the blood.
- *Respiratory control* is a fitting capstone to previous discussions.
- Emphasize that the respiratory organs have some nonrespiratory functions such as coughing, sneezing, speech, yawning, regulation of pH, and sense of smell.

Classroom and Laboratory Enrichment

- Show that measurement of oxygen consumption can be used to determine rate of metabolism by using a simple respirometer (see a biological supply catalog), or use a computer simulation of the experiment.
- Ask for volunteers to debate the "nonsmokers versus smokers" rights. Ask them to provide data, rather than emotional attacks.
- Have a person who has visited a high altitude location report on the breathing discomforts she/he experienced.
- Construct or purchase a working model of the lungs, chest cavity, and diaphragm.
- Demonstrate the Heimlich maneuver on a class volunteer; then ask classmates to demonstrate (gently!) on one another.
- Have the students use a spirometer to measure respiratory air volumes. Compare the results of smokers and nonsmokers.
- Invite a scuba diver to discuss the special gas composition and gas exchange problems associated with deep dives.
- Use a chart or dissectible mannequin to locate major organs of the human respiratory system.
- Exhibit a model of a human larynx and trachea.
- Display a freeze-dried preparation of a sheep's lungs.

Ideas for Classroom Discussion

- Which of the following *respiratory surfaces* does not require participation of the blood for delivery of oxygen to the tissues? (a) integument; (b) gills; (c) tracheae; (d) lungs.
- What is the singular advantage of the one-way flow of air in the *parabronchi* of the bird lung?
- What is happening in the condition we call "hiccups"? What causes it? What are the best short-term remedies? On what physiological principles (if any) are they based?
- Why is the best body position for public speaking and singing a standing, or "sitting tall," position?
- The air at high altitudes is sometimes described in everyday language as "thin." How does this translate in technical terms?
- Can you die from holding your breath? Explain the neural mechanisms that are operating here.
- What are the functions of the sinuses?
- "Respiratory distress syndrome" (hyaline membrane disease) is the primary cause of respiratory difficulty in immature newborns. What are its symptoms and cause? How is it treated?

Term Paper Topics, Library Activities, and Special Projects

- Vertebrate muscular activity is dependent on an oxygen supply carried by the blood (specifically red cells). When demand exceeds supply, an *oxygen debt* is incurred and activity slows or stops. Such is not the case with insects with their indefatigable flight muscles. Investigate why this difference exists.
- Certain woodwind musicians have perfected a breathing technique that allows the production of continuous sound from the instrument. Investigate reports of this technique, and interview a practitioner. Arrange a demonstration if possible.
- Check a book of world records for the longest continuous bout of hiccups. What is the "ultimate" cure for a persistent case?
- How many people in the United States die of simple choking each year? What age is most affected? What is the most commonly lodged object?
- Using cigarette ads gathered from magazines published during the past forty years, show the change in public attitude toward smoking.
- How might the anatomy and physiology of persons who were born and raised at very high altitudes be different from those who were born and raised in the lowlands and have only recently become acclimated to high elevations?
- How does the new "smokeless" cigarette work? Why is the American Medical Association so adamantly opposed to its approval by the Food and Drug Administration?
- Investigate the effects of air pollution on respiratory functions.

Films, Filmstrips, and Videos

- *The Lungs and Respiratory System.* EBEC, 1975, 16 minutes, 16 mm, color. Warns of the hazards of smoking and air pollution. Employs models, scanning electron microscopy, graphics, and demonstrations to illustrate the anatomy and functions of lungs and the respiratory system.
- *Respiration in Man.* EBEC, 1969, 26 minutes, 16 mm, color. A great deal of the respiratory system is seen in this film. An isolated animal lung is examined to demonstrate the role of blood in the lung. Demonstrations with unique laboratory equipment provide visual proof of the exchange of carbon dioxide for oxygen.
- *Respiratory Systems in Animals.* IU, 1971, 14 minutes, 16 mm, color. The interactions of transport systems with respiratory organs are shown in earthworms, axolotls, clams, crayfish, fish, frogs, fetal pigs, and humans. X-ray photography, experimentation, and dissections are presented.

32

DIGESTION AND ORGANIC METABOLISM

Revision Highlights

Only minor editing.

Chapter Outline

TYPES OF DIGESTIVE SYSTEMS AND THEIR FUNCTIONS

HUMAN DIGESTIVE SYSTEM: AN OVERVIEW

- **Components of the Digestive System**
- **General Structure of the Gastrointestinal Tract**
- **Gastrointestinal Motility**

CONTROL OF GASTROINTESTINAL ACTIVITY

STRUCTURE AND FUNCTION OF GASTROINTESTINAL ORGANS

- **Mouth and Salivary Glands**
- **Pharynx and Esophagus**
- **Stomach**
- **Small Intestine**
- ***Commentary:* Human Nutrition and Gastrointestinal Disorders**
- **Large Intestine**
- **Enzymes of Digestion: A Summary**

HUMAN NUTRITIONAL NEEDS

- **Energy Needs**
- **Carbohydrates**
- **Fats**
- **Proteins**
- **Vitamins and Minerals**
- **Objective and Subjective Views of Obesity**

ORGANIC METABOLISM

- **The Vertebrate Liver**
- **Absorptive and Post-Absorptive States**
- **Controls Over Organic Metabolism**
- **Case Study: Feasting, Fasting, and Systems Integration**

SUMMARY

Objectives

1. Know the various ways in which a digestive system can be structured. Realize the behavioral limitations of organisms with incomplete digestive systems.
2. Understand the structure and function of the human digestive system.

3. Summarize the daily nutritional requirement of a 25-year-old man who works at a desk job and exercises very little. State what he needs in energy, carbohydrates, proteins, and lipids and name at least six vitamins and six minerals that he needs to include in his diet every day.
4. Explain how the human body manages to meet the energy and nutritional needs of the various body parts even though the person may be feasting sometimes and fasting at other times.

Key Terms

nutrition
systems integration
digestive system
extracellular fluid
incomplete digestive system
complete digestive system
lumen
discontinuous feeding habit
ruminants
motility
secretion
digestion
absorption
gastrointestinal tract
mucosa
submucosa
muscle layer
serosa
pyloric sphincter
peristalsis
segmentation
sphincters
mouth
tooth
molars
incisors
cuspids
saliva
salivary glands
salivary amylase
mucin
pharynx
esophagus
stomach
pepsinogen
pepsins
peptic ulcer
small intestine
villi
microvillus
pancreas
trypsin
chymotrypsin
carboxypeptidase
aminopeptidase
insulin
glucagon
liver
bile
gallbladder
micelle
colon, large intestine
feces
rectum
anus
appendix
appendicitis
bulk
essential amino acids
net protein utilization
vitamins
minerals
trace elements
obesity
anorexia nervosa
bulimia
absorptive state
post-absorptive state
beta cells
insulin
alpha cells
glucagon

Lecture Outline

I. Introduction
   A. Nutrition considers how the body takes in, digests, absorbs, and uses food.
   B. Systems integration considers how systems interact to meet the metabolic needs of the organism.
   C. When a bear eats a salmon, nutrition includes more than what happens in the gut.
   D. The integration of digestive, circulatory, and respiratory systems must be considered.

II. Types of Digestive Systems and Their Functions
   A. A digestive system is a body cavity or tube where food is reduced to molecules that can cross the lining of the gut.
   B. An incomplete digestive system (for example, in a planarian) has one opening.
   C. A complete digestive system is a tube with two openings where food moves in one direction through the lumen.
      1. Regions may be specialized for one function, such as transport.
      2. Animals with discontinuous feeding habits may have organs for storage.
      3. Ruminants (e.g., cows) eat grass continuously and have multiple stomachs to digest cellulose. **(Fig. 32.4, p. 452; TM 135)**
   D. Four functions of the digestive system include:
      1. Motility, that is, the breaking down, mixing, and transporting of ingested nutrients, and the elimination of undigested residues.
      2. The secretion of needed enzymes and hormones.
      3. The digestion of nutrients by reducing food to molecules small enough to cross the gut lining.
      4. The absorption of digested nutrients into the blood or lymph.

III. Human Digestive System: An Overview
   A. Components of the Digestive System
      1. We have discontinuous feeding habits and our gastrointestinal tract has many specialized regions.
      2. Subdivisions include the mouth, pharynx, esophagus, stomach, small intestine, colon, rectum, and anus.
      3. Accessory glands include the salivary glands, liver, gall bladder, and pancreas.
   B. General Structure of the Gastrointestinal Tract
      1. It is divided into mucosa (epithelium + connective tissue), submucosa, muscle layer, and serosa. **(Fig. 32.5, p. 452; TA 97)**
   C. Gastrointestinal Motility
      1. Peristalsis moves food when rings of circular muscles contract behind it and relax in front of it.
      2. Segmentation is a back-and-forth movement in the intestine that helps to mix the contents of the lumen. **(Fig. 32.7, p. 454; TM 136)**
      3. Sphincters prevent backflow.

IV. Control of Gastrointestinal Activity **(Fig. 32.8, p. 454; TM 137)**
   A. Controls respond to the volume and composition of food in the lumen.
   B. Receptors in the gut wall trigger reflexes in response to breakdown products and secretions.
   C. Endocrine hormones help regulate the digestion and absorption.

V. Structure and Function of Gastrointestinal Organs
   A. Mouth and Salivary Glands
      1. A mammal uses teeth to start the mechanical reduction of food.
      2. Each tooth consists of enamel, dentine, and an inner pulp.
      3. Molars grind, incisors bite, and cuspids grasp and tear.
      4. Saliva contains salivary amylase, bicarbonate ions, and mucins that help form the lubricated bolus.
   B. Pharynx and Esophagus
      1. Voluntary muscle contractions move the bolus into the pharynx.
      2. This stimulates mechanoreceptors that trigger contractions in the pharynx and esophagus—a reflex that moves the food into the stomach.
   C. Stomach
      1. The stomach is a muscular sac that stores and mixes food, secretes substances, and controls the rate that food enters the small intestine.
      2. Components of Gastric Fluid
         a. In the mucosa, exocrine cells secrete HCl, pepsinogens, and mucus.
         b. Endocrine cells release hormones that contribute to the secretion of gastric fluid.
         c. HCl dissolves food to produce chyme, kills microbes, and converts pepsinogens to pepsin.
         d. Pepsins break proteins into peptide fragments.
         e. HCl secretion is stimulated when the stomach is distended and receptors are triggered.
         f. Peptide fragments and caffeine stimulate gastrin release, which in turn stimulates HCl secretion.
         g. Reflexes triggered by food in the mouth or stress can stimulate HCl secretion.
         h. When gastric fluid digests part of the stomach wall, a peptic ulcer results.
      3. Stomach Emptying
         a. Peristaltic waves mix chyme and strong waves cause the sphincter to close. **(Fig. 32.7, p. 454; TM 136)**
         b. The larger the meal, the faster the stomach empties.
         c. Increases in acidity, osmotic pressure, and fat content trigger the release of hormones that inhibit motility.
         d. Fear and depression also inhibit motility.
   D. Small Intestine. The small intestine has three regions: the duodenum, jejunum, and ileum.

1. Intestinal Villi **(Fig. 32.10, p. 457; TM 138)**
    a. Halfway through the small intestine, the chyme contains amino acids, fatty acids, and monosaccharides.
    b. These small organic molecules move across the epithelium.
    c. The mucosa is folded into absorptive villi, and epithelial cells of each villus have crowns of microvilli—both increase the surface area. **(Fig. 32.9, pp. 456–457; TA 97)**
2. Absorption Mechanisms
    a. The uptake of many monosaccharides (glucose) and amino acids requires active transport.
    b. Fatty acids and monoglycerides cross by diffusion.
    c. Glucose and amino acids diffuse into the blood.
    d. In the epithelium, fatty acids and monoglycerides combine into triglycerides and aggregate into chylomicrons. These fat droplets enter lymph vessels.
    e. The small intestine also absorbs water and salts.
3. The Role of the Pancreas in Digestion
    a. Ducts from the pancreas and liver join and empty into the duodenum.
    b. Exocrine cells secrete enzymes in response to hormones and neural signals.
    c. Enzymes digest carbohydrates, fats, proteins, and nucleic acids.
    d. Secretions include trypsin, chymotrypsin, carboxypeptidase, aminopeptidase, and bicarbonate ions.
4. Role of the Liver in Digestion
    a. The liver secretes bile, a solution containing bile salts and pigments, cholesterol, and lecithin.
    b. Bile is stored in the gall bladder between meals.
    c. Bile salts emulsify fats and provide pancreatic lipase with access to triglycerides.

E. Large Intestine
1. The large intestine stores and concentrates feces—undigested and unabsorbed material, water, and bacteria.
2. Na^+ is actively transported into the colon and water follows.
3. The appendix is a projection of the cecum and contains lymphatic tissue.

F. Enzymes of Digestion: A Summary **(Table 32.2, p. 461; TM 139)**

VI. Human Nutritional Needs

A. Early humans ate fruits, seeds, and plants.
B. Today, we eat more fats, sugars, and salts and less fiber.
C. Colon and breast cancer, cardiovascular disorders, kidney stones, and obesity result.
D. Energy Needs
1. To maintain weight, caloric intake must balance energy output.

E. Carbohydrates
1. Complex carbohydrates are the main energy source.
2. Fruits, vegetables, and grains should be at least 55% to 58% of daily caloric intake.
3. The average American eats 128 pounds of sucrose per year.

F. Fats
1. Fats make up 40% of the American diet—they should be less than 30%.
2. The body requires about one tablespoon of polyunsaturated oil per day.

G. Proteins
1. Eight of the twenty common amino acids are essential amino acids.
2. The net protein utilization (NPU) compares different proteins to an ideal of 100.
3. Eating cereal grains with beans increases the overall NPU.

H. Vitamins and Minerals
1. Humans require at least fourteen vitamins from food.
2. Calcium and magnesium are needed for many enzymes to function.
3. Potassium is needed for protein synthesis and muscle and nerve function.
4. Iron is needed for cytochromes and hemoglobin.

I. Objective and Subjective Views of Obesity
1. Obesity involves an excess of fat in the body's adipose tissues.

2. Obese persons have a greater risk of high blood pressure, atherosclerosis, and diabetes.
3. In the United States anorexia nervosa and bulimia are not uncommon.

VII. Organic Metabolism **(Fig. 32.12, p. 466; TM 140)**
   A. The vertebrate liver is central to the storage and interconversion of absorbed carbohydrates, fats, and proteins.
   B. Absorptive and post-absorptive states of organic metabolism.
      1. In the absorptive state, ingested molecules enter the blood from the GI tract.
         a. Pools of organic molecules are built up.
         b. Excess carbohydrates and other molecules are transformed into fats (adipose tissue) and glycogen.
      2. In the post-absorptive state, the GI tract draws from internal pools of organic molecules.
         a. Glycogen and amino acids in the liver are converted to glucose and secreted into the blood.
         b. Most cells use fats as an energy source.
   C. Controls Over Organic Metabolism **(Fig. 32.13, p. 467; TM 141)**
      1. In the islets of Langerhans of the pancreas, beta cells can secrete insulin and alpha cells can secrete glucagon.
   D. Case Study: Feasting, Fasting, and Systems Integration **(Fig. 32.13, p. 467; TM 141)**
      1. Suppose that you take a walk in the woods after you eat a large breakfast.
      2. At first, glucose will begin to rise in your blood.
      3. In response, beta cells will secrete insulin and blood glucose will fall.
      4. After a while, the blood glucose level will fall below normal and glucagon output will increase. **(Fig. 32.14, p. 469; TM 142)**
      5. Then you climb up a high mountain, and you breathe faster and deeper.
      6. You forgot lunch, and your adrenal medulla produces hormones that break down fats.
      7. After days of fasting, ACTH would break down proteins for an energy source.

Suggestions for Presenting the Material

- The topics of *respiration, circulation,* and *excretion* can be correlated with *digestion* by the use of Figure 32.2. This figure is useful no matter what sequential variation of these four systems you use.
- After a brief discussion of *incomplete* versus *complete* digestive systems, the chapter focuses on two main topics: (a) human digestive organs and function and (b) human nutritional needs and metabolism.
- The traditional and most logical method of presentation is to follow a mouthful of food as it passes from mouth to anus. Along the way, you can be as detailed as your course requires. For example, you may ask students to know structures, general secretions, and main functions as presented in Table 32.1 but add only a few, all, or none of the enzymes in Table 32.2.
- Similarly, you will need to inform students as to the amounts of material from Tables 32.4 and 32.5 you require them to learn. Certainly, the authors of the text did not intend for these tables to be memorized entirely but provided them for reference and completeness.
- As each digestive organ is discussed, reference to the overhead transparency of Figure 32.6 should be made. It would also be very helpful if students were following along in Table 32.1 as you lecture.
- You may wish to delay the "Control of Gastrointestinal Activity" section until the entire system has been described.
- At some point in your discussion, be sure to emphasize that *digestion* and *absorption* are inseparable in the total function of providing nutrition to body cells.
- Students will enjoy your lectures more if you include brief notes on the various "problems along the way" that cause us minor, and occasionally major, distress.

- Nutrition is receiving increasingly more emphasis in our lives. You may wish to devote an entire lecture to this timely topic. This is also an excellent opportunity to review the contents of Chapter 4 and especially Table 4.2.
- Although it occupies only a few pages of the text, the discussion of "organic metabolism" and the accompanying figures present a wealth of information. You may need to streamline this material to accommodate the needs of your students.
- This is an ideal time to discuss some of the fad weight-loss diets as well as the more legitimate ones. Students are extremely interested in this topic, and it is a beneficial one for them to discuss.
- When discussing obesity, introduce the concept of body mass index (BMI) as an indicator of the degree of obesity. Consult a nutrition text for details.

Classroom and Laboratory Enrichment

- Because the digestive organs lie cramped in a small space, one upon the other, drawings such as Figure 32.6 are not as useful as life-size models. If your department has one, use it throughout your journey through the G-I tract.
- Demonstrate peristalsis by placing a ball of suitable size inside a flexible tube (such as a section of old bicycle tire innertube) and squeezing to move it along.
- The action of a digestive enzyme (salivary amylase) can be demonstrated using the procedure outlined in the "Enrichment" section for Chapter 7 of this resource manual.
- Show a film depicting the consequences of vitamin and mineral deficiencies in the human diet.
- Ask if any volunteers from within the class, or outside, would be willing to tell the class about his/her digestive or eating disorder.
- Use molecular models to demonstrate the process of digestion.
- Mix oil and water together in a flask to show their immiscibility. Then add soap to the mixture to illustrate emulsification.

Ideas for Classroom Discussion

- What is meant by "heartburn"? Is its use in television antacid advertising misleading especially for young viewers? What would you propose as a better term?
- In an incomplete digestive system, what types of foods would be avoided due to elimination difficulties?
- How is the stomach of a cow like the gut of a termite? How could antibiotics given a ruminant animal for a blood infection interfere with digestive function?
- Millions of people around the world suffer from deficiencies in amount and nutritional composition of food. Surprisingly, dietary disorders of persons in affluent countries result from "too much of good things." What are the consequences of the American diet?
- Diet plans for weight reduction are numerous and proliferating daily. How can the wary consumer recognize a plan that could be dangerous?
- What do you think of the programs that call for regimes of fasting and "purification of body fluids"? Are they biologically sound?
- Give some of the reasons that dietary fiber, such as bran, is so important in our diet.
- Why do some adults, who could drink milk as infants without difficulty, experience intestinal pain (due to gas) and/or dehydrating diarrhea when they drink milk?

Term Paper Topics, Library Activities, and Special Projects

- Survey your class to collect data on height and weight as in Figure 32.11. Does your class data fall within the ranges given here?
- From the library, obtain a "calorie chart." Monitor and record your caloric intake for a week. Prepare a report.
- Ulcers are regarded by most people as a badge of success for bank presidents and young stockbrokers. Investigate the serious nature of this medical problem including the ages of the sufferers and what treatments are prescribed.
- *Anorexia* and *bulimia* seem to have appeared only recently as eating disorders. Trace the history of what is known about these conditions.
- The role of vitamins in human health is a fascinating story. Select one or two vitamins and report on the history of discovery surrounding each.
- Investigate how the drug cimetidine (Zantac) suppresses the secretion of acid by the parietal cells of the stomach.
- Find out why drinking coffee or other caffeine-containing beverages increases the sensation of hunger.

Films, Filmstrips, and Videos

- *Digestion and the Food We Eat (The Mechanics of Life).* BFA, 1971, 9 minutes, 16 mm, color. Presents the process of digestion and examines where the principal processes occur, which enzymes are involved, and how absorption takes place.
- *The Digestive System.* NGS, 1988, 17 minutes, 16 mm, color. See images of chewing and swallowing, of bile being released from the gallbladder, of cilia lining the walls of the small intestine as you journey through the digestive system. Photography by Lennart Nilsson.
- *Digestive Systems in Animals.* IU, 1971, 15 minutes, 16 mm, color. Shows a variety of the ways that animals obtain food and the types of digestive cavities they have. Peristalsis and the action of villi in the small intestine are shown as they occur in a pigeon. Dissections of various anesthetized animals are shown and explained.
- *Human Body: Nutrition and Metabolism.* CORF, 1962, 14 minutes, 16 mm, color.
- *Nutrition: A–Z, Part I.* Cincon, 1975, 25 minutes, 16 mm, color. Presents the American diet: calories, fats, carbohydrates, proteins, minerals, food supplements, vitamins E, C, and B complex. Discusses what is wrong with what the average American eats and explains the process of digestion.
- *Nutrition: A–Z, Part II.* Cincon, 1975, 25 minutes, 16 mm, color. Deals with vitamins A and D, dietary disorders, beef, fish and poultry, vegetables and fruits, milk, food poisoning. Also discusses how infants and children eat and how that affects their health.
- *Nutrition: A–Z, Part III.* Cincon, 1975, 25 minutes, 16 mm, color. Discusses how food is altered by additives. Talks about fad diets and food quackery. Focuses on dietary problems of the aged and of overweight people and discusses "natural" and organic foods.

33

TEMPERATURE CONTROL AND FLUID REGULATION

Revision Highlights

Rewritten for greater clarity. Pilomotor response and peripheral vasoconstriction and vasodilation added. Consequences of extreme sweating (as during marathon races) described. Updated information on fever. Refined picture of water and solute balances in extracellular fluid. New illustration and better explanation of nephron structure and associated blood vessels. Refined description of countercurrent multiplication and control of sodium reabsorption. New section on acid-base balance, using bicarbonate/carbon dioxide buffer system as the example.

Chapter Outline

CONTROL OF BODY TEMPERATURE
- **Temperatures Suitable for Life**
- **Heat Gains and Heat Losses**
- **Classification of Animals Based on Temperature**
- **Temperature Regulation in Mammals**
- ***Commentary:* Falling Overboard and the Odds for Survival**

CONTROL OF EXTRACELLULAR FLUID
- **Water Gains and Losses**
- **Solute Gains and Losses**
- **Urinary System of Mammals**
- **Overview of Urine Formation**
- **A Closer Look at Filtration**
- **A Closer Look at Reabsorption**
- **Acid-Base Balance**
- ***Commentary:* Kidney Failure, Bypass Measures, and Transplants**
- **Case Study: On Fish, Frogs, and Kangaroo Rats**

SUMMARY

Objectives

1. Understand the degrees to which ectotherms, endotherms, and heterotherms can control their body temperatures. Be able to explain how heat gain and loss occur in birds and mammals and how these animals maintain a steady body temperature.
2. Explain how the chemical composition of extracellular fluid is maintained by mammals.

Key Terms

radiation	convection	ectotherms	behavioral temperature regulation
conduction	evaporation		

endotherms
heterotherms
shivering
hypothermic
hyperthermic
fever
excretion
nutrients
mineral ions
waste products
ammonia
urea
uric acid
urine
kidneys
cortex
medulla
nephrons
collecting ducts
renal pelvis
ureter
urinary bladder
urethra
urinary system
urination
glomerulus
glomerular capillaries
Bowman's capsule
proximal tubule
loop of Henle
distal tubule
peritubular capillaries
filtration
bulk flow
reabsorption
secretion
glomerulonephritis
kidney dialysis machine
hemodialysis
peritoneal dialysis
countercurrent multiplication
antidiuretic hormone
aldosterone
hypertension

Lecture Outline

I. Introduction
   A. Animals first evolved in shallow seas where they adapted their tissues to a salty environment and stable temperatures.
   B. As animals invaded the land, their internal environment approximated the seas they left behind.
   C. To maintain a hospitable internal environment, animals make controlled adjustments to the changing environment.

II. Control of Body Temperature
   A. Temperatures Suitable for Life
      1. Most enzymes function within the 0°–40° range.
      2. Above 40°, many enzymes denature.
      3. Enzyme activity is halved for each 10° drop from optimal temperature.
      4. How can animals keep their body temperature constant? They must balance heat gain and loss.
   B. Heat Gains and Heat Losses
      1. Change in body heat = heat produced + heat gained – heat lost.
      2. Radiation is energy lost as infrared and other wavelengths when the environment is cooler than the animal.
      3. Conduction is the transfer of heat from one object to another.
      4. Convection is heat transfer because of moving air or liquids.
      5. Evaporation is heat loss due to sweat changing from a liquid to a gas.
      6. Animals gain heat by metabolism and they gain or lose heat by radiation, conduction, convection, and evaporation.
   C. Classification of Animals Based on Temperature
      1. Ectotherms, such as lizards, gain the most heat from the environment.
         a. Ectotherms use behavioral temperature regulation, such as by basking in the sun when they are cold.
         b. They crawl under rocks when cold at night.
      2. Endotherms and Heterotherms (birds and mammals)
         a. Their body temperature is controlled by metabolism and behavioral adjustments.
         b. Most have an active life style—a foraging mouse uses thirty times more energy than a foraging lizard.
         c. Animal shape and insulation (e.g., fur, feathers, and fat) are adaptations for this life style.
         d. Heterotherms like the hummingbird drop their body temperatures at night to avoid running out of energy.
      3. Advantages of Ectothermy Versus Endothermy
         a. Ectotherms have an advantage in the tropics, and reptiles exceed mammals in these regions.

b. Endotherms have an advantage in moderate to cold environments—there are no lizards in polar regions.

D. Temperature Regulation in Mammals
1. Rapid temperature changes can raise or lower the core temperature slightly.
2. Normal core temperature is restored by feedback control mechanisms.
3. Responses to Cold Stress
   a. Skin thermoreceptors → hypothalamus → shivering.
   b. Behavioral responses include a cat curling into a ball or a human clapping hands.
   c. Vasoconstriction of peripheral blood vessels is another response.
   d. When all these measures are inadequate, hypothermia results.
4. Responses to Heat Stress
   a. Peripheral blood vessels dilate.
   b. Skeletal muscle activity slows down.
   c. The hypothalamus stimulates sweating, which results in evaporation.
5. Fever
   a. The hypothalamus resets the body's thermostat to a higher core temperature.
   b. Fever seems to enhance the effectiveness of the body's immune response.

III. Control of Extracellular Fluid

A. Water Gains and Losses
1. Normally, water losses balance water gains.
2. Gains in water result from absorption in GI tract and as a product of metabolism.
3. Thirst is controlled by the hypothalamus.
4. Losses in water occur from excretion, evaporation from epithelia, sweating, and elimination by the GI tract.

B. Solute Gains and Losses
1. Solutes are added to the internal environment by absorption, secretion, and metabolism.
2. Waste products include CO_2, ammonia (deamination), urea, and uric acid (from nucleic acid degradation).

C. Urinary System of Mammals
1. Urine (forms in kidneys) = water + ions + organic waste.
2. Kidneys regulate the volume and composition of the internal sea.
3. Each kidney is composed of a renal capsule, cortex, and medulla. **(Fig. 33.5, p. 478; TA 98)**
4. Urine forms in nephrons, enters the renal pelvis, travels through the ureter to the urinary bladder, and leaves through the urethra. **(Fig. 33.5, p. 478; TA 98)**
5. Urinary system = two kidneys, two ureters, urinary bladder, and urethra.
6. Kidney stones are deposits of uric acid that collect in the renal pelvis or lodge in the urethra.
7. The capacity for voluntary control of urination develops at two years of age as the nervous system develops.
8. Nephrons: glomerulus with glomerular capillaries in Bowman's capsule → proximal tubule → loop of Henle → distal tubule → collecting ducts → renal pelvis.
9. Blood vessels at the nephron: renal arteries → arterioles → glomerular capillaries → efferent arterioles → peritubular capillaries (which surround proximal and distal sections of the nephron) → renal veins.

D. Overview of Urine Formation **(Fig. 33.6, p. 479; TA 99)**
1. Filtration—blood pressure forces filtrate out of the glomerular capillaries into Bowman's capsule, then into the proximal tubule.
2. Reabsorption—occurs in tubules; water and usable solutes return to the capillaries.
3. Secretion—peritubular capillaries secrete H^+, K^+, foreign substances (e.g., drugs), uric acid, and other wastes into nephrons.
4. Concentrated urine results such that the quality of the extracellular fluid is maintained.

E. A Closer Look at Filtration
1. Rate of Filtration
   a. Filters nearly one-fourth of the cardiac output per minute—45 gallons per day.
   b. Arterioles delivering blood to glomeruli have a wider diameter than most arterioles.

c. Glomerular capillaries are highly permeable to water and small solutes—but not to platelets and blood cells.

2. Factors Influencing Filtration
   a. The rates are influenced mainly by the rate of water reabsorption.
   b. The rates are also influenced by hormonal and neural controls; for example, during exercise, blood is diverted from the kidneys to muscles.

F. A Closer Look at Reabsorption
1. The daily minimum obligatory water loss is 400 ml; it increases or decreases with water intake.
2. Large volumes of water are reabsorbed; this is controlled by the concentration gradient between the nephron and the interstitial fluid.
3. Transport Processes at the Proximal Tubule
   a. Sodium ions are pumped out of the filtrate into the interstitial fluid.
   b. Water and some other substances passively follow and diffuse into peritubular capillaries.
4. Countercurrent Multiplication
   a. Countercurrent multiplication permits water to be reclaimed by maintaining a concentration gradient between the nephron and the interstitial fluid.
   b. It involves flow in opposite directions in the loop of Henle.
5. Control of Water Reabsorption **(Fig. 33.9, p. 482; TM 143)**
   a. Antidiuretic hormone (ADH) secretion is under hypothalmic control and helps to conserve water when needed.
   b. ADH makes the walls of distal tubules and collecting ducts more permeable to water, and thus the urine becomes more concentrated.
6. Thirst Mechanism—when needed the hypothalamus stimulates ADH secretion and water-seeking behavior.
7. Control of Sodium Reabsorption
   a. When too much Na^+ is lost, extracellular fluid volume is reduced, and pressure receptors are triggered.
   b. In response, a part of the kidney—the juxtaglomerular apparatus—secretes an enzyme, renin.
   c. Renin changes angiotensin I (a blood protein) into angiotensin II—this causes the adrenal cortex to secrete aldosterone, which stimulates Na^+ reabsorption in the distal tubule and collecting ducts.
   d. Hypertension can adversely affect kidney function.

G. Acid-Base Balance
The kidneys help to buffer the blood by secreting excess H^+ and restoring HCO_3^-.

H. Case Study: On Fish, Frogs, and Kangaroo Rats
1. Bony fish in seawater live in a hypertonic environment and tend to lose water. They continually drink and pump out excess solutes through fish gills.
2. Freshwater fish tend to gain water; their kidneys pump out excess water as dilute urine.
3. Kangaroo rats in the desert gain most water through metabolic oxidation of seeds; their urine and feces are very concentrated.

Suggestions for Presenting the Material

- This chapter expands the concept of homeostasis introduced in Chapter 23 by focusing on temperature and fluid regulation.
- Your lectures on "control of body temperature" will cover familiar topics such as *heat gain and loss, ectotherms and endotherms,* and *response to temperature stress.* The latter topic is well presented in Figure 33.2 and Table 33.2.
- Be sure to explain the inadequacy of the layman's terms *warmblooded* and *coldblooded.*

- The "Control of Extracellular Fluid" section contains material that is less familiar to students, especially the intricacies of the kidney. Therefore, this section (sometimes referred as the "excretory system") will deserve some time and effort.
- The authors begin the chapter with an excellent prelude—water and solute gains and losses. These paragraphs make clear the necessity for the urinary system described next.
- Again the authors have presented the material in ever-increasing detail as the discussion progresses. Some instructors may find the "Closer Look" sections too detailed for their classes; others may enjoy these expanded explanations.
- If you use the word *excretion* in your lectures, clearly distinguish it from the word *elimination,* which is the voiding of undigested waste via the anus.
- Discussion of the nephron structure is meaningless without constant reference to the transparency of Figure 33.5. Be certain to distinguish between structures carrying *blood* and those carrying *filtrate.*
- An analogy to the common drip-type coffee brewing machine is helpful (see the "Enrichment" section).
- Reinforcement of ADH and aldosterone functions (Chapter 26) should be made during your presentation of kidney function.
- Observation of Table 33.4 will show that large quantities of fluid are filtered from the blood, but nearly all fluid is replaced. Use the "storeroom-cleaning" analogy (see the "Enrichment" section) as a possible clue to why this is the body's method.
- In your lecture include some examples of larger desert mammals, such as the eland, and the special adaptations they have for conserving water and for regulating their body temperature.

Classroom and Laboratory Enrichment

- The measurement of metabolic rate using a respirometer (see the "Enrichment" section of Chapter 31 in this manual) can be varied to show the effects of temperature on an endotherm (mouse) and ectotherm (frog).
- To aid the students' conceptualization of the nephron function, the following comparison to a drip-type coffee maker is made:

Nephron Part	*Coffee Brewer Part*
a. Afferent arteriole carrying blood in	a. Hot water
b. Blood with wastes	b. Coffee grounds
c. Glomerulus	c. Filter paper
d. Bowman's capsule	d. Filter holder
e. Proximal and distal ducts, Henle's loop, collecting ducts	e. Carafe

- The storeroom-cleaning analogy reveals a possible explanation for why the kidney removes from the blood much more than it will eventually excrete. Pose this question based on the two scenarios that follow: Which of the following, (a) or (b), results in a more efficient cleaning of the storeroom?
  a. Carefully removing and disposing only those few selected items that are in plain view and identifiable as "no longer needed"
  b. Removing *all* items from the storeroom; then sweeping, dusting, mopping; finally replacing only those items selected as "still worth keeping"
- The complete analysis of urine can reveal a wealth of information concerning the status of body metabolism. Ask a clinical lab technician to speak on modern analysis techniques.
- If possible, invite a kidney transplant or dialysis patient to discuss his or her condition with the class. Some kidney centers have a speaker's bureau for just this purpose. They may also provide dialysis equipment for observation.
- Use a mannequin to show the locations of the urinary organs.
- Exhibit a model of a kidney to illustrate its parts and the blood vessels associated with it.
- Display a model of a nephron.

- Use a model of a renal corpuscle to illustrate its structure.
- Provide microscope slides of the kidney for student viewing.

Ideas for Classroom Discussion

- Of these three processes—filtration, reabsorption, secretion—which is (are) accomplished by a kidney dialysis machine? Explain any limitations of the device.
- Why do high-protein diet supplements for increasing muscle mass or losing weight include warnings saying that water intake must be increased when consuming the product?
- In the storeroom-cleaning analogy (see the "Enrichment" section), which scenario do you think results in a better cleaning of the storeroom? Do you think the same would apply to kidney function?
- When asked what the kidney does, most people would probably respond that it filters the blood. Why is this answer not a complete statement of kidney function?
- Obviously, humans can survive using only one kidney. Why then do we have two?
- We humans are a fashion-conscious lot. We spend perhaps hundreds of dollars covering our midsections and arms with sweaters and coats, while leaving heads and legs (females, hopefully) bare. Is this biologically sensible? What areas of the body are the greatest dissipators of heat?
- Is it possible to overload our kidneys' capacity by excessive intake of water?
- Why does eating salty foods make you thirsty?
- Why does eating salty foods make you temporarily gain weight?

Term Paper Topics, Library Activities, and Special Projects

- Investigate the workings of a kidney dialysis machine. Include historical perspectives and recent technological advances.
- In recent years, there have been scattered reports of persons (mostly children) surviving a plunge into icy waters. See if you can locate these reports and the follow-up stories that may give indications of any permanent damage.
- Although Western cultures find the practice bizarre, the consumption of one's own urine is practiced in Eastern cultures (India, for example). Report on the supposed benefit of such actions and the possible dangers.
- Popular wisdom reasons that because incidence of colds and flu increases in fall and winter, it is the drop in temperature and human exposure to it that are to blame. Find out if this is really true. Are there more compelling causes?
- What are "kidney stones"? What are some of the factors responsible for their formation?
- Investigate the reasons why humans cannot meet their water needs by drinking seawater exclusively.

Films, Filmstrips, and Videos

- *Excretory Systems in Animals.* IU, 1971, 16 minutes, 16 mm, color. Paramecia, amoebae, hydra, and planaria are shown as examples of animals that excrete carbon dioxide and ammonia by diffusion. The function of contractile vacuoles is shown also. Nephridia, Malpighian tubules, and green glands are shown, and their function is discussed. Excretory organs of frogs and a dissected dog kidney are examined. Animation is used to show nephron function.
- *Regulating Body Temperature.* EBEC, 1972, 22 minutes, 16 mm, color.
- *The Work of the Kidneys.* EBEC, 1972, 20 minutes, 16 mm, color. This film explains the physiological mechanisms used by vertebrate kidneys to control the volume and composition of body fluids. The anatomy of a kidney is described and its importance as a regulator is emphasized.

34

PRINCIPLES OF REPRODUCTION AND DEVELOPMENT

Revision Highlights

Major rewrite and reorganization into two parts: first a description of main developmental stages, then details of mechanisms underlying development. This affords a clear path through complex material; it also affords more flexibility in assigning the basics alone. Updated picture of cell differentiation and morphogenesis. More straightforward description of pattern formation; SEMs of human hand formation provide one of the examples. Refined summary.

Chapter Outline

Objectives

1. Understand how asexual reproduction differs from sexual reproduction. Know the advantages and problems associated with having separate sexes.
2. Describe early embryonic development and distinguish each: oogenesis, fertilization, cleavage, gastrulation, and organ formation.
3. Explain how a spherical zygote becomes a multicellular adult with arms and legs.

Key Terms

sexual reproduction
asexual reproduction
fission
budding
clones
parthenogenesis
hermaphrodites, -ditism
viviparous
ovoviviparous
oviparous
synchronous
external fertilization
internal fertilization
testes
penis
vagina
ovary
yolk
gametogenesis
fertilization
cleavage
gastrulation
organogenesis
growth
tissue specialization
oocyte
animal pole
vegetal pole
gray crescent
cleavage
blastula
blastocoel
blastodisk
extraembryonic membranes
blastocyst
gastrulation
germ layers
endoderm
mesoderm
ectoderm
gastrula
archenteron
neural tube
incubate
adult
larva
metamorphosis
direct development
indirect development
regeneration
genetically equivalent
cell differentiation
morphogenesis
stem cells
identical twins
active cell migrations
chemotaxis
adhesive cues
neural plate
neural tube
pattern formation
ooplasmic localization
embryonic induction
Drosophila
limited division potential
imaginal disks
homeotic mutation
controlled cell death
brain neurohormone
juvenile hormone
ecdysone
aging

Lecture Outline

I. Introduction
   A. How is a single-celled zygote transformed into the specialized cells and tissues of the adult form?

II. The Beginning: Reproductive Modes
   A. Asexual Reproduction
      1. In fission, the parent divides into two equivalent parts; examples include certain flatworms.
      2. In budding, new individuals develop as outgrowths of the parent (*Hydra* is an example).
      3. The offspring are clones, or genetically identical copies of the parent, and are well-adapted forms in stable environments.
   B. Sexual Reproduction
      1. Sexual reproduction usually involves the fusion of gametes from a male and female parent.
      2. In parthenogenesis, an unfertilized egg develops into an adult (this occurs for example in beetles, frogs, turkeys, and lizards).
      3. In hermaphrodites, such as earthworms and tapeworms, one individual has male and female reproductive organs.
   C. Some Strategic Problems in Having Separate Sexes
      1. Male and female reproductive cycles must be synchronous with gametes released at the same time.
      2. Energy outlays are required for males and females to recognize each other.
      3. External fertilization in water requires the production of large numbers of gametes.
      4. Internal fertilization requires testes, penis, vagina, and ovary, and gametes need to be protected from harsh conditions.
      5. Yolk or attachment to the mother is needed to nourish the embryo.

III. Stages of Development (**Fig. 34.3, p. 492; TM 144**)
   A. In animal gametogenesis, the sperm and egg mature.
   B. After fertilization, cleavage produces the blastula.
   C. In gastrulation simple tissues form and organogenesis follows.
   D. Gametogenesis
      1. Sperm cells contain mostly parental DNA.
      2. Oocytes contain specialized proteins, enzymes, mRNA, and parental DNA.
      3. Frog eggs show polarity—an animal pole near the nucleus and a vegetal pole with yolk.
   E. Fertilization (**Fig. 34.5, p. 493; TM 145**)
      1. Sperm penetration causes changes in the cortical granules of the egg.
      2. A gray crescent forms opposite the penetration site and the body axis is established.
   F. Cleavage
      1. Mitotic divisions are not accompanied by an increase in cell size—the cells get smaller and form the blastula.
      2. In sea urchins, the blastula contains a blastocoel.
      3. In amphibians, the yolk impedes cleavage near the vegetal pole and the blastocoel forms near the animal pole.
      4. In higher vertebrates, the yolk restricts cleavage so much that a blastodisk forms.
   G. Gastrulation
      1. In sea urchins, cell migrations form the archenteron.
      2. In vertebrates, a neural tube forms following embryo elongation.
      3. The three germ layers—endoderm, mesoderm, and ectoderm—form.
   H. Organogenesis
      1. Following gastrulation, cell types emerge that eventually form organs.
   I. Post-Embryonic Pathways of Development
      1. In some animals, such as nematodes, the transition from embryo to adult only takes an increase in size and gonad maturation.
      2. In higher vertebrates, body proportions must also change.
      3. Indirect development in insects involves a larval form and metamorphosis into an adult form.
      4. In regeneration, animals (such as crabs) can replace a limb.

IV. Mechanisms of Development
   A. DNA replication and mitosis assure that all cells are genetically equivalent.
   B. During gastrulation, different cell lineages follow different patterns of gene expression.
   C. Unequal distribution of cytoplasmic substances leads to cell differentiation.
   D. Cell interactions result in morphogenesis.
   E. Developmental Information in the Egg
      1. Before fertilization, mRNA, proteins, and ribosomal subunits are stored.
      2. Microtubules and other cytoskeletal elements are oriented in specific directions for cell divisions.
      3. Many mRNA transcripts accumulate in different cytoplasmic regions.
      4. Yolk may be asymmetrically distributed.
   F. Cell Differentiation
      1. Cell differentiation occurs through controls over gene expression.
      2. A differentiated cell has the same genes as the fertilized egg.
   G. Mechanisms Underlying Morphogenesis
      1. Cell Migrations
         a. Active cell migration by pseudopods involves chemotaxis (it occurs in the nervous system, for instance).
         b. Recognition proteins on cells or the extracellular matrix can provide adhesive cues (for example, pigment cells move on blood vessels, not axons).
         c. Adhesive cues let cells know where to stop.
         d. The folding of sheets of cells involves microtubules and microfilaments.
      2. Localized Growth and Cell Death
         a. Morphogenesis depends on localized growth, which is linked to regulatory genes.

b. Controlled cell death transforms paddlelike appendages into hands and feet.
c. Adult chimpanzees and humans differ considerably in proportions of body parts. **(Fig. 34.12, p. 501; TM 146)**

H. Pattern Formation
1. Pattern formation refers to the mechanisms responsible for the differentiation of tissues and their positioning in space.
2. Ooplasmic localization causes the cells to differentiate according to what cytoplasmic substances they inherited.
3. Embryonic induction causes a body part to differentiate in response to signals from an adjacent body part.
a. The optic cup induces the epidermis to form a lens.
b. Inductions involve the diffusion of a chemical (e.g., a hormone).
c. Homeotic mutations affect regulatory genes that lead to defects in inducer substances (e.g., an "antenna" disk may develop into a leg).

V. Aging and Death
A. After the growth and differentiation of complex animals, cells deteriorate and the organism ages.
B. In humans, skin wrinkles, fat deposits, muscle mass decreases, kidney function decreases, and so on.
C. Moorhead and Hayflick found cell lines that divide fifty times and die, which implies that cells have a limited division potential.
D. DNA may gradually lose the capacity for cell repair.
E. Defects in collagen may adversely affect the exchange of materials to and from cells.
F. The genes that code for self-markers on the membranes may deteriorate, and autoimmune responses may intensify.

Suggestions for Presenting the Material

- This chapter and the next cover reproduction in general, and human reproduction specifically. While Chapter 35 (human) focuses on the details of one organism, the present chapter introduces *general terms of reproduction* (sexual versus asexual), *development* (fertilization, cleavage, gastrulation, organogenesis, etc.), *growth, metamorphosis, differentiation, aging and death.*
- The progression of topics within the chapter is logical, but topics can be abbreviated or omitted if necessary.
- Embryonic development is best demonstrated by use of a film or videotape (see the "Enrichment" section); but if these are not available, 2 x 2 transparencies of Figure 34.6 will substitute, albeit not as well. If you are the photographer, prepare "overall" slides of each page and individual shots (macro lens needed) of stages (a) through (l).
- Figure 34.4 is a good condensation of embryonic comparisons, but without adequate background it may be too "abstract" for beginning, nonscience students. It would best be used as a display photo on the screen while the lecture material is presented elsewhere.
- In the "Morphogenesis and Growth" section the material is quite unfamiliar, especially in "pattern formation."
- More familiar is the topic of *metamorphosis,* especially as it appears in insects. This is a good place to "reclaim" your young audience if you feel they have been plowing through some thick waters with you.
- If you feel comfortable in presenting it, the story of neuroendocrine control in insects (Figure 34.17) is fascinating. One analogy that will aid understanding is to compare juvenile hormone to the brake on an automobile, that is, both *retard* forward progress; and to compare ecdysone to the accelerator, that is, both *promote* forward progress.

- Aging is something your youthful class will not find of much interest now. But perhaps you can find some recent research tidbit on which to base your presentation of what is really a puzzling process.

Classroom and Laboratory Enrichment

- If at all possible, show a videotape or film depicting development of some animal. Because of the dynamic nature of this process and the rapid changes, static photographs are woefully inadequate.
- Although it is not always convenient to do so, the demonstration of live chick embryos is a real attention-arresting sight. Don't neglect to place an embryo under a stereomicroscope to see the heartbeat and blood flow.
- The early development of sea urchin embryos is not as difficult to demonstrate as that of the chick. Biological supply houses sell demonstration kits. Timing is a critical factor for viewing all the stages, so you should plan to videotape the sequence.

Ideas for Classroom Discussion

- What advantage(s) does asexual reproduction have over sexual union?
- Is there such a creature as a "female earthworm"? Explain your answer in anatomical terms.
- Some persons think the "yolk" of the chicken egg corresponds to the "nucleus." Is this true?
- Death is an unpleasant subject but can you think for a moment, exactly what is death? How is death officially defined by doctors and coroners? Would their definition be different from that of a developmental biologist?
- The Old Testament records human life spans ranging in the hundreds of years. How can you rationally explain this?
- When an insect is in the pupal stage, seemingly there is no activity. Some people have even called it the "resting stage"—erroneously! Biochemically and histologically what is happening during the pupal stage?
- All animals reproduce at rates sufficient to maintain their populations. Humans are the only ones whose proliferation seems to be under very few limiting factors. Is this so? Why?

Term Paper Topics, Library Activities, and Special Projects

- Exciting progress is being made in research related to the aging process. Prepare a report on recent developments.
- Investigate the use of juvenile hormone mimic chemicals to control insects. What are the successes, failures, limitations, advantages, and prognoses for the future?
- Embryologists can transplant imaginal disks of insect larvae to produce *very* unusual adults, for example, legs where antennae should be. Investigate the procedures used and the results produced. Could this be done in invertebrates?

Films, Filmstrips, and Videos

- *Amphibian Embryo.* EBEC, 1963, 16 minutes, 16 mm, color. Uses time-lapse and microphotography with animated cross-section diagrams to show how a single-celled amphibian egg is transformed into a multicellular organism.
- *Cell Differentiation: The Search for the Organizer.* CORT, 1983, 15 minutes, 16 mm, color. Hans Spemann, a Nobel Prize recipient and an embryologist, explains experiments that he calls

"conversations with an embryo" that led to discovery of the organizer, a group of cells responsible for induction.

- *The Chick Embryo: From Primitive Streak to Hatching.* EBF, 1960, 13 minutes, 16 mm, color. Demonstrates the development of a chick embryo from cleavage to hatching. Uses time-lapse photography and microphotography to show the principal processes that occur in development.
- *Development and Differentiation.* CRMP, 1974, 20 minutes, 16 mm, color. Discusses the roles of DNA and the cytoplasm in development and follows a frog embryo through gastrulation.
- *Protist Reproduction.* MLA-Wards, 1977, 13 minutes, 16 mm, color. Shows the reproductive styles of protists, including fission, with replication of organelles and subcellular particles. Also examines forms of budding, conjugation, and reproduction using gametes.
- *Reproductive Systems.* NGS, 1988, 25 minutes, 16 mm, color. Fertilization and development of a new human life. Photography by Lennart Nilsson.

35

HUMAN REPRODUCTION AND DEVELOPMENT

Revision Highlights

Text updated; for example, peptide hormones from Sertoli cells that modulate control of male reproductive function are now covered. More logical illustration of hormonal effects on ovarian and uterine function (Figure 35.8). New section on endometriosis. New illustrations of placenta relative to fetal circulation (Figure 35.14) and of embryonic developmental stages most sensitive to damage from cigarette smoke, alcohol, viral infection, etc. (Figure 35.21). New Commentary on breast cancer and testicular cancer. Updated text and illustration on birth control methods. Updated Commentary on sexually transmitted diseases, with more material on AIDS and new micrographs of causative agents of gonorrhea and syphilis.

Chapter Outline

Objectives

1. Describe the structure and function of the male and female reproductive tracts.
2. Outline the principal events of prenatal development.
3. Know the principal means of controlling human fertility.
4. Know how breast and testicular cancer can be detected by self-examination.

Key Terms

gonads
testes (sing.: testis)
ovaries
sperm
secondary oocytes
scrotum
seminiferous tubules
interstitial cells
Sertoli cells
acrosome
epididymis
vas deferens
seminal vesicle
prostate gland
bulbourethral glands
semen
testosterone
follicle-stimulating hormone, FSH
luteinizing hormone
GnRH
estrogen
progesterone
oviduct
uterus
myometrium
endometrium
cervix
vagina
vulva
clitoris
granulosa cells
primary follicle
zona pellucida
secondary oocyte
ovulation
ovum
estrous cycle
menstrual cycle
corpus luteum
follicular phase
antrum
midcycle surge of LH
endometriosis
coitus
orgasm
mature ovum
capacitation
blastocyst
trophoblast
chorionic gonadotrophin
embryonic disk
extraembryonic membranes
yolk sac
allantois
amnion
chorion
umbilical cord
placenta
first trimester
fetus
second trimester
third trimester
lactation
thalidomide
fetal alcohol syndrome, FAS
mammography
mammogram
biopsy
radical mastectomy
metastasis
abstinence
rhythm method
withdrawal
douching
spermicidal foam
spermicidal jelly
diaphragm
condoms
IUD
the Pill
vasectomy
tubal ligation
abortion
miscarriages
in vitro fertilization
gonorrhea
syphilis
chlamydia

Lecture Outline

I. Primary Reproductive Organs
   A. Reproductive system = pair of gonads (testes or ovaries) + accessory glands and ducts.
   B. It functions to produce gametes and sex hormones, which affect reproduction and secondary sexual traits.
   C. Male gametes are sperm and female gametes are eggs or secondary oocytes.
   D. All gonads look alike in the embryo until the seventh week of development.

II. Male Reproductive System **(Fig. 35.2, p. 510; TA 100)**
   A. The testes are divided into 250–300 lobes, each containing two or three seminiferous tubules where sperm form.
   B. Interstitial cells, a connective tissue, secretes hormones including testosterone.
   C. Spermatogenesis **(Fig. 35.3, p. 511; TA 101)**
      1. From puberty onward, continuous spermatogenesis occurs; it takes nine to ten weeks.
      2. Sperm development requires a temperature cooler than the body core; hence, the testes are suspended in the scrotum.
      3. Diploid spermatogonia—mitosis → primary spermatocytes—meiosis I → haploid secondary spermatocytes → meiosis II → haploid spermatids → mature sperm.

4. The developing cells are nourished by Sertoli cells.
5. Each sperm has a head (nucleus and acrosome), midpiece (mitochondria), and tail (microtubules).

D. Sperm Movement Through the Reproductive Tract
1. Testis → epididymis → vas deferens → duct from seminal vesicle → prostate gland → urethra.
2. Fluids added to sperm from the prostate gland (buffers vaginal pH), seminal vesicles (fructose + prostaglandins), and bulbourethral glands (mucus—lubricates penis) = semen. **(Table 35.1, p. 512; TM 147)**

E. Hormonal Control of Male Reproductive Functions
1. Testosterone, produced by interstitial cells, stimulates spermatogenesis, the formation of reproductive organs and secondary sex characteristics, and helps to develop and maintain normal sexual behavior.
2. Follicle-stimulating hormone (FSH) and luteinizing hormone (LH) are gonadotropins produced by the anterior pituitary.
3. Testosterone production is controlled by a negative feedback loop.

III. Female Reproductive System **(Table 35.2, p. 513; TM 148)**

A. Two ovaries produce and release oocytes monthly and secrete estrogen and progesterone.
1. Oviducts channel oocytes into the uterus.
2. The pear-shaped uterus houses the developing embryo during pregnancy; it has a myometrium (muscle layer) and an endometrium (lining).
3. External genitalia are the vulva (labia majora and labia minora), clitoris, and urethral opening.

B. Oogenesis **(Fig. 35.7, p. 514; TA 102)**
1. A primary oocyte with surrounding cell layer is a primary follicle.
2. Just before ovulation, the primary oocytes complete meiosis I.
3. At ovulation, meiosis II begins and will not be complete unless fertilization occurs.

C. Menstrual Cycle: An Overview **(Table 35.3, p. 515; TM 148)**
1. Most female mammals come into "heat" or estrus.
2. Humans and other primates have a menstrual cycle (there is no relatrionship between heat and fertility), and the uterine lining is sloughed at the end of each cycle of twenty-eight days.

D. Control of Ovarian Function **(Fig. 35.8, p. 516; TA 103)**
1. Follicular Phase
The follicular phase lasts about thirteen days.
   a. One follicle fills an antrum with estrogen secretions; then the follicle balloons and ruptures at ovulation.
   b. At the start of a cycle, because FSH and LH levels increase, the follicle secretes estrogen.
   c. Estrogen then causes a midcycle surge in LH that triggers meiosis I to resume, estrogen secretion to fall, ovulation to take place, and the formation of the corpus luteum from the remains of the follicle.
2. Luteal Phase
   a. The corpus luteum secretes progesterone and estrogen and persists for twelve days if fertilization does not occur.
   b. High levels of progesterone and estrogen cause FSH and LH levels to fall and this prevents follicle formation.
   c. If fertilization does not occur, the corpus luteum fails, FSH and LH levels rise, and a new cycle begins.

E. Control of Uterine Function
1. Estrogen and progesterone cause the endometrium to develop and prepare for pregnancy.
2. Without fertilization occurring, progesterone and estrogen levels fall, and the endometrium disintegrates.

IV. Sexual Union and Fertilization
   A. If sperm ejaculation occurs with ovulation, sperm enter the oviducts within sixty minutes and pregnancy can result.
   B. The secondary oocyte is stimulated into completing meiosis II when sperm reach its surface; this forms a mature ovum.
   C. Following capacitation, a sperm nucleus enters the egg cytoplasm and fuses with the egg nucleus.

V. Prenatal Development
   A. First Week of Development
      1. Before the first week ends, the blastocyst contacts and adheres to the uterine lining.
      2. The trophoblast secretes chorionic gonadotropin that maintains the corpus luteum.
      3. The corpus luteum continues to secrete hormones that maintain the endometrium.
      4. Cell divisions of the trophoblast result in cells that become implanted in the mother's tissues.
   B. Extraembryonic Membranes
      1. Shelled eggs contain membranes that function in nutrition, respiration, and excretion—the yolk sac, the allantois, the amnion, and the chorion.
      2. Humans also have a yolk sac (which forms part of the GI tract); the allantois contributes blood vessels, the amnion protects the embryo with fluid, and the chorion contributes to the placenta.
      3. Humans have an umbilical cord which connects to these extraembryonic membranes.
   C. The Placenta **(Fig. 35.14, p. 522; TA 104)**
      1. The placenta is composed of the endometrium and extraembryonic membranes (especially chorion).
      2. The embryo receives nutrients and oxygen and disposes of wastes.
      3. The blood vessels of the mother and embryo remain separate.
      4. The placenta secretes hormones to maintain pregnancy.
   D. Embryonic and Fetal Development
      1. First Trimester
         a. The embryonic disk forms in the blastocyst.
         b. Gastrulation results in germ layers.
         c. Heart, nervous system, and respiratory structures form.
         d. Segmentation becomes obvious.
         e. By the ninth week, a fetus exists. **(Fig. 35.16, p. 523; TA 104)**
         f. By the twelfth week, all the major organs are formed and the arms and legs move.
      2. Second Trimester
         a. The lanugo covers the body and the eyes open.
      3. Third Trimester
         a. The fetus is ready for birth by the middle of the third trimester.
   E. Birth
      1. Birth begins with contractions of the uterine muscles; the cervical canal dilates, and the amniotic sac ruptures.
      2. The fetus is expelled before the placenta, and the umbilical cord is severed.
   F. Lactation
      1. During pregnancy, estrogen and progesterone stimulate the development of the mammary glands.
      2. After birth, colostrum is secreted and is rich in proteins and lactose.
      3. Then prolactin from the pituitary stimulates milk production.
      4. Suckling stimulates the release of oxytocin that causes the contraction of the milk glands and uterine tissue.
   G. Case Study: Mother as Protector, Provider, Potential Threat **(Fig. 35.21, p. 528; TM 149)**
      1. Some Nutritional Considerations
         a. During pregnancy, the mother needs increased vitamin and food intake.
         b. A poor maternal diet can damage the fetus—especially the brain.
      2. Risk of Infections

   a. Antibodies can cross the placenta and protect the fetus from most bacterial infections.
   b. Some viral infection, such as German measles, can cause malformations.
3. Effects of Prescription Drugs
   a. Thalidomide, a tranquilizer, caused many birth defects.
4. Effects of Alcohol
   a. Fetal alcohol syndrome can include mental retardation.
5. Effects of Smoking
   a. Smoking during the second half of pregnancy can result in low birth weight and other problems.

VI. Control of Human Fertility
   A. Some Ethical Considerations
   Consider that 9,900 infants are born each hour—how can human fertility be controlled?
   B. Possible Means of Birth Control (**Fig. 35.22, p. 530; TM 150**)
      1. Approaches to birth control include fertility control, implantation control, and abortion.
      2. Behavioral interventions include abstinence, the rhythm method, and withdrawal.
      3. Other methods include spermicides, a diaphragm, condoms, an IUD, the Pill, vasectomy, tubal ligation, and abortion.
   C. In Vitro Fertilization
      1. About 15% of American couples are infertile.
      2. A two- to four-day-old embryo implants less than 20% of the time.

Suggestions for Presenting the Material

- Most students approach this chapter with a sigh of relief and an attitude that says "finally, something I know everything about." The intuitive instructor will build upon what *accurate* information the students already know and will gently, but authoritatively, correct misinformation. This also prohibits any belittling of incomplete, or inaccurate, "folklore."
- The sequence of topics is logically presented in the chapter: male and female structures, menstrual cycle, sexual union, prenatal development, fertility control.
- The overhead transparencies of male and female reproductive systems should be in view of the students during your presentation. Call attention to the summary tables (35.1 and 35.2).
- When comparing the male and female systems, it is helpful to note that the male produces and delivers sperm, much as the female produces and delivers eggs. However, the female also provides: (a) a site for fertilization, and (b) a site for embryonic and fetal development. This provides a convenient lead into the discussion of the menstrual cycle.
- The details of the menstrual cycle are of interest to both sexes but especially to females, of course. Because events are happening simultaneously in the pituitary, ovary, and uterus, it is almost a necessity to keep Figure 35.8 in constant view. The critical feature of this figure is the time line along the bottom, which puts all events into perspective.
- These points need emphasis in your lecture on female reproduction:
  a. Females are born with a finite number of eggs, which means eggs grow old (implications for causing genetic defects in offspring).
  b. Menstrual cycle can be defined as "the monthly release of an egg and all the preparations for it."
  c. Retention or sloughing of endometrium is determined by progesterone levels at the end of the menstrual cycle.
- You may be able to surprise your students with the fact that only a few days of each month comprise the "fertile period." But hasten to inform your young "experimenters" that the fertile days can "move around" depending on a variety of nutritional, psychological, and health factors.
- Table 35.4 presents a good overview of human development. The specifics of each development in each time interval can lead to a "cataloging" approach, which can be alleviated by using the videotape referred to in the "Enrichment" section.

- The most practical aspect of the chapter is the section on "control of human fertility." Students are eager for this information, especially when it can be presented with special reference to the mode of action of each device. Be sure to emphasize the effectiveness as shown in Figure 35.22.
- The topic of sexual disease is presented in this chapter under "Commentary." Individual instructors may, or may not, wish to present this. But certainly some reference, even without details, should be made to the toll these diseases exact on society.

Classroom and Laboratory Enrichment

- Because you probably teach students with a great range of sexual knowledge and experience, adjust your enrichment activities accordingly. For example, many students will appreciate observing birth control devices, or slide photos of them. Others may be intimately familiar with their use, or at least give that impression.
- The topic of prenatal development will be greatly enhanced by the use of a videotape such as *The Miracle of Life*, distributed by Crown Video through retail outlets.
- If you decide to present details of the physical manifestations of sexually transmitted diseases, use discretion in which slides you show.
- Students enjoy bragging about their sexual knowledge. Perhaps you can bring some perspective to this by preparing a true/false quiz of common facts and fallacies associated with human reproduction. Allowing the students to remain anonymous, administer the quiz prior to your lecture. Tabulate the results and report to the class as you give the accurate information.

Ideas for Classroom Discussion

- In Chapter 14, the incidence of Down syndrome was said to increase with maternal age, especially for mothers over 40. Based on the information in the present chapter, can you explain why?
- Using anatomical terms, explain why men who have had a vasectomy operation are still able to expel normal amounts of semen but no sperm.
- Explain the events that result in the production of identical and fraternal twins.
- What propels sperm from their point of origin to the opening where they exit the body?
- What mechanisms prevent the entry of more than a single sperm into the egg?
- The placenta supplements, or completely replaces, the activity of three organ systems in the fetus. What are they?
- Many communities and even states restrict the teaching of human reproduction. Why do you think this body system is singled out over, say, digestion or respiration, for such a prohibition?

Term Paper Topics, Library Activities, and Special Projects

- It is a curious, but real, fact that most sex crimes are committed by males. Investigate what psychologists and physicians say about the underlying cause(s) for this behavior.
- Sexual dysfunction is a frustrating situation for those involved, whether male or female. Prepare a list, based on your reading, of the most prevalent disorders and their treatment.
- We hear about "sperm banks" now and then. Do such repositories actually exist? Where are they? How do they function?
- Any interruption in the menstrual cycle, whether temporary or permanent, is cause for concern. Report on the causes and effects of the cessation of menstruation in females who drastically reduce their body weight (anorexia) or body fat (as in body building).
- Search for the physiological explanation for the cessation of menstruation (menopause)—usually when a women is between the ages of 40 and 50. Is there any comparable phenomenon in men?

- It is known that overcrowding and stressful conditions reduce reproductive behavior in rodents. Is there any published evidence of such a phenomenon in humans?

Films, Filmstrips, and Videos

- *A Child Is Born.* BFA, 25 minutes, 16 mm, color. Covers pregnancy and shows a normal delivery; slightly sentimental.
- *Human Reproduction* (2nd ed.). MGHT, 20 minutes, 16 mm, color.
- *Life before Birth.* Time-Life, filmstrip. Excellent photographs by Nilssen follow developmental progress of humans from fertilization to birth.
- *Menstrual Cycle.* Lilly, 1971, 12 minutes, 16 mm, color. Explains the basic hormonal and histological changes that occur in the ovaries, uterus, and other organs and tissues during the normal menstrual cycle.
- *Prenatal Development.* CRMP, 1974, 20 minutes, 16 mm, color. Describes fertilization and the subsequent development of human embryos and fetuses. Describes the appearance of the embryo at four and eight weeks, and the fetus at twelve, sixteen, twenty, and twenty-eight weeks.
- *When Life Begins.* CRM/MGHT, 1972, 14 minutes, 16 mm, color. A sequence of events in the growth and development of the human fetus from ovulation to birth is recorded. The rapid growth of organs in the early stages of embryonic development is clearly shown.

36

POPULATION GENETICS, NATURAL SELECTION, AND SPECIATION

Revision Highlights

Very well-received chapter; only minor modification. For example, random mating is more clearly identified as a condition required for genetic equilibrium; clarification of headings and illustration on genetic drift and other factors related to population size; vivid photos instead of line art for disruptive selection among *Papilio dardanus* populations.

Chapter Outline

POPULATION GENETICS
- **Sources of Variation**
- **The Hardy-Weinberg Baseline for Measuring Change**
- **Factors Bringing About Change**

MUTATION

FACTORS RELATED TO POPULATION SIZE
- **Genetic Drift**
- **Founder Effect**
- **Bottlenecks**

GENE FLOW

NATURAL SELECTION
- **Modes of Natural Selection**
- **Stabilizing Selection**
- **Directional Selection**
- **Disruptive Selection**
- **Sexual Selection**
- **Selection and Balanced Polymorphism**

EVOLUTION OF SPECIES
- **Divergence**
- **When Does Speciation Occur?**
- **Reproductive Isolating Mechanisms**
- **Modes of Speciation**

SUMMARY

Objectives

1. Understand how variation occurs in populations and how changes in allele frequencies can be measured.
2. Know how mutations, gene flow, and population size can influence the rate and direction of population change.
3. Describe four kinds of selection mechanisms that help shape populations.
4. Describe three modes of speciation.

Key Terms

typological
genotype
phenotype
Hardy-Weinberg principle
genetic equilibrium
mutation
alleles
population
genotypic frequencies
Hardy-Weinberg equilibrium
allele frequencies
genetic drift
gene flow
natural selection
founder effect
bottlenecks
stabilizing selection
directional selection
disruptive selection
polymorphism
mark-release-recapture
mimicry
differential fertility
balanced polymorphism
differential mortality
sexual dimorphism
speciation
species
divergence
reproductive isolating mechanism
mechanical isolation
behavioral isolation
hybrid inviability
allopatric speciation
parapatric speciation
sympatric speciation
polyploidy
hybridization

Lecture Outline

I. Introduction
  A. Carl von Linné's approach to classification was typological—he made use of perfect types.
  B. Many similar organisms show extreme variation in certain traits.
  C. Variation among individuals of a species is the raw material for evolution.

II. Population Genetics
  A. Sources of Variation
    1. Gene mutations create new alleles.
    2. Independent assortment and crossing over are other sources of variation.
    3. Chromosomal aberrations also cause variation.
    4. Fertilization between genetically varied gametes is another source of variation.
    5. More genetic variation is possible than is expressed at one time.
  B. The Hardy-Weinberg Baseline for Measuring Change
    1. Populations, not individuals, evolve.
    2. A population is a group of individuals of the same species with random mating.
    3. Gene pool = sum of all genes in a population.
    4. Variation is expressed as allele frequencies.
    5. The Hardy-Weinberg principle states that in the absence of disturbing factors, the frequencies of different genotypes in a population will reach an equilibrium and remain stable from generation to generation—this is genetic equilibrium.
    6. It is never reached; it is a reference point from which to measure evolutionary change.
    7. Factors Bringing About Change
      a. Mutation
      b. Genetic drift
      c. Gene flow
      d. Natural selection

III. Mutation
  A. Mutation is the original source of the genetic variation.
  B. It occurs randomly and may be harmful or beneficial depending upon the environment where the mutation is expressed.

IV. Factors Related to Population Size
  A. Genetic Drift **(Fig. 36.3, p. 543; TM 151)**
    1. Genetic drift is the chance increase or decrease in the abundance of different alleles.
    2. It is more rapid in small populations.
  B. Founder Effect

1. Founder effect is a case of genetic drift when a population is founded by a small number of individuals.
2. It is common in oceanic islands.

C. Bottlenecks
1. Bottlenecks occur when a large population is drastically reduced in size.
2. For example, hunters reduced the elephant seal population to about 20, but it is now 30,000.

V. Gene Flow
A. Gene flow is the physical flow of alleles into and out of populations.
B. Few populations of a species are isolated from other populations of the same species.
C. Gene flow tends to decrease genetic variation between populations of a species.

VI. Natural Selection
A. Introduction
1. Genetically based advantageous phenotypes tend to increase in frequency.
2. More offspring are born than can survive.
3. Members of a population vary, and much of this variation is heritable.
4. Some heritable traits are more adaptive than others.
5. Those with more adaptive traits have a greater chance of leaving more offspring; this is differential reproduction.
6. Natural selection is the result of differential reproduction.

B. Modes of Natural Selection **(Fig. 36.4, p. 545; TM 152)**
1. Stabilizing selection
2. Directional selection
3. Disruptive selection (this results in polymorphism)

C. Stabilizing Selection **(Fig. 36.4, p. 545; TM 152)**
1. Human birth weight averages 7 pounds.
2. Horseshoe crabs show little change in 250 million years.

D. Directional Selection **(Fig. 36.4, p. 545; TM 152)**
1. The Peppered Moth
   a. Before the mid-1800s the light-gray form was prevalent; by 1898, the dark form increased in frequency.
   b. The mark-release-recapture method demonstrated that industrial pollution kills lichens and darkens the tree trunks; this verified the hypothesis that natural selection accounted for an increase of dark forms in industrial areas.
2. Pesticide Resistance
   a. An example of directional selection.

E. Disruptive Selection **(Fig. 36.4, p. 545; TM 152)**
1. With mimicry, one species is the mimic, another is the model.
2. The model may be inedible and the mimic may be very tasty.

F. Sexual Selection
1. Larger, more conspicuous and aggressive males may be selected by females; this is sexual dimorphism.

G. Selection and Balanced Polymorphism
1. Two or more alleles of a gene persist at a higher frequency than the mutation rate.
2. For example, the HbS allele is common in areas where malaria is common.

VII. Evolution of Species
A. Introduction
1. Species originate by speciation.
2. A species is an interbreeding population that produces fertile offspring.

B. Divergence
1. Divergence occurs when a barrier isolates parts of a population.
2. A buildup of differences in allele frequencies may result in populations that can't interbreed under natural conditions—the populations become separate species.

C. When Does Speciation Occur? (Fig. 36.10, p. 551; TM 153)

D. Reproductive Isolating Mechanisms

1. Reproductive isolating mechanisms prevent the exchange of alleles between populations.
2. Mechanical Isolation
   Differences in the structure or function of reproductive organs may prevent individuals of different populations from producing hybrid zygotes.
3. Isolation of Gametes
   Sperm and egg may have incompatibilities (this is true of different sea urchin species).
4. Isolation in Time
   One species of cicada reproduces every thirteen years, another species every seventeen years.
5. Behavioral Isolation—courtship rituals may precede mating.
6. Hybrid Inviability and Infertility

E. Modes of Speciation

1. Allopatric speciation takes place when species form as a result of geographic separation.
2. Parapatric speciation occurs in populations that live side by side in different environments.
3. Sympatric Speciation
   a. Sympatric speciation occurs as a result of ecological, behavioral, or genetic boundaries that arise *within* the boundaries of a single population.
   b. It can be caused by mutation and polyploidy and may involve hybridization.

Suggestions for Presenting the Material

- If you are following the sequence of chapters in the book, it has been some time since genetics was discussed. Therefore, brief parenthetical reviews may be necessary during the presentation of this chapter.
- Because of the number of topics in this chapter, it would be a good idea to present an outline that shows the main and subordinate themes.
- The main points of the chapter are as follows:
  Variation is the result of several factors.
  The Hardy-Weinberg equation provides baseline for calculating gene frequencies under unrealistic conditions.
  Several factors yield change in the real world:
    mutation
    genetic drift (founder, bottleneck)
    gene flow
    natural selection (stabilizing, directional, disruptive)
  Sufficient change will yield different species.
    There are different mechanisms of speciation.
    There are various modes of speciation.
- Work a Hardy-Weinberg problem in the manner suggested in the "Enrichment" section below.
- Several of the topics in the chapter do not have visual material in the textbook. Check other texts for photo examples of founder effects, bottlenecks, isolating mechanisms, and so on.
- Pesticide resistance in insects is perhaps the best and most recent example of natural selection. It should be presented clearly!

Classroom and Laboratory Enrichment

- Choose an easy-to-see trait governed by one gene with two alleles such as tongue-rolling (the ability to roll one's tongue into a U-shape) or free earlobes (earlobes whose bases are not attached to the jawline), and ask students to determine their own phenotype. Determine the number of homozygous recessive individuals in the class (those who are non-tongue-rollers or have attached earlobes). Use the Hardy-Weinberg principle to calculate the frequencies of the dominant allele and the recessive allele.
- Demonstrate genetic drift by tracing changes in allele frequency throughout time. In small hypothetical populations, select a trait governed by two alleles and calculate the frequency of each allele. Different groups of students could be assigned populations of different sizes. Follow each population throughout several generations as some of its members (selected by coin tosses) succumb to disease, predation, and other random causes of early death. How does population size affect genetic variation over time?
- Show 35 mm slides or films about endangered species that are threatened by sharp reductions in population size and subsequent loss of genetic variability.
- What happens to the genetic variability of small, isolated populations of laboratory organisms after many generations without the introduction of new organisms? Design and implement an experiment using any organism with a short generation time and several easy-to-see traits that can be followed from one generation to the next.
- How does artificial selection by humans affect gene frequencies of domestic plants and animals? Pursue this question with experiments or demonstrations.
- Use films or overhead transparencies to discuss examples of mimicry. Discover and observe, if possible, local examples of mimicry among insect or animal populations.
- Prepare an overhead transparency on which you have written the solution to a Hardy-Weinberg problem (check a genetics or majors biology text). This method is more convenient than blackboard presentation, and you will feel more confident. Don't disclose the entire page at once. Instead, arrange the problem stems and solutions in such a way that one can be uncovered while the other is hidden.
- Select a well-known example of the founder effect in a human population, perhaps a religious group. Using slides show how a phenotypic characteristic (for example, polydactyly) "spreads" from one generation to the next.
- Explain the development of insect resistance to DDT as a modern-day example of natural selection. Point out that DDT was introduced to the world in the early 1940s and in just ten years resistant strains were reported in many countries. By the time it was banned in the United States in 1973, virtually every housefly was resistant to its effects.

Ideas for Classroom Discussion

- How did Darwin's observation of variation among species help him to develop his principle of evolution?
- What is the difference between a theory and a principle? Why does your text refer to the principle of evolution rather than the theory of evolution?
- How does phenotypic variation arise? Ask students to list as many sources of phenotypic variation as they can. They should be able to remember how genetic variation comes about from their earlier study of genetics.
- What is the difference between natural selection and artificial selection?
- How are new alleles created? Is the creation of new alleles an important source of genetic change? Why or why not?

- What are some phenotypic variations that might have assisted the success of *Homo sapiens*? Ask your students to think of some imaginative examples of variations that might be useful in the future evolution of our species.
- How representative of the human population is your class? Discuss the importance of sample size with reference to determining allele frequencies. Ask your students to think about gene pool size and the founder effect if they were stranded forever on an uninhabited island. Would certain alleles be over- or underrepresented? What would happen to allele frequencies after many generations?
- Think of examples of human alleles whose frequencies vary from one global region to the next.
- What did Darwin's study of the different finch species among the Galapagos Islands tell him about speciation? What conclusions can you make about the evolutionary histories of the different species of Galapagos finches, given what you now know about the process of speciation?
- Why are conservationists concerned when the genetic variation within a population of rare or endangered organisms begins to decrease?
- How does sexual selection benefit a species? Would the introduction of alleles from a similar but different species introduce variety and thus help the species? Why or why not?
- How do reproductive isolating mechanisms help a species?
- Is population size a critical factor for success of mimics? What might happen to a mimic species if it began to greatly outnumber its inedible model?
- Why is the statement "She has evolved into a fine pianist" *not* biologically accurate?
- Of the five sources of phenotypic variation, why is *mutation* the only one that *creates* new alleles? How do the others yield variation?
- Which of the factors that cause changes in allelic frequencies could be under conscious human control?
- How could the results of various scientific experiments be influenced by using strains of laboratory mice that are continuously inbred?
- In the instance of resistance to insecticides, it was pointed out that the *resistant* strains gradually replaced the *susceptible* ones. Are these two strains different species? How could you prove/disprove your answer?

Term Paper Topics, Library Activities, and Special Projects

- Read about research of the founder effect in human populations isolated by geography or custom.
- What is the frequency of the allele for Tay-Sachs disease among Ashkenazic Jews? How do scientists explain the high frequency of this allele in this segment of the population?
- How has the loss of genetic diversity (possibly resulting from a population bottleneck at some time in the past) affected cheetahs? Report on recent research efforts in this area.
- Describe how artificial selection in the genus *Brassica* has resulted in several very different vegetable varieties.
- What is the frequency of the allele for cystic fibrosis in the United States? Does the frequency of this allele differ among different segments of the population? Is the allele frequency changing over time?
- How do commercial plant breeders and agricultural biologists maintain genetic variability among the plants they raise?
- Why is inbreeding harmful to a species? Select a species or a group of organisms (a dog breed, for example), and discuss the results of inbreeding.
- Examine the role of geographic barriers (such as high elevations, mountaintops, isolated stream drainages, islands) in the development of a group of closely related regional species or genera.
- Prepare brief biographies of G. H. Hardy and W. Weinberg. How is it that they discovered their principle "independently"? Did they ever collaborate?

- Among invertebrates there are several hermaphroditic (both sexes in the same animal) species such as the common earthworm. How is self-fertilization avoided when these animals reproduce? (Answer: anatomical features and differential timing of gamete maturation)
- It is known that HbS/HbA (sickle-cell heterozygote) persons in Africa are more likely to survive a malarial infection. Is the percentage of heterozygous individuals in nonmalaria areas (such as the United States) different from that in malaria areas?

Films, Filmstrips, and Videos

- *Darwin and the Theory of Natural Selection.* CORF, 12 minutes, 16 mm, color.
- *Darwin's Bulldog.* Time-Life, 50 minutes, 16 mm, color. A dramatization of the 1860 Oxford debate in which Huxley brilliantly defended Darwin's theory of natural selection.
- *Did Darwin Get It Wrong?* BBC, PBS, TLF, 1983, 57 minutes, 16 mm/videocassette, color. *Nova Series.* Examines the controversy over Darwin's theory of natural selection that has raged and waned for over a century. Challenges to the theory continue to emerge from scientific laboratories, fossil evidence, and a vocal, powerful minority called "Creationists."
- *The Galapagos: Darwin's World within Itself.* EBEC, 1971, 19 minutes, 16 mm, color.
- *Genetic Polymorphisms and Evolution.* MIFE, 1977, 16 minutes, 16 mm, color. Two case studies are presented that show how genetic mutations evolved as a result of the need for survival in a changing environment. Sickle-cell anemia is discussed in detail.
- *Inheritance in Populations.* MGHT, 16 minutes, 16 mm, color.
- *Interactions in Heredity and Environment.* MGHT, 16 minutes, 16 mm, color.
- *Linkage-Population Genetics.* MIFE, 1976, 20 minutes, color. Genetic linkage is discussed and some illustrative matings are demonstrated. The analysis of linkage data is explained, and the Hardy-Weinberg law is clearly demonstrated. Genetic drift, gene flow, mutation, and selection are also discussed.
- *Natural Selection.* EBEC, 1963, 16 minutes, 16 mm, color.

37

PHYLOGENY AND MACROEVOLUTION

Revision Highlights

Minor changes; for example, corrected caption to Figure 37.7.

Chapter Outline

A QUESTION OF PHYLOGENY
- **Classification Schemes**
- **Speciation and Morphological Change**

RECONSTRUCTING THE PAST
- **The Fossil Record**
- **Comparative Morphology**
- **Divergence and Convergence in Form**
- **Comparative Biochemistry**

GEOLOGIC TIME

MACROEVOLUTIONARY PATTERNS
- **Evolutionary Trends**
- **Adaptive Radiation**
- **Origin of Higher Taxa**
- **Extinction and Replacement**

SUMMARY

Objectives

1. Understand the various classification schemes and realize the difficulty in determining from fossil evidence where the limits of one species end and the limits of another begin.
2. Be able to cite what biologists generally accept as evidence that supports their belief in evolution. Explain how observations from comparative morphology and comparative biochemistry are used to reconstruct the past.
3. Know the time boundaries of the five geologic eras and identify the principal organisms associated with each era.
4. Understand the factors that encourage increased rates of speciation and the formation of larger taxonomic groups. Know also the factors that bring about extinction and replacement of species.

Key Terms

species
category of relationship
higher taxa
macroevolution
phenotype
phylogeny
evolutionary taxonomy
lineage
gradualism
punctuation
comparative morphology
morphological divergence
convergence
neutral mutation
molecular clock
DNA hybridization studies
Archean
Proterozoic
Paleozoic
Mesozoic
Cenozoic
microevolutionary units of evolution

Lecture Outline

I. Introduction
  A. Species are the real entities in the tree of descent; the higher taxa are categories of relationship. **(Fig. 37.1, p. 555; TM 154)**
  B. Macroevolution considers trends among groups of species.

II. A Question of Phylogeny
  A. Classification Schemes
    1. Classification schemes are based upon phenotypic similarities—morphological, physiological, and behavioral traits.
    2. Phylogeny refers to evolutionary relationships.
    3. In a tree of descent, each twig is a line of descent, or lineage.
    4. Descent may be represented by evolutionary taxonomy or cladistics. **(Fig. 37.2, p. 556; TM 155)**
  B. Speciation and Morphological Change **(In-text art, p. 556; TM 156)**
    1. Morphological "gaps"separate even closely related species.
    2. Gradualism holds that change occurs within species by processes like genetic drift and directional selection. **(Fig. 37.3, p 557; TM 156a)**
    3. Punctuation suggests that morphological change occurs during rapid speciation, with little change occurring once a species forms. **(Fig. 37.3, p 557; TM 156b)**

III. Reconstructing the Past
  A. The Fossil Record
    Completeness varies with the type of organism and environment.
  B. Comparative Morphology
    Comparative morphology is used to create hypotheses; for example, early stages of development are highly conserved.
  C. Divergence and Convergence in Form
    1. Evolution proceeds by the modification of existing structures.
    2. Morphological divergence leads to departures from the ancestral form.
    3. Convergence occurs when at least two species of dissimilar lineages living in a similar environment come to resemble each other.
  D. Comparative Biochemistry
    1. Introduction
      a. All organisms are fundamentally the same in many aspects of their biochemistry.
      b. Neutral mutations accumulate through random processes and can be used as a molecular clock.
    2. Immunological Comparisons **(Fig. 37.7, p. 560; TM 157)**
    3. DNA Hybridization Studies **(Fig. 37.8, p. 561; TM 158)**

IV. Geologic Time
  A. The geologic time scale is a biological one in that the eras correspond to major evolutionary events.
  B. The ages of geologic time were established through radioactive dating methods.

V. Macroevolutionary Patterns
  Species are the units of evolution.
  A. Evolutionary Trends
  B. Adaptive Radiation
    Adaptive zones are ways of life often opened by a key innovation like wings. **(Fig. 37.11, p. 565; TM 159)**
  C. Origin of Higher Taxa
  D. Extinction and Replacement
    Radiations may follow extinctions but are rarely the cause of extinctions.
    1. Background Extinction

   a. Background extinction is the steady rate of species turnover in lineages through time.
2. Mass Extinction
   a. Mass extinction is an abrupt change in extinction rates affecting several higher taxa simultaneously.
   b. Five mass extinctions were global events of catastrophic proportions; some may have been caused by comets.

Suggestions for Presenting the Material

- This chapter is a "bridge" from the previous chapter, which examined mechanisms of change and speciation, to the following chapters, which detail some of the results of evolution.
- This chapter builds on the brief introduction to classification given in Chapter 2. You may want to briefly review Linnaeus and his categories of classification.
- Emphasize that humans construct classification schemes and therefore these systems are subject to change and interpretation.
- Stress the tenuous conclusions that are in constant revision when scientists attempt to reconstruct the past. Just as a good medical diagnosis is not based on one examination or one lab test, a good analysis of past evolutionary history is not based on any one line of evidence, but rather, several lines of corroborating evidence.
- It is probably not crucial that students learn to reproduce every detail of Figure 37.9 (the geologic time scale), but they should be able to arrange the major eras and periods in sequence and provide approximate dates.
- Some instructors may require the calculation of dates using radioactive half-lives. Certainly, emphasis should be placed on *how* these calculations are used.
- Unfortunately, a thorough look at macroevolution and adaptive radiation necessitates an understanding of eras and periods as well as the names of taxonomic groups as Figure 37.11 shows. This will require "adjustment" to your audience.

Classroom and Laboratory Enrichment

- Discuss methods of fossil preservation. Examine actual fossils or films, videos, or 35 mm slides of fossils. Visit collections of fossils in nearby museums.
- Perform exercises in which student groups each devise their own classification system for an array of similar objects, for example, buttons or bolts. Groups should share their results when finished. Did each group invent different classification schemes? Discuss the merits and shortcomings of each system. What are the attributes of a good classification system?
- Discuss the historical background of the classification of one group of organisms. Compare early views of biologists regarding the evolutionary history of this group to those of modern biologists. Discuss how new research techniques have shed new light upon the group's history.
- Show overhead transparencies of phylogenetic trees of familiar organisms, such as horses, vertebrates, or mammals.
- Look for examples of divergence and convergence among groups of organisms.
- Use data from immunological or DNA hybridization studies to reconstruct the phylogenetic branch points during the evolutionary history of a group of organisms.
- Examine the primate phylogenetic tree. What morphological traits have been used to construct this phylogeny?
- Prepare a summary table of the "tools" used in "reconstructing the past." For each tool, list the procedure, reliability, advantages/disadvantages, accuracy, and so forth.
- See if you can obtain some electrophoresis patterns (starch gel, or photos of same) that show actual "runs." Interpret the similarities and differences in the proteins from different sources (see the "Discussion" below). If you can arrange for a researcher to present this information, so much the better.

Ideas for Classroom Discussion

- What evidence was available to Darwin at the time he formulated his principle of evolution?
- What types of conditions favor fossil formation? What conditions are poor for fossil formation?
- What kinds of organisms are well represented in the fossil record? What types of organisms have left little or nothing in the fossil record?
- How does microevolution differ from macroevolution? Can biologists actually see microevolution occurring? Give some examples.
- How do worldwide distribution patterns of fossils suggest a common evolutionary origin for many organisms?
- Why do many evolutionary biologists believe that the punctuational model of evolution explains the gradual development of different organisms better than does the gradual model?
- What triggered the rapid adaptive radiation of mammals beginning at the start of the Cenozoic Era?
- Which evolutionary model is supported more fully by the fossil evidence—gradualism or punctuation?
- Of all the methods used to reconstruct the past, which is most conclusive? Least conclusive? Which of these was (were) not available to scientists of Darwin's time?
- One of the methods used to show evolutionary relationships is *comparative biochemistry.* Of course it is the DNA that we wish to compare, but it is easier to compare the products of DNA, namely proteins (see the "Enrichment" section below). Using the knowledge gained from Chapter 16, show how the comparisons of proteins can reveal information about DNA.

Term Paper Topics, Library Activities, and Special Projects

- Describe famous fossil finds. You may wish to include famous fossils that were later found to be hoaxes, such as the Piltdown man discovered in England in 1912.
- Discover what fossil evidence has revealed about the past history of your region of the United States.
- Discuss modern examples of adaptive radiation.
- Write a report about recent theories attempting to explain the demise of the dinosaurs.
- Describe radioactive dating.
- What is thought to be the current rate of species extinction today? Is the rate of extinction today higher than can be accounted for by background extinction? What areas of the earth are experiencing the highest rates of species extinction?
- When reconstructing the past evolutionary history of the earth and its living forms, vast periods of time are necessary to account for the amount of change. How do strict creationists view the geologic time scale as presented in Figure 37.9?
- What are the current theories for the great mass extinctions of the past? Which is (are) the most plausible?
- Research the causes of extinction *today.* Are all, or most, of the causes the result of human intervention?

Films, Filmstrips, and Videos

- *Adaptive Radiation: The Mollusks.* EBF, 1968, 18 minutes, 16 mm, color. Good photography and a good example of adaptive radiation.
- *Darwin's Finches: Clues to the Origin of Species.* BFA, 1961, 11 minutes, 16 mm, color.
- *Evolution in Progress.* BFA, 1968, 18 minutes, 16 mm, color.
- *Stephen Jay Gould: This View of Life.* Time-Life/NOVA, 1985, 57 minutes, 16 mm, color. Gould lends the liveliest voice in science to his views of evolutionary theory.

38

ORIGINS AND THE EVOLUTION OF LIFE

Revision Highlights

New illustrations of fossils and reconstructions of early plants (*Cooksonia*, *Psilophyton*, and *Archaeopteris*, Figure 38.13).

Chapter Outline

ORIGIN OF LIFE
- **The Early Earth and Its Atmosphere**
- **Spontaneous Assembly of Organic Compounds**
- **Speculations on the First Self-Replicating Systems**

THE AGE OF PROKARYOTES
- **Prokaryotic Metabolism and a Changing Atmosphere**
- **Divergence Into Three Primordial Lineages**

THE RISE OF EUKARYOTES
- **Origin of Mitochondria**
- **Origin of Chloroplasts**
- **Beginnings of Multicellularity**

FURTHER EVOLUTION ON A SHIFTING GEOLOGICAL STAGE
- **Life During the Paleozoic**
- **Life During the Mesozoic**
- **The Cenozoic: The Past 65 Million Years**

PERSPECTIVE

SUMMARY

Objectives

1. Describe how life might have spontaneously arisen on Earth approximately 3.5 billion years ago.
2. Understand how ancient prokaryotes are thought to have changed the primeval atmosphere of Earth and diverged into three primordial lineages.
3. Outline the theory that accounts for the origin and rise of eukaryotes.
4. Describe the shifty shenanigans of tectonic plates in the Paleozoic, Mesozoic, and Cenozoic Eras.

Key Terms

Stanley Miller
microspheres
liposomes
Desulfovibrio
eubacteria
archaebacteria
urkaryotes
mesosomes
symbiosis
plate tectonic theory
Gondwana
Laurasia
Pangaea
Tethys Sea
Cambrian
Ordovician
Silurian
Devonian
Carboniferous
Permian
Mesozoic Seas
archosaurs

Lecture Outline

I. Introduction
   A. By the late 1860s, the "fixity of species" was crumbling as a scientific concept.
   B. By the middle of the twentieth century, the "fixity of continents" theory crumbled.
   C. The evolution of life is linked to the evolution of the earth.

II. Origin of Life
   A. The Early Earth and Its Atmosphere
      1. About 3.8 billion years ago, the primitive atmosphere of the earth formed without oxygen.
      2. Because of its size and distance from the sun, liquid water formed on its surface.
   B. Spontaneous Assembly of Organic Compounds **(Fig. 38.2, p. 572; TM 160)**
      All the building blocks of living systems can form under abiotic conditions.
   C. Speculations on the First Self-Replicating Systems
      1. Organic material accumulated in shallow seas over a 300-million-year period.
      2. Templates for Protein Synthesis **(Fig. 38.3, p. 573; TM 161)**
         a. Clay particles may have first served as templates to assemble amino acids into proteins with enzymatic activity.
         b. Ribonucleotides may have then stuck to the clay or amino acids.
         c. DNA may have evolved later.
         d. This was not pure chance; some reactions are more probable than others.
      3. Models of the First Plasma Membranes
         a. Chemical control requires chemical isolation.
         b. Abiotically, amino acids form proteins that assemble into microspheres, and the microspheres can acquire a lipid-protein film.
         c. Abiotically, lipids may form liposomes that have many of the properties of biological membranes.
         d. The first cells were probably membrane-bound sacs containing nucleic acids that served as templates for proteins.

III. The Age of Prokaryotes
   A. Prokaryotic Metabolism and a Changing Atmosphere
      1. The first cells were prokaryotes—anaerobic and heterotrophic.
      2. Fermentation evolved first, then electron transport systems—photosynthesis and aerobic respiration.
   B. Divergence into Three Primordial Lineages
      1. These are the eubacteria, archaebacteria, and urkaryotes.

IV. The Rise of Eukaryotes
   A. Introduction
      1. The nucleus evolved from an infolding of the plasma membrane around the DNA.
   B. Origin of Mitochondria
      1. Mitochondria evolved from symbiosis between urkaryote and aerobic bacteria.
      2. Aerobic bacteria and mitochondria have a similar size, structure, and biochemistry.
   C. Origin of Chloroplasts
      1. Chloroplasts arose from symbiosis of aerobic urkaryote and photosynthetic bacteria.
      2. Chloroplasts have similar metabolism and DNA as cyanobacteria.
   D. Beginnings of Multicellularity
      The first fossils are 700 million years old.
      1. Regulatory genes began to control activities of tissues and then organs.
      2. Division of labor permits larger size and efficiency.

V. Further Evolution on a Shifting Geologic Stage **(Fig. 38.10, p. 579; TM 162)**
   A. Introduction
      1. According to plate tectonic theory, the continents changed position and orientation.

2. When the land masses separated, speciation proceeded; when the land masses collided, diversity declined.

B. Life During the Paleozoic

1. The Cambrian **(Fig. 38.14 left, p. 582; TA 105)**
   a. Most organisms lived on or just below the seafloor (examples include trilobites, mollusks, and echinoderms).
2. The Ordovician **(Fig. 38.14 left, p. 582; TA 105)**
   a. Cephalopods and armored fish evolved.
3. The Silurian **(Fig. 38.14 left, p. 582; TA 105)**
   a. Armored fish evolved jaws and plants began to invade the mud at the margin of land.
4. The Devonian **(Fig. 38.14 left, p. 582; TA 105)**
   a. Plants continued their invasion of the land and fish dominated the seas and some began to breathe air.
5. The Carboniferous **(Fig. 38.14 left, p. 582; TA 105)**
   a. Gymnosperms, reptiles, and insects evolved.
6. The Permian **(Fig. 38.14 right, p. 583; TA 106)**
   a. Dinosaurs, birds, snakes, and lizards emerged, and 90% of marine species became extinct.

C. Life During the Mesozoic **(Fig. 38.14 right, p. 583; TA 106)**

1. Mesozoic Seas
   a. Adaptative radiation of marine species increased diversity—a process that is still continuing.
2. Mammals, Dinosaurs, and Flowering Plants
   a. Mammals evolved from therapsids.
   b. The "ruling reptiles" evolved from thecodonts and disappeared at Cretaceous-Tertiary boundary.
   c. Birds emerged (*Archaeopteryx* and *Protoavis*)
   d. Flowering plants began radiation.

D. The Cenozoic: The Past 65 Million Years **(Fig. 38.14 right, p. 583; TA 106)**

1. At first, mammals were rodentlike and opossumlike creatures.
2. Then, adaptive radiation increased the number of genera from 100 to more than 3,000. **(Fig. 38.18, p. 586; TM 163)**
3. The Cenozoic was a period of great diversity and dispersal of land masses.

VI. Perspective

A. C, H, N, and O are the primary elements that can spontaneously form all the basic molecules of life.

B. Chemical evolution led to biological evolution.

C. All species are products of interactions of the environment and with each other.

Suggestions for Presenting the Material

- You may want to begin this topic by surveying the class for the explanations most frequently given to account for the origin of life on planet Earth. (Responses from my classes are, in frequency order: 1. special creation, 2. arrival from distant planets, 3. "just happened.")
- Based on the above survey, you may find that your students know the *least* about *spontaneous generation* and are indeed skeptical that "life could come from nonliving matter." Be sure you include some mention of the difference in possibilities of spontaneous generation in the *past* and *today* (see the "Enrichment" section).
- Use a summary table of contributing scientists (see the "Enrichment" section) to guide your presentation of the origin of life.

- The second major topic of this chapter is the transition of *prokaryotes* to *eukaryotes*. You will need to redefine these two types of cells, but further characterization is just around the corner in Chapter 39.
- Emphasize the role of free oxygen in the transition mentioned above.
- In the final portion of the chapter, the student is again presented with the geologic time scale (similar to Figure 37.9). However, this time more "events" are assigned to the eras and periods. There is much information in just a few pages. You may need to specify what you require here.

Classroom and Laboratory Enrichment

- Show actual fossils or photographs of fossils representing some of the life forms prevalent during the Paleozoic, Mesozoic, and Cenozoic Eras.
- Discuss the similarities and differences between cyanobacteria and chloroplasts and between aerobic bacteria and mitochondria. Students could make wet mounts in lab of several primitive cyanobacteria and cells of photosynthetic eukaryotes with chloroplasts.
- Demonstrate the events of continental drift using an overhead transparency to represent the earth and pieces of transparency film to represent the land masses. Films and videotapes can also be used to effectively explain continental drift.
- Assign one group of students to each of the geologic periods. Ask each group to report on the climate, geologic events, and dominant groups of organisms of that period.
- Prepare an overhead summarizing the work of Redi, Spallanzani, and Pasteur in disproving spontaneous generation under "recent" conditions on earth.
- Construct a summary table listing the *scientist(s)* and *major contributions* to the origin of life question:

Scientist	*Contribution*
a. Miller and Urey	a. *Amino acids* from inorganics
b. Cairns-Smith	b. *Protein assembly* on clay templates
c. Lawless	c. Clay crystals attract only *left-handed molecule* forms
d. Fox	d. Lipid-protein *microspheres*
e. Hargreaves and Deamer	e. *Liposome* formation

- Retrieve your overheads of *chloroplasts* and *mitochondria* to show comparison of their structure. Briefly review the similarities and differences in their functions.
- Obtain a set of 2 x 2 slides that picture reconstructions of the geologic time periods. These will provide some visualization of the rather "abstract" information in the text.

Ideas for Classroom Discussion

- Why is that body of thought known as "scientific creationism" not a science? List the types of evidence supporting the principle of evolution.
- Evolutionary biologists believe that life on earth first evolved in the absence of oxygen. Where did the oxygen in our atmosphere come from?
- How have extinctions facilitated the development of new species?
- What role did climate changes during the Permian Period play in the development of modern land plants?
- What are some features that might account for the spectacular success of the mammals since their origin in the Mesozoic?
- Diagram the apparatus used by Stanley Miller to study the synthesis of organic compounds in an atmosphere like that of the early earth. Ask students what they think of Miller's findings.

- Do you think it is possible for scientists studying the evolution of life to also believe that God played a role in the creation of life? Why or why not?
- What distinguishes a nonliving lipid-bound sphere containing nucleic acids and amino acids from a living cell?
- Describe the metabolic pathways used by the first living cells to obtain energy.
- Why was aerobic respiration necessary for the evolution of eukaryotes?
- What are some of the advantages of multicellularity?
- Why did Charles Darwin refer to the evolution of flowering plants as "an abominable mystery"?
- Name some organisms living today that are virtually unchanged from their earliest appearances in the fossil record millions of years ago. Why do some organisms fail to change significantly throughout geologic time?
- Are the continents still moving today?
- Why do you think people believed as recently as 100 years ago that spontaneous generation still occurs on the earth?
- What are the objections you could raise to the possibility of life arriving here millions of years ago from distant planets?
- On what evidence do scientists reconstruct a primordial earth that resembles Figure 38.1?
- Why was no free oxygen (O_2) included in Miller's experiment? What did the electrical spark simulate?
- In the hypothetical series of events (Figure 38.3) leading from spontaneous formation of molecules to living cells, which step is most puzzling and the one about which little is known?
- Could any of the "creations" resulting from the experiments listed in the table in the "Enrichment" section be described as "living"? Why or why not?

Term Paper Topics, Library Activities, and Special Projects

- Trace the evolutionary history of one group of plants or animals throughout geologic time.
- Prepare a time line tracing the geologic and biologic history of your section of the United States.
- Discuss the varying degrees of success of some of the groups of organisms whose evolutionary histories are shown in Figure 38.18. Describe ways in which the radiation or extinction of one group of organisms may have influenced that of another. For example, do you think it is a coincidence that both the insects and the flowering plants have greatly increased in number during the Cenozoic Era? Did a decrease in lycopods and horsetails contribute in any way to the expansion of the gymnosperms? How did the climate change in North America after lycopods and horsetails evolved?
- Discuss the history of the experiments by Miller and Urey.
- Collect fossils or photographs or descriptions of fossils from your region of the United States. What do they reveal about the past history of your area?
- Discuss the formation of coal and oil. Prepare a map showing areas of coal and oil concentration in the United States.
- Discuss recent controversial studies that use human mitochondrial DNA to draw conclusions about human evolution.
- Describe the effects of the plate tectonic theory on organismal diversity.
- Where did the names of the eras and periods used in the geologic time scale come from? Discuss the invention and history of the geologic time scale.
- See if you can locate the original article by Miller and Urey in which they reported their famous experiment. Read it carefully to see what speculative application they made for their experiment.
- Although the idea of life originating on earth from forms traveling from distant planets is highly imaginative, serious proposals have been advanced. Locate some of these and evaluate their merit.

Films, Filmstrips, and Videos

- *Molecular Biology.* CORT, 1981, 15 minutes, 16 mm, color. *Biological Sciences Series.* Animation illustrates the origin and evolution of life beginning with the primitive Earth four billion years ago.
- *Origin of Life.* Wiley, 1971, 25 minutes, 16 mm, color. Uses cinemicrography, narration, and animation to illustrate the organization and development of the living cell. Examines the structure of the major organic compounds involved in living systems.
- *The Origin of Life: Chemical Evolution.* EBEC, 1969, 10 minutes, 16 mm, color. *Heredity and Adaptive Change Series.* Examines experiments that pertain to theories of the origin of life in the seas of primitive Earth, including the work of Oparin, Ponnamperuma, Miller, and Urey.
- *Origin of Life: Biochemical Evolution.* MG, WILEY, 1971, 26 minutes, 16 mm, color. Explains the conditions of the primitive Earth that could have given rise to amino acids, carbohydrates, lipids, and nucleic acids; then to macromolecules and primitive life forms.
- *Plate Tectonics: A Revolution in the Earth Sciences.* PBS, 1985, 28 minutes, videocassette, color. *Earth Explored Series.* The series explores evolution on Earth. This program illustrates how mountains build, affecting migration of animals. It includes animation, scenes in the Alps, and visits to active volcano areas.
- *The Search for Life.* Time-Life, 30 minutes, color. Shows Miller's experiment, the formation of proteinoid microspheres, and DNA.

39

VIRUSES, MONERANS, AND PROTISTANS

Revision Highlights

Reorganized, updated, rewritten. Viruses now described before monerans and protistans, with more text and new illustrations on viral structure, replication cycles, and new photo of effect of viral infection on tulips. Update on prions. Clearly organized section on bacteria, with emphasis on metabolic diversity. New text and illustrations on general characteristics and body plan of bacteria. New text and photo on Gram staining. Distinguishes between "characterizing" and "classifying" bacteria and explains why latter is so difficult. New description and illustration of archaebacteria, then of eubacteria. Examples of eubacteria selected to illustrate basic mode of nutrition (photosynthetic autotrophs, chemosynthetic autotrophs, heterotrophs). New table of some major bacteria groups, using *Bergey's Manual* as a guideline. Examples used are ones described in this or other chapters. New photos of cyanobacteria, better description of heterocyst formation. Lyme disease covered. Revised section on bacterial behavior, including bacteria that cooperates in predatory behavior and fruiting body formation. Updated classification of protistans (Table 39.4); slime molds placed here rather than with fungi, with a note about difficulty in classifying them. *Giardia* is among the pathogens described. Tighter concluding section on evolution of multicellularity. Revised summary.

Chapter Outline

VIRUSES
- **General Characteristics of Viruses**
- **Viral Infectious Cycles**
- **Animal Viruses**
- **Plant Viruses**
- **Viroids and Prions**

MONERANS
- **Characteristics of Bacteria**
- **Classification of Bacteria**
- **Archaebacteria**
- **Eubacteria**
- **A Final Word on the "Simple" Bacteria**

PROTISTANS
- **Slime Molds**
- **Euglenids**
- **Chrysophytes**
- **Dinoflagellates**
- **Protozoans**

ON THE ROAD TO MULTICELLULARITY

SUMMARY

Objectives

1. List five specific viruses that cause human illness and describe how the viruses do their dirty work.
2. Describe the principal moneran forms and ways of living.
3. Describe the four categories of protistans. Tell how protistans differ from monerans, viruses, and multicellular eukaryotes. Give some common names of protistans.

Key Terms

microbes
virus
viral capsid
lytic pathway
lysogenic pathway
bacteriophage
influenzaviruses
pandemic
herpesviruses
poxviruses
rhinoviruses
retroviruses
viroids
prions
scrapie
kuru
Creutzfeldt-Jacob disease
photosynthetic autotrophs
chemosynthetic autotrophs
heterotrophs
binary fission
peptidoglycan
capsule
slime layer
pili
bacterial flagella
archaebacteria
methanogens
extreme halophiles
thermoacidophiles
eubacteria
photosynthetic bacteria
cyanobacteria
heterocysts
green bacteria
purple bacteria
chemosynthetic bacteria
nitrifying bacteria
heterotrophic bacteria
Lyme disease
endospores
magnetotactic bacteria
myxobacteria
Myxococcus
fruiting bodies
continuum of diversity
cellular slime molds
plasmodial slime molds
euglenids
pellicle
dinoflagellates
red tides
protozoans
flagellate protozoans
trypanosomes
foraminiferans
heliozoans
radiolarians
sporozoans
ciliated protozoans

Lecture Outline

I. Viruses
   A. General Characteristics of Viruses **(Fig. 39.2, p. 591; TA 107)**
      1. Viruses are noncellular infectious agents consisting of a nucleic acid core and a protein coat.
      2. They replicate only within a host cell.
      3. The genetic material of a virus is DNA or RNA, and it may be a single- or double-stranded molecule.
      4. The coat, or capsid, may be enclosed within a lipid bilayer from the cell membrane.
   B. Viral Infectious Cycles **(Fig. 39.3, p. 592; TA 108)**
      1. Virus binds to the host cell, enters the cytoplasm, and then creates viral proteins and nucleic acid.
      2. The viruses called bacteriophage can follow a lytic or lysogenic pathway of infection.
   C. Animal Viruses
      1. Flu pandemics are caused by genetic recombinants between human viruses and the viruses of other animals.
   D. Plant Viruses
      1. Plant viruses are transmitted by invertebrates and may reproduce in these animal vectors.
   E. Viroids and Prions
      1. Viroids consist of single-stranded RNA without a protein coat; they can infect plants and may cause cancer.
      2. Prions appear to be proteins that can cause fatal diseases (kuru is an example).

II. Monerans
   A. Introduction

1. Bacteria, unlike viruses, are able to live and reproduce as independent units.

B. Characteristics of Bacteria **(Fig. 39.6, p. 595; TA 109)**
   1. They are prokaryotes that are photosynthetic autotrophs, chemosynthetic autotrophs, or heterotrophs.
   2. Bacteria reproduce by binary fission.
   3. Their chromosome is circular DNA and, in addition, may have plasmids.
   4. The wall usually contains peptidoglycan and may have a sticky capsule or slime layer.
   5. They have pili for attachment during conjugation.
   6. Bacterial flagella make motility possible.
   7. Coccus, bacillus, or spirillum shapes occur commonly.

C. Classification of Bacteria
   1. They may be Gram-negative or Gram-positive.

D. Archaebacteria
   1. Archaebacteria live in extreme environments reminiscent of ancient environments.
   2. They have unique ribosomes and membrane lipids.
   3. All have similar DNA sequences and lack peptidoglycan in their cell walls.
   4. They include methanogens, halophiles, and thermoacidophiles.

E. Eubacteria
   1. Eubacteria form a diverse group divided mainly by Gram-positive walls, Gram-negative walls, and lack of cell walls; they also differ in nutritional modes.
   2. Some photosynthetic eubacteria use oxygen as a final electron acceptor and may fix nitrogen in heterocysts.
   3. Chemosynthetic eubacteria include nitrifying bacteria that help to cycle nitrogen.
   4. Heterotrophic eubacteria include many human pathogens, such as *Treponema* and *Rickettsia*.
   5. Endospore-forming heterotrophs (e.g., *Clostridium*) produce endospores that are resistant to heat.
   6. A strain of *E. coli* is the leading cause of infant mortality.

F. A Final Word on the "Simple" Bacteria
   1. Some bacteria can sense a stimulus such as light or oxygen and react (i.e., move away).
   2. Cells of *Myxococcus Xanthus* can move as a unit.

III. Protistans

A. Protistans are a kingdom of single-celled eukaryotes; kingdom boundaries are poorly defined. **(Fig. 39.15, p. 601; TM 164)**

B. Slime Molds
   1. Each cellular slime mold is composed of an aggregation of amoebalike cells.
   2. Each plasmodial slime mold is a multinucleate mass of cytoplasm.

C. Euglenids
   1. Euglenids are flagellated, photosynthetic cells that have a pellicle and an eyespot.

D. Chrysophytes
   1. These include "yellow-green algae," "golden algae," and diatoms with certain xanthophylls and beta-carotene.

E. Dinoflagellates
   1. Dinoflagellates are marine flagellates with grooved cellulose plates; they move by flagella.
   2. Some produce neurotoxins and cause red tides.

F. Protozoans
   1. Flagellated Protozoans
      a. Flagellated protozoans include trypanosomes, which cause sleeping sickness.
   2. Amoeboid Protozoans
      a. Amoeboid protozoans use pseudopods to capture prey or for motility.
      b. Amoebas are "naked," and *Entamoeba histolytica* causes amoebic dysentery.
      c. Foraminiferans extend pseudopods through tiny holes.
      d. Heliozoans have fine pseudopods that radiate like sun rays.
      e. Radiolarans resemble heliozoans but have a silica skeleton.

3. Sporozoans include *Plasmodium,* some of which cause malaria.
4. Ciliated protozoans show great structural complexity.

IV. On the Road to Multicellularity
  A. Many protistans remain attached after division and form colonies.
  B. Colonies may be quite complex; *Volvox* may be an experiment in multicellularity.

Suggestions for Presenting the Material

- Students will be extremely interested in the viruses and unicellular organisms covered in this chapter, especially those that are capable of causing well-known diseases.
- Most students are unfamiliar with the structure and function of the viruses, bacteria, and protists. Emphasize the nonliving nature of the viruses, a point that students often have trouble grasping.
- Briefly review the five kingdoms, reminding students that this chapter covers organisms from two different kingdoms (in addition to the viruses). Use diagrams or overheads that show the viruses, monerans, and protistans drawn to actual scale (Figure 39.1 is a good example). Ask students again to recall the differences between prokaryotic and eukaryotic cells.
- This chapter offers an excellent opportunity to discuss epidemiology and to show how scientists use the scientific method when attempting to find the causative agent of a disease. The recent widespread public awareness of AIDS also opens an avenue for discussion of the accuracy with which scientific stories are covered by the media.

Classroom and Laboratory Enrichment

- Visit an electron microscope in a research laboratory where research on viruses is being done. Ask the technician to discuss preparation techniques for viral specimens and to demonstrate the operation of the microscope. If possible, allow students to prepare their own viral suspensions of a plant virus (such as tobacco mosaic virus) in the laboratory for subsequent examination under the electron microscope.
- Use overhead transparencies or films or videos to show the steps of a viral infectious cycle.
- Prepare an exhibit of living organisms or portions of organisms, such as leaves, affected by viruses.
- Use the tip of a toothpick to collect bacteria from sources such as plaque from the surface of a tooth or yogurt with live cultures. Prepare a simple bacterial smear, stain with crystal violet, and observe at 1000x.
- Prepare gram stains of bacterial species of varying morphology and gram status from cultures obtained from biological supply houses.
- Collect and examine cyanobacteria from stagnant ponds, tree trunks, and greenhouse flowerpots.
- View protozoa in a hay infusion.
- Look for slime molds on moist logs and fallen trees in damp forest areas.
- Collect diatoms from the edges of quiet ponds and slow-moving streams.
- Gather information from local health authorities about any of the diseases caused by bacteria or protists that are found in your area. Sexually transmitted diseases such as gonorrhea, syphilis, chlamydial infections, and trichomoniasis would be good examples. What is the number of cases per month within your city or state? Is the number of new cases rising or falling?
- Survey your environment for the presence of bacteria. This can be done with help from a microbiologist who can prepare media and plates. Then you and your students can (a) take swabs of various surfaces; (b) allow dust to settle on a plate; (c) allow insects to land; (d) apply samples of food, and so on.
- Remind students of the beneficial aspects of bacteria including their usefulness in producing products through genetic engineering.

- Many protists are highly mobile creatures and fun to watch. If you have a small class, secure some cultures and allow students to make their own preparations for microscopic observation. If your class is large, use a videotape (several biological supply houses offer these).
- Seek class members or faculty colleagues who may have traveled to a tropical area where diseases unknown in the United States are present. Ask them to report on any precautions taken before, during, and after their trip.

Ideas for Classroom Discussion

- How would you classify the viruses? Would you consider them to be nonliving? If the viruses are nonliving, why do you think they are covered in the study of biology?
- What is a pathogen? What is the Greek word from which this term is derived?
- What is epidemiology? Discuss the roles played by the World Health Organization and the Center for Disease Control in the study and treatment of outbreaks of disease.
- How does the human body respond to a sudden onslaught of viral invaders?
- What are viroids and prions? How are they similar to/different from viruses?
- Why is the kingdom Monera said to be "the most metabolically diverse" of all of the five kingdoms? Why are organisms of such varied metabolisms all placed in the same kingdom?
- Why are the methanogens, halophiles, and thermoacidophiles placed in their own subkingdom (the Archaebacteria) of the kingdom Monera?
- What are some steps that tropical countries can take to reduce the rate of malaria?
- In older classification schemes, bacteria were included with the plants. For what reason was this a rather poor "match"?
- Normally, we would think of "autotrophic" as synonymous with "green" if we were considering members of the plant kingdom. What two sources of energy are available to *bacterial* autotrophs?
- Why is the "Gram" of Gram stain written with a capital letter?
- What do the prefixes "eu-" and "archae-" mean when attached to the root "-bacteria"?
- What is the role of disease in limiting the growth of human and animal populations? (Remember Malthus!)

Term Paper Topics, Library Activities, and Special Projects

- Where did viruses come from? Discover what scientists know about the evolutionary origin of the viruses.
- How are viruses destroyed? What kinds of chemicals and filters are used in the laboratory to ensure that an experimental liquid is virus-free?
- Research the history, mode of action, and treatment of several economically important viral diseases of crops.
- Discuss the role of the immune system in fighting viral outbreaks or bacterial infections. How does the immune system "recognize" viral strains or bacterial species that have caused disease within the individual on previous occasions?
- Discuss the worldwide Spanish flu epidemic of 1918–1920.
- Collect stories about AIDS from magazines and newspapers. Analyze the stories for scientific accuracy and completeness. In what areas do such stories effectively cover AIDS? In what ways do they mislead or fail to inform the reader?
- Report on the latest scientific research in the treatment of any one of the diseases caused by viruses discussed in your text.

- Are public swimming areas (pools, lakes, springfed ponds, rivers) in your area required to routinely check for evidence of pollution from sewage effluent? Discover the maximum *E. coli* count allowed by law before the facility must be closed to public swimming. How is this figure determined? What does it mean? Why is it unsafe to swim in waters that have a relatively large number of *E. coli* bacteria?
- How do antibiotics such as penicillin and tetracycline kill bacteria? Discuss the modes of action of several commonly used antibiotics.
- Describe the role of certain species of bacteria in producing food products such as buttermilk, sour cream, yogurt, and sauerkraut.
- Describe the morphology and physiology of any one of the pathogenic genera of bacteria. Discuss the diseases caused by the members of the genus and their treatments.
- Discuss the geographic distribution and economic impact of "red tides" in the United States.
- Write a report on any one of the diseases caused by protozoans. Discuss the historical roles played by these diseases in situations such as the colonization of new lands, wars, and the building of the Panama Canal.
- The five kingdom system of classification (see Table 2.2) was proposed by Robert H. Whittaker in 1963. Locate the original article to see if his original version has changed much up to the present time.
- Your text indicated that flu epidemics (and pandemics) tend to occur in ten- to forty-year cycles. Check an epidemiology or medical textbook to locate possible explanations for this periodicity.
- How does a retrovirus differ from other viruses in its operation?
- The following are diseases caused by spirochetes and transmitted in different ways:
  a. Lyme disease by ticks
  b. Relapsing fever by lice
  c. Syphilis by human sexual contact

  Are the different spirochete species able to interchange vectors? Explain why.

Films, Filmstrips, and Videos

- *Bacteria and the Ecology of Planet Earth.* MLA, 1980, 13 minutes, 16 mm, color. Bacteria recycle nutrients through decomposition and "fix" atmospheric nitrogen for use by producers. This film follows the role of bacteria in the nitrogen cycle thoroughly.
- *Microorganisms That Cause Disease.* MLA-Wards, 1977, 11 minutes, 16 mm, color. A clear, comprehensive showing of the five principal groups of pathogenic microorganisms: bacteria, viruses, fungi, rickettsiae, and protozoa. Stresses that microorganisms that are pathogenic cause infectious diseases through destruction of cells. Electron micrographs, photomicrographs, and animated diagrams show the structure of various microorganisms.
- *Microscopic Plants.* MGHT, 30 minutes, 16 mm, color.
- *A New Look at Bacteria.* MLA, 1980, 16 minutes, 16 mm, color. Shows locomotion patterns and behavioral responses in bacteria. Gliding forms, spirochetes, and flagellated bacteria are shown moving through their own environments.
- *A New Look at Leeuwenhoek's "Wee Beasties."* MLA, 1977, 12 minutes, 16 mm, color. Beautiful photography of protistans through microphotography.
- *Plant Life: Bacteria.* EBEC, 1962, 19 minutes, 16 mm, color.
- *Protist Behavior.* MLA, 1977, 11 minutes, 16 mm, color. Shows the behavior of protistans responding to various stimuli.
- *Protist Ecology.* MLA, 1977, 12 minutes, 16 mm, color. The protistan role in ecological relationships is richly illustrated.
- *Protist Physiology.* MLA, 1977, 13 minutes, 16 mm, color. Shows several techniques that demonstrate specific physiological characteristics of protistans.

- *Protist Reproduction.* MLA, 1977, 10 minutes, 16 mm, color. Shows the reproductive methods of protists, including fission, with replication of organelles and subcellular particles; specialized forms of budding; sexual processes in ciliates and algae; and the production of eggs and sperm of colonial protists.
- *Viruses.* Wards, 1980, 15 minutes, 16 mm, color. This film studies the culture of viruses and research in virology. Hidden viruses may play a role in cancer, aging, and inheritable diseases. A Bruce Russell film.
- *Viruses—Threshold of Life.* CORF, 1967, 14 minutes, 16 mm, color.
- *World of Microbes.* MGHT, 30 minutes, 16 mm, color.

40

FUNGI AND PLANTS

Revision Highlights

Major revision. Better introduction to chapter. *Part I, Fungi:* Expanded coverage, with better balance among all important groups, including chytrids and water molds. New photos of powdery mildew, apple scab, *Amanita,* budding cells of yeast, stinkhorn fungus, and conidia of *Penicillium;* new life cycles for *Rhizopus* and for typical mushroom. New Commentary on fungi and human affairs. *Part II, Plants:* Expanded coverage, simplified tables, and new life cycles for *Chlamydomonas,* moss, fern, pine, and flowering plant. Tightened introduction on evolutionary trends among plants. More commercial applications of red algae and brown algae. New photos of sea palms, green algae (including sexual reproduction in *Spirogyra*), and *Marchantia;* expanded coverage of ferns, including a new SEM of sporangia; and a new two-page photo spread on the radiation of flowering plants into diverse environments. Improved summary.

Chapter Outline

PART I. KINGDOM OF FUNGI
- The Fungal Way of Life
- Fungal Body Plans
- Overview of Reproductive Modes
- Major Groups of Fungi
- Chytrids
- Water Molds
- Zygospore-Forming Fungi
- Sac Fungi
- Club Fungi
- Imperfect Fungi
- Mycorrhizae
- *Commentary:* A Few Fungi We Would Rather Do Without
- Lichens

PART II. KINGDOM OF PLANTS
- Evolutionary Trends Among Plants
- Classification of Algae
- Red Algae
- Brown Algae
- Green Algae
- The Land Plants
- The Bryophytes
- Lycopods, Horsetails, and Ferns
- Existing Seed Plants
- The Gymnosperms
- Angiosperms—The Flowering Plants

SUMMARY

Objectives

1. Describe the various types of fungal body plans, patterns of reproduction, and natural history.
2. Name at least one specific example of each of the five groups of true fungi.

3. Outline the evolutionary advances that converted marine algal ancestors into forms that could exist on wet land. Then state the advances that converted primitive homosporous marsh plants into dry-land flowering plants.

Key Terms

saprobes
parasites
mycelium, -lia
hypha, hyphae
fungal spores
dikaryotic stage
chytrids
rhizoids
water molds
resting spore
zygospore
sac fungi
ascus, asci
club fungi
ergotism
mycorrhiza, -izae
lichen
saprophytic
parasitic
binary fission
budding
fragmentation
xylem
phloem
isogamy
oogamy
plant spores
homosporous
heterosporous
megaspores
pollen grains
agar
algin
holdfasts
kelps
archegonium, -ia
antheridium, -ia
sporangium, -ia
rhizoids
lycopods, Lycophyta
strobilus, strobili
horsetails, Sphenophyta
rhizomes
frond
sorus, sori
gymnosperms
angiosperms
cycad
ginkgos
gnetophytes
pollen tube
embryo sac

Lecture Outline

I. Part I. Kingdom of Fungi
  A. The Fungal Way of Life
    1. Fungi are heterotrophs that are saprobes or parasites.
  B. Fungal Body Plans
    1. Fungal spores give rise to gametes or multicellular haploid bodies.
    2. A mycelium is composed of hyphae, which may form sporangia or gametangia.
  C. Overview of Reproductive Modes
    1. Asexual reproduction occurs by binary fission, budding, spores, or fragmentation of the hyphae.
    2. Sexual reproduction may require a dikaryotic stage.
  D. Major Groups of Fungi
    1. The existing fungi are eukaryotic, their nutrition is extracellular and absorptive, and they possess chitinous walls.
  E. The chytrids or Chytridiomycetes are mostly single-celled saprobes or parasites of aquatic plants and animals.
  F. The water molds or Oomycetes have extensive mycelia and produce flagellated spores and resting spores; some hypha become modified to produce gametangia.
  G. In the zygospore-forming fungi or Zygomycetes, sexual reproduction results in zygospores; mycelial growth is common from asexual spores. **(Fig. 40.6, p. 614; TA 110)**
  H. Sac fungi or Ascomycetes produce haploid spores in asci; multicellular sac fungi produce ascocarps which bear or contain asci.
  I. Club fungi or Basidiomycetes include mushrooms, shelf fungi, bird's nest fungi, stinkhorns, and puffballs; haploid spores are produced by meiosis on basidia which appear on basidiocarps. **(Fig. 40.12, p. 618; TA 111)**
  J. Imperfect fungi or Deuteromycetes have no sexual forms; examples are *Candida albicans* which infects humans and *Penicillium* used to produce cheese and the antibiotic penicillin.
  K. In mycorrhizae there are symbiotic relationships between club fungi and plant roots; the fungus regulates ion flow into the plant.
  L. Lichens are structures produced by a symbiotic relationship between sac fungi and a cyanobacterium or green alga. **(Fig. 40.15, p. 621; TM 165)**

II. Part II. Kingdom of Plants
   A. Evolutionary Trends Among Plants **(Table 40.2, p. 623; TM 166)**
      1. From Nonvascular to Vascular Plants
         a. Land plants evolved xylem and phloem to transport water, sugars, and other nutrients.
      2. Toward a Dominant Sporophyte
         a. There is an alternation of generations between a diploid sporophyte and a haploid gametophyte.
      3. Isogamy to Oogamy
      4. The Evolution of Seeds
         a. Homosporous—some algae and simple vascular plants fall into this category.
         b. Heterosporous—many vascular plants produce megaspores and microspores that develop into gametophytes.
   B. Classification of Algae
   C. Red Algae
      Almost all of the 4,000 or so species of red algae are marine organisms.
      1. Rhodophyta have no flagellated motile cells in their life cycle.
      2. Phycobilins contribute to photosynthesis, especially in deep waters.
   D. Brown Algae
      The brown algae include about 1,500 species, many of which are large seaweeds.
      1. Phaeophyta contain xanthophylls and often live in the intertidal zone attached to rocks by holdfasts.
      2. Kelps may grow 50 meters and have complex blades, stipes, and floats.
   E. Green Algae
      There are at least 7,000 species of green algae.
      1. Green algae have the same photosynthetic pigments as land plants.
      2. They have diverse patterns of reproduction. **(Figs. 40.22 and 40.24, pp. 626, 627; TA 112, TM 167)**
   F. The Land Plants
      1. By 400 million years ago, bryophytes and vascular plant lineages were established.
      2. A layer of sterile cells surrounds the reproductive cells.
      3. The embryo sporophyte experiences early development within the female gametophyte.
   G. Bryophytes
      There are about 16,000 species of mosses, liverworts, and hornworts.
      1. Bryophytes are small, nonvascular plants that live in moist places.
      2. There are independent gametophyte and dependent sporophyte generations. **(Fig. 40.25, p. 628; TA 113)**
   H. Lycopods, Horsetails, and Ferns
      1. Some lycopods were once tree-sized; now they are small plants with true roots, stems, and leaves.
      2. Horsetails were once treelike; only *Equisetum* has survived (it has a sporophyte with rhizomes and aerial branches).
      3. Ferns **(Fig. 40.30, p. 631; TA 114)**
         a. Ferns have underground stems and aerial leaves.
         b. Gametophytes lack vascular tissue; hence they must live in moist places.
      4. Lycopods, horsetails, and ferns are the amphibians of the plant kingdom.
   I. Existing Seed Plants
      1. These include gymnosperms and angiosperms.
   J. The Gymnosperms
      1. Introduction
         a. In both gymnosperms and angiosperms, gametophytes are parasitic upon sporophyte tissues.
         b. Gymnosperm seeds are not completely covered by protective tissue layers.
      2. Cycads were once abundant, but about 100 species remain that produce massive cones.
      3. Ginkgos were once diverse; one species survives and is commonly grown in cities.

4. Among the gnetophytes, *Welwitschia* has a deep taproot and a small exposed part with cones and leaves.
5. Conifers
   a. Conifers contain pine, spruce, fir, hemlock, juniper, cypress, and redwoods.
   b. They are cone-bearing, woody trees and shrubs, most with needlelike or scalelike leaves.
   c. The pine life cycle is representative. **(Fig. 40.34, p. 636; TA 115)**

K. Angiosperms: The Flowering Plants
There are about 250,000 known species of angiosperms.
1. In angiosperms, the seed is carried within the mature ovary.
2. There are two classes: monocots and dicots.
3. The diploid sporophyte dominates the life cycle.
4. Embryos are nourished by the endosperm.
5. Seeds are packaged in fruits.
6. The life cycle of *Lilium* is representative. **(Fig. 40.35, p. 637; TA 116)**
7. Flowering structures coevolved with insects.

Suggestions for Presenting the Material

- Ask students to recall the classification scheme as presented in Table 2.2 by showing the overhead transparency.
- Emphasize that it is only tradition—not biological logic—that places the fungi and the plants together in the same chapter. Once students understand the major structural and nutritional differences that separate these two kingdoms, the rationale for placing fungi in their own kingdom instead of with the plants will become clear.
- Be sure to spend ample time discussing the parasitic and saprophytic life styles found among the fungi; students will be much more familiar with the autotrophic mode of nutrition among photosynthetic green plants.
- Ask students to name examples of fungi and use these examples when discussing fungal diversity. Students will be surprised at the variety of organisms found in this kingdom.
- When students think of plants, almost all of the examples that come to mind will be angiosperms. Ask for examples of plants that do not have flowers.
- Review alternation of generations; if students still find this confusing, they will have trouble comparing one plant life cycle to another. Highlight evolutionary hallmarks (summarized in Table 40.4) such as development of vascular tissue, dominant sporophyte, heterospory, oogamy, nonmotile gametes, and seeds that distinguish simple plants from those that are more complex.

Classroom and Laboratory Enrichment

- Present a flow chart depicting the role of fungi in the carbon cycle.
- Use overhead transparencies to show life cycles of species of fungi and plants. If the life cycle is unlabeled, you can ask your students to find the points at which meiosis and fertilization occur. Ask them which part of the life cycle is haploid and which part is diploid.
- Demonstrate variety within the kingdom Fungi by growing cultures of representative fungal species or ordering them from a biological supply house. Use 35 mm slides to survey the fungi. Use prepared microscope slides to look at vegetative hyphae and reproductive structures of representative fungi.
- Collect samples of plant tissues exhibiting signs of pathogenic fungi. Some examples include leaves infected with powdery mildew, rusts, or black spot, young seedlings killed by "damping off" (caused by the fungus *Pythium*), and citrus fruits covered with blue-green *Penicillium*.

- Ask the produce manager of a local supermarket to save rotting or damaged fruit for one or two days. Place the fruits on display in lab, and allow students to look at fungi with a dissecting scope or to make wet mounts and view them under the compound scope.
- Start a culture of bread mold *(Rhizopus)* by wiping a small piece of preservative-free bread across a dusty floor or cabinet top, misting with water, and placing in a covered petri dish for several days.
- Grow *Pilobolus* on extremely fresh horse manure that has been placed on filter paper in a culture dish covered with an upside-down beaker. Soon after *Pilobolus* has started to appear, sporangia ejected by *Pilobolus* stalks will appear on the inside of the beaker.
- Collect fungi from forests and fields after cool, damp weather conditions. Lichens can be collected from a wide variety of locations regardless of weather conditions.
- Prepare a yeast culture in lab by dissolving an envelope of baker's yeast and a small amount of sugar in a beaker of warm water. Keep the beaker warm. Make a wet mount and observe budding yeast cells under the microscope.
- Demonstrate the ability of yeast to perform fermentation by making beer or wine in the laboratory.
- Use fresh grocery store mushrooms *(Agaricus)* as an example of a basidiomycete. Pass them around the classroom or lab, and ask students to observe the basidiocarp, stalk, cap, and gills.
- Collect green algae in freshwater ponds and streams. *Spirogyra* is a common green alga whose twisting, ribbonlike chloroplasts are beautiful to observe under the microscope. Collect coastal brown algae if access to the shoreline is possible. Cultures ordered from biological supply houses or prepared slides may also be used.
- Prepare and serve Japanese foods containing red algae.
- Look at fossils of ancient lycopods, horsetails, ferns, and gymnosperms. Discuss how changing climates influenced the geographic distributions and the sizes of these plants.
- Prepare a small display of portions of bryophytes, lycopods, horsetails, ferns, and gymnosperms from local areas where plant collection is allowed. Prepared slides and live materials ordered from biological supply houses can supplement your collection.
- Collect reproductive structures of gymnosperms (use photos, drawings, or models to represent those taxonomic divisions for which structures are unavailable). Compare them to the reproductive structures of angiosperms (students can dissect flowers).
- Use models, photos, drawings, or overhead transparencies to discuss the life cycle of pine (a good representative gymnosperm because it is familiar to students). Compare it to the life cycle of a typical angiosperm.
- Because of the numerous classification categories included within this chapter (see Tables 40.1, 40.3, 40.4) it is easy to get lost in a "phylum forest." To alleviate this difficulty, alert the students to the comprehensive classification scheme found in Appendix I. You may wish to make a transparency (enlarged slightly for easier viewing) on your copy machine.
- Purchase a set of 2 x 2 transparency slides that will provide a survey of the plant world including algae.
- Redraw several of the life cycle figures in this chapter but omit several key labels. Photocopy these and distribute as a labeling exercise.

Ideas for Classroom Discussion

- Why do some biologists classify the chytrids and the water molds in the kingdom Protista?
- Describe the similarities and differences between fungal spores and spores found among members of the plant kingdom.
- What do we mean when we say a fungus is "imperfect"?
- What is "ich"? Students who keep tropical fish should know.

- Wildlife biologists have occasionally observed wild bears who appear to stagger and walk unsteadily after consuming large quantities of fallen fruit in late summer and early autumn. What has happened to them?
- What features had to evolve in plants to enable survival on land?
- What are the only land plants with a dominant, independent gametophyte and a dependent sporophyte?
- What happened to the treelike lycopods and horsetails that were dominant during the Carboniferous Period? How did the Carboniferous Period get its name?
- What are some differences and similarities between a pine cone and a fruit?
- Which type of pollination is more efficient—wind pollination, as seen among conifers, or insect pollination, as seen among some of the angiosperms?
- The relationship of the algae and fungi that compose a lichen is mutualism. What is the symbiotic relationship of the entire *lichen* to the *tree* to which it is attached (see Chapter 44 for categories)?
- What feature is the basis for classification of the fungi?
- What is the status of the word *algae*? Organisms that carry that designation may be included in which kingdoms and on what basis?
- What major structural difference separates the bryophytes from the ferns?
- Is it true that lichens grow only on the north side of trees? What is the basis of this? Is this also true of lichens in the Southern Hemisphere?

Term Paper Topics, Library Activities, and Special Projects

- Discuss the economic importance of fungi, both as agents of decay and disease and as organisms important in research and industry.
- Discuss the diagnosis and treatment of fungi that grow on the human skin and mucous membranes.
- Describe the role of yeasts in the making of beer and wine.
- Write a report about the use of lichen species as indicators of pollution.
- Compile a list of wild mushrooms found in your area.
- Discuss the possible future role of algae as a human food source.
- Read package labels of dairy products such as ice cream and chocolate milk, and look for food additives made from red algae.
- Prepare a vascular plant flora for your campus or some other local area.
- Find vegetation maps showing the worldwide distributions of major vegetation associations of gymnosperms and angiosperms (for example, grasslands, temperate forests, coniferous forests, and so on).
- Consult a textbook on insect control methods to investigate the use of fungi as biocontrol agents for insects.
- Obtain a listing of the active ingredients in an antifungal medication for athlete's foot. Consult the Merck Index to learn how these chemicals inhibit fungal growth.
- Obtain a field guide to the edible mushrooms. What easily seen features warn us of poisonous varieties?

Films, Filmstrips, and Videos

- *Algae.* IU, 1965, 17 minutes, 16 mm, color.
- *Angiosperms—The Flowering Plants.* EBEC, 1962, 21 minutes, 16 mm, color. Describes the structural and reproductive characteristics that distinguish angiosperms from other plants. Animated and

time-lapse photography are used to trace the processes of pollination, formation of seeds and fruits, seed dispersal, and plant growth in angiosperms. Illustrates a wide variety of angiosperms and their importance to man.

- *The Evolution of Vascular Plants: The Ferns.* EBF, 1962, 17 minutes, 16 mm, color. Describes the evolution of transport systems in terrestrial plants. Discusses the adaptive advantages of transport systems in ferns and shows how botanists determined the story of plant evolution by studying both fossils and living plants.
- *Fungi.* EBEC, 1960, 16 minutes, 16 mm, color. Uses time-lapse photography, photomicrography, and animation to show how fungi grow, how they obtain food, their importance as decomposers, and their economic significance as agents in food processing and causing disease.
- *Fungi.* MLA, 1977, 16 minutes, 16 mm, color. This film examines the major groups of fungi, showing how certain molds have evolved chemical substances (antibiotics) that serve to reduce competition with other microorganisms, while other fungi produce poisons that are extremely toxic to humans.
- *Fungi.* BOUH, 1983, 19 minutes, 16 mm, color. Defines and traces the similar life cycles of a variety of fungi. Uses time-lapse and micrographic photography with animation to examine reproduction, growth, and feeding. Contrasts the feeding habits of saprophytic and parasitic fungi.
- *Gymnosperms.* EBEC, 1961, 17 minutes, 16 mm, color.
- *How Pine Trees Reproduce.* EBEC, 1963, 16 mm, color.
- *Life of the Molds.* 21 minutes, 16 mm, color. Includes the types, characteristics, and life cycles of molds, using time-lapse photography to show growth.
- *A New Look at Algae.* MLA, 1977, 15 minutes, 16 mm, color. Striking examples of adaptation in freshwater and marine algae are shown in this beautifully photographed film.
- *The Origin of Land Plants: Liverworts and Mosses.* EBEC, 1963, 14 minutes, 16 mm, color. Traces the evolution of land plants and illustrates the structural characteristics, reproductive processes, and adaptive mechanisms of mosses and liverworts. Shows their relationships to higher land plants. Microphotography, macrophotography, and animated drawings are used to illustrate the adaptive and reproductive processes.

41

ANIMAL DIVERSITY

Revision Highlights

Major revision and expansion to provide balanced coverage of invertebrate groups and better coverage of vertebrates. Simple introduction to general characteristics of animals, including a less abstract comparison of types of body cavities that students can relate to (the human body is one of the examples). The table summarizing major animal phyla is now at the end of the chapter, where its details will make more sense. Additional new photos of corals, sea anemones, hydrozoans (including *Physaria*), planula stage, comb jellies, *Schistosoma japonicum*, tapeworm scolex, proglottid, nematode, rotifer, chitons, *Aplysia*, octopus, leech, onychophoran, spiders, millipede, brittle stars, sea urchins, sea stars, acorn worm, coelacanth, lamprey, sea horse cave salamander, American toad, green sea turtle, Galapagos tortoise, crocodile, coral snake, tuatara, and chameleon; two-page spread on insect diversity. New diagrams of radial vs. bilateral symmetry, medusa vs. polyp, flatworm anatomy, tapeworm life cycle, nematode anatomy, rotifer anatomy, book lung, sea star anatomy, and amniote egg. Better coverage of cnidarians, new section on comb jellies, more examples of parasitic nematodes (e.g., pinworms, hookworms, *Trichenella, Wuchereria*). Better comparison of protostomes with deuterostomes. Section on mollusks now precedes sections on annelids/arthropods. Chitons covered now, and better treatment of cephalopods. Improved section on arthropods, especially insects. Better picture of echinoderm diversity, including a close look at sea star anatomy. Revised organization for chordates (invertebrate chordate/vertebrate distinction clearer). Tightened discussion of tunicates. Simple history of vertebrate evolution, followed by expanded text description of fishes, amphibians, reptiles, birds, and mammals.

Chapter Outline

GENERAL CHARACTERISTICS OF ANIMALS
- Body Plans
- Representative Animal Phyla

SPONGES

CNIDARIANS

COMB JELLIES

FLATWORMS
- Turbellarians
- Trematodes
- Cestodes

NEMERTEANS

NEMATODES

ROTIFERS

TWO MAIN EVOLUTIONARY ROADS

MOLLUSKS
- Chitons
- Gastropods
- Bivalves
- Cephalopods

ANNELIDS
- Earthworms
- Leeches
- Polychaetes

Annelid Adaptations

ARTHROPODS

Arthropod Adaptations

Chelicerates

Crustaceans

Insects and Their Kin

ECHINODERMS

CHARACTERISTICS OF THE CHORDATES

Urochordates

Cephalochordates

THE VERTEBRATES

From Jawless to Jawed Fishes

Key Developments in the Transition to Land

Modern Fishes

Amphibians

Reptiles

Birds

Mammals

SUMMARY

Objectives

1. Describe the major advances in body structure and function that made invertebrates and vertebrates increasingly large and complex.
2. Discuss the relationship between segmentation and the development of paired organs and paired appendages.
3. List the functions of a coelom and describe the role coelomic development played in animal evolution.
4. Be able to reproduce from memory a phylogenetic tree that expresses the relationships between the major groups of animals.

Key Terms

metazoans
vertebrates
invertebrates
multicellular
coelom, coelomate
hemocoel
acoelomate
pseudocoelomate
pseudocoel
segmentation
uncephalized
cephalized
collar cells
amoeboid cells
gemmule
hydrozoans
sea anemones
medusa
polyp
nerve net
mesoglea
nematocysts
planula
tapeworms
pharynx
flame cells
hermaphroditic system
flukes
schistosomiasis
cestodes
tapeworms
scolex
proglottids
protostomes
deuterostomes
radial cleavage
spiral cleavage
mantle
radula
ctenidia
torsion
filter feeders
siphons
seta, setae
parapodium, -dia
nephridium, -dia
exoskeleton
book lungs
carapace
mandibles
maxillae
antennae
insect wings
tube feet
water-vascular system
ampulla
chordates, Chordata
urochordates
cephalochordates
ostracoderms
scales
amniotic egg

Lecture Outline

I. Introduction
   A. The multicelled animals, or metazoans, are divided into invertebrates and vertebrates.

II. General Characteristics of Animals
   A. Most animals are multicellular, heterotrophic, diploid, have tissues that are arranged into organs and organ systems, and are motile at some point in their life cycle.

B. Animal life cycles include a period of embryonic development; germ tissue layers, the ectoderm, endoderm, and in most species, mesoderm, give rise to adult organs.

C. Body Plans

1. Body Symmetry **(Fig. 41.2, p. 642; TA 117)**
   Animals usually show radial or bilateral symmetry.
2. Type of Gut
   The animal gut may have one or two openings.
3. Body Cavities **(Fig. 41.3, p. 642; TA 118)**
   a. The coelom is a type of body cavity lined by epithelium.
   b. The coelom in insects is filled with blood; this type of body cavity is a hemocoel, and is part of an open circulation system.
   c. Acoelomates are animals like flatworms that lack a body cavity.
   d. The pseudocoelom occurs in nematodes and other animals that have a coelom not lined with a peritoneum.
4. With segmentation, bodies consist of a series of body units (earthworms and insects are examples).
5. Cephalization means having a head and usually also a brain; flatworms are the simplest example.

D. Representative Animal Phyla
These represent trends in evolution. **(Fig. 41.4, p. 643; TM 168)**

III. Sponges

A. Sponges have an asymmetric body organized into pores, canals, and chambers. **(Fig. 41.7, p. 645; TM 169)**
B. They reproduce sexually or asexually by fragments or gemmules.

IV. Cnidarians

Cnidarians include hydras, jellyfishes, sea anemones, and corals. They are characterized by

A. True tissues and radial symmetry
B. A nerve net
C. Nematocysts **(Fig. 41.12, p. 647; TM 171)**
D. Some have both medusa and polyp forms in their life history. **(Figs. 41.8 and 41.9, pp. 645–646; TM 170)**
E. An example is *Obelia*, a colonial form. **(Fig. 41.13, p. 648; TA 119)**

V. Comb Jellies

A. Ctenophora are predatory marine animals with eight rows of comblike structures composed of cilia.

VI. Flatworms

A. Flatworms are bilateral, cephalized animals with three germ layers. **(Fig. 41.17, p. 650; TA 120)**
B. Turbellarians are free-living carnivores with protonephridia. **(Fig. 41.17, p. 650; TA 120)**
C. Trematodes are flukes and include important human parasites like *Schistosoma.* **(Fig. 41.18, p. 651; TA 121)**
D. Cestodes are tapeworms and are intestinal parasites of vertebrates. **(Fig. 41.21, p. 653; TA 122)**

VII. Nemerteans

A. Nemerteans are ciliated externally and have a complete gut and circulation system and a unique proboscis.

VIII. Nematodes

A. Round worms are covered by a cuticle and have a complete gut and a pseudocoelom. **(Fig. 41.22, p. 654; TM 172)**

B. There are free-living forms and important human parasites, such as hookworms and *Wuchereria*.

IX. Rotifers

A. Rotifers are small aquatic animals with a crown of cilia. **(Fig. 41.24, p. 655; TM 173)**

X. Two Main Evolutionary Roads **(In-text art, p. 656; TM 174)**

A. In protostomes, the first embryonic opening becomes the mouth; they include mollusks, annelids, and arthropods.

B. In deuterostomes, the first embryonic opening becomes the anus; they include echinoderms and chordates.

XI. Mollusks

A. All have a head, a foot, and a visceral mass. **(Fig. 41.26, p. 657; TA 123)**

B. Chitons are elongate animals, covered with plates that use their radula to graze.

C. Gastropods include snails and slugs that often undergo torsion during development. **(In-text art, p. 658; TM 175)**

D. Bivalves such as clams and scallops have a large foot modified for burrowing. **(Fig. 41.31, p. 659; TA 124)**

E. Cephalopods such as squids and octopuses have large brains and are fast-swimming predators.

XII. Annelids

A. Annelids are segmented protostomes.

B. Earthworms aerate the soil because each one ingests its own weight in soil per day.

C. Leeches are predators of invertebrates and parasites of vertebrates.

D. Polychaetes are burrowing, crawling, free-swimming, and attached species.

E. Annelid Adaptations

1. Annelids have bristle-like setae for traction or they are modified into paddles for swimming; they may also serve for respiration.
2. Coelomic chambers act as a hydrostatic skeleton.
3. Nephridia regulate the composition of body fluids.

XIII. Arthropods

A. Arthropod Adaptations

1. The arthropod exoskeleton is a hard cuticle that provides protection, prevents water loss, and provides support needed for land animals; it requires molting to accommodate growth.
2. Segments become specialized to form the head, thorax, and abdomen.
3. Jointed appendages are derived from annelid parapodia.
4. Respiratory systems include insect tracheas.
5. Specialized sensory structures include the compound eye.

B. Chelicerates include horseshoe crabs, ticks, mites, spiders, and scorpions.

1. Ticks are blood-sucking parasites that can serve as vectors of many diseases.
2. Spiders use chelicerae to kill prey with venom.

C. Crustaceans include shrimps, lobsters, barnacles, isopods, copepods, and water fleas.

1. The body parts of crustaceans include mandibles. **(Fig. 41.43, p. 667; TM 176)**
2. They are important components of marine and freshwater food webs.

D. Insects and Their Kin

1. Millipedes and centipedes have paired legs along the trunk; they are scavengers and carnivores.
2. Insects
   a. Insects have a head, thorax, and abdomen, and often wings.
   b. They also have a tracheal system, Malpighian tubules, and modified mouth parts.

XIV. Echinoderms

A. Echinoderms include "spiny-skinned" sea stars, sea urchins, sand dollars, brittle stars, and sea cucumbers.
B. They are deuterostomes with radial symmetry. **(Fig. 41.52, p. 674; TA 125)**
C. They have a water-vascular system with tube feet.

XV. Characteristics of the Chordates

A. Most are vertebrates with a brain and nerve cord protected by a backbone of cartilage or bone.
B. Others lack a backbone but have slits in the pharynx, a dorsal tubular nerve cord, and a notochord.

XVI. The Invertebrate Chordates

A. Urochordates are commonly called tunicates or sea squirts; they lack a coelom.
   1. The notochord is limited to the tail.
   2. The pharynx collects food and serves as a respiratory organ.
B. Cephalochordates
   Cephalochordates are often called lancelets or "amphioxus" because their body is sharply tapered at both ends.

XVII. The Vertebrates

A. From Jawless to Jawed Fishes
   1. Ostracoderms were the first fish (they were jawless).
   2. Jaws evolved from gill supports. **(Fig. 41.58, p. 679; TM 177)**
B. Key Developments in the Transition to Land
   1. Evolution of Gills, Lungs, and the Heart
      a. When rasping and sucking replaced filter-feeding, gill tissue became more important in respiration.
      b. Closed circulatory systems evolved in lampreys and fishes (two-chambered hearts) as well as in amphibians (three-chambered hearts). **(Fig. 41.59, p. 679; TM 178)**
      c. The high metabolic rate of land animals required the evolution of the four-chambered heart and true lungs.
   2. Evolution of Limbs—paired fins → lobed fins → limbs. **(Fig. 41.60, p. 680; TM 179)**
   3. Evolution of Nervous and Sensory System
      a. Evolution of land vertebrates occurred with evolution of visual and auditory areas of the brain.
      b. Mammals first revealed the potential for the cerebral cortex.
C. Modern Fishes
   1. Modern fishes developed streamlined bodies to move through water; fins propel and stabilize.
   2. Of jawless fishes, only lampreys and hagfishes remain; they have eel-like bodies.
   3. Cartilaginous fishes include sharks, skates, and rays with endoskeletons of cartilage.
   4. Most bony fishes are ray-finned fishes.
D. Amphibians include salamanders, frogs, toads, and apodans (tiny, legless, tropical forms).
   1. Amphibians have a bony endoskeleton and four legs.
   2. They lay their eggs in wet environments.
E. Reptiles
   1. Reptiles evolved in the Late Carboniferous, as vertebrates followed insects inland; jaws and limbs evolved.
   2. They have internal fertilization and amniotic eggs. **(Fig. 41.65, p. 684; TA 126)**
   3. Adaptive radiation occurred in the Mesozoic; dinosaurs dominated the land.
   4. Reptiles have bony endoskeletons, lungs, and skin that resists drying out.
F. Birds
   1. Birds evolved from reptiles during the Jurassic.
   2. Wings have feathers, powerful muscles, and lightweight bones. **(Fig. 41.66, p. 686; TM 180)**

G. Mammals
   1. There are egg-laying mammals (Monotremata), pouched mammals (Metatheria), and placental mammals (Eutheria).
   2. Mammals have milk-secreting glands and hair.

Suggestions for Presenting the Material

- This chapter is as diverse as its name implies. It begins with definition of some basic terms used in describing the animal phyla. Then the "march through the phyla" begins. This can be a tedious and overwhelming experience unless you exercise special care to add those little extras that will hold student interest.
- Perhaps in previous biology courses, students have considered protozoans as animals. Recall that these one-celled organisms are now placed in the kingdom Protista. Therefore, the present chapter begins with true multicellular animals—the sponges.
- Table 41.1 is a comprehensive summary placed at the conclusion of the chapter, but you may wish to refer to it as you complete each phylum.
- In a survey that reduces volumes of information to about thirty pages, there must be a brevity that will certainly shortchange someone's favorite animal group. However, the authors have been fair in presenting the chief characteristics for which each phylum is noted. Even so, the few features may be too many and further reduction may need to be made by the individual instructor.
- If the instructor has a good grasp of animal diversity, especially invertebrates, the chapter may seem oversimplified, but to a botanist it may be quite the opposite. No matter where you find yourself in this spectrum, remember the naive student and have compassion.
- One way to enliven your presentation is to relate the members of each phylum to the students' daily lives. This may be more difficult for sponges than arthropods, but with a little forethought it can be done. See samples in the "Enrichment" section.
- Done right, this chapter presents one of the greater challenges in the book!

Classroom and Laboratory Enrichment

- There is no scarcity of visual material depicting the diversity of animal life. It ranges from professionally made 2 x 2 transparencies, to single-phylum films, to full-length video surveys of the animal kingdom such as David Attenborough's *Life on Earth* (available at retail outlets).
- If you are fortunate to live in or near a large city with a zoo, arrange a field trip, but this time with a difference. Prepare the students for note-taking on various topics such as classification, body symmetry and cavities, body system development, ecological niche, and so on. Comment on the lack of animal representatives from the invertebrate phyla (at least ones that are plainly visible).
- Hopefully most, or perhaps all, of your students will have an accompanying laboratory experience where they can see, touch, and dissect representatives of the various phyla. If not, it will be even more crucial that you present adequate visual material—at least 2 x 2 transparencies if you have to shoot them yourself.
- As mentioned in the "Presentation" section, relating animals to daily life will enhance your lectures. Below is a sampling that you can build upon:

  PORIFERA: Natural *sponges* (expensive) are still the best for cleaning purposes.

  CNIDARIA: We wouldn't want to tangle with a *jellyfish,* but a valuable piece of *coral* on your ring finger is OK.

  PLATYHELMINTHES: The *tapeworms* and *flukes* have disgusting parasitic habits but are all a part of nature's balance.

  NEMATODA: Everyone is familiar with *roundworms* in pets, but do you know that humans can get them too?

ANNELIDA: The familiar *earthworm* is a very beneficial tiller of the soil, but is its relative the *leech* so well respected?

ARTHROPODA: The largest phylum contains pesky *insects* as well as tasty *lobsters*.

MOLLUSCA: *Snails* and *clams* are quite familiar—we even eat them! But what about *squid* and *octopus*?

ECHINODERMATA: *Sea stars* seem exotic because they and their relatives are found only in the sea.

CHORDATA: Finally, "familiar" animals, but they constitute less than 5% of all the animals on earth.

Ideas for Classroom Discussion

- Why are protozoans no longer included in the kingdom Animalia?
- Give the literal meaning of each phylum name in the chapter.
- Even if you cannot remember all the details concerning the members of each phylum, can you tell the major niches that each phylum seems to fill?
- The blood fluke, *Schistosoma*, is an important human parasite. What role do parasites play in nature? Are they necessarily villains, as we often think?
- When a pet owner says "My animal has worms," what worm is most likely present? How could you distinguish between a flatworm and a roundworm, other than by shape?
- Some people don't consider an organism as a true animal unless it bleeds red when injured. Which animals presented in this chapter would fit this "definition"?
- Arthropods as a group have more unique features than perhaps any other group. Name as many of these features as you can.
- Why do persons you know refuse to eat insects when close relatives such as crayfish and lobsters and near relatives such as snails and oysters are gourmet items?
- Are all chordates, vertebrates? Are all vertebrates, chordates? Explain.

Term Paper Topics, Library Activities, and Special Projects

- In the early days of biological investigation, sponges and even some cnidarians were classified as plants. Search some older textbooks to see why scientists changed their minds.
- Prepare a report on the extent of platyhelminth parasitic infestation in humans. List the condition, number of cases, and economic loss to society.
- The phylum Echinodermata is unique in not possessing any freshwater representatives. The reasons for this should be found in an invertebrate biology textbook; see if you can locate the details.
- Chordates dominate our thinking when the word *animal* is used. But they constitute about 3% of the total species (based on the numbers in Table 41.1). Relate this fact to the concept of pyramids (energy, biomass, numbers) as presented in Chapter 45.

Films, Filmstrips, and Videos

- *Annelids.* Wards-MLA, 1976, 10 minutes, 16 mm, color. Beautiful photography of diverse annelids.
- *Arthropods.* MLA, 1976, 13 minutes, 16 mm, color. Examines the diversity of arthropods and points out their principal distinguishing features.
- *The Chordates: Diversity in Structure.* CORF, 1969, 13 minutes, 16 mm, color.
- *Coelenterates.* MLA, 1976, 10 minutes, 16 mm, color. Beautiful photography of living cnidarians.
- *Echinoderms.* MLA, 1976, 10 minutes, 16 mm, color. Presents the principal features of each of the five classes of echinoderms. Beautifully photographed.

- *Flatworms.* MLA, 1976, 10 minutes, 16 mm, color. Beautiful photography of diverse flatworms.
- *Mollusks.* MLA, 1976, 10 minutes, 16 mm, color. Shows the main characteristics, feeding methods, locomotory patterns, and breeding habits of specific mollusks.
- *Parasitism—Parasitic Flatworms.* EBEC, 1962, 17 minutes, 16 mm, color. Traces the development of parasitic flatworms and shows several life cycle stages that are rarely photographed. Defines predation, mutualism, and internal and external parasitism, and gives examples of each of these relationships. Reveals the complicated life cycle of the blood flukes and lung and liver flukes.
- *Segmentation: The Annelid Worms.* EBEC, 1961, 16 minutes, 16 mm, color. Describes the structure and functions of the segmented worm body system and explains the evolutionary implications of segmentation. The major classes of the annelids are illustrated, and the digestive and reproductive systems are shown in animation.
- *What Is a Reptile?* EBEC, 1961, 18 minutes, 16 mm, color. Illustrates representative kinds of the four orders of reptiles and describes their physical characteristics, reproductive processes, and evolutionary development. Discusses reasons for the reptiles' successful evolution and survival: the evolution of the hard-shelled egg, the waterproof skin, and the improved limbs. Describes the characteristics and behavior of various reptiles.

42

HUMAN ORIGINS AND EVOLUTION

Revision Highlights

Updated, major revision. Updated primate and hominid classification. Expanded introduction on mammalian characteristics that relate to human evolution (especially dentition, extended period of infant dependency and learning, flexibility in response in novel situations). Revised section on trends in primate evolution (including new illustration comparing skeletal organization and stance of monkeys, apes, and humans). New section on modifications that occurred during evolution of hands, and behavioral evolution. New photo of tree shrew. Updated treatment of fossil evidence of evolution of early anthropoids, hominoids. Updated picture of "bushy" early period of hominid evolution; new illustrations comparing size and stature of australopiths with modern humans. More recent proposed phylogenetic trees for hominids; new two-page illustration comparing hominid skull shapes and photos of Olduvai stone tools. Evidence of early social and cultural development described; four-color reconstruction of early hominids in the Olduvai habitat. Revised summary.

Chapter Outline

THE MAMMALIAN HERITAGE

PRIMATE CLASSIFICATION

TRENDS IN PRIMATE EVOLUTION

- **From Quadrupeds to Bipedal Walkers**
- **Modification of Hands**
- **Enhanced Daytime Vision**
- **Changes in Dentition**
- **Brain Expansion and Elaboration**
- **Behavioral Evolution**

PRIMATE ORIGINS

THE HOMINIDS

- **Australopiths**
- **Stone Tools and Early *Homo***
- ***Homo erectus***
- ***Homo sapiens***

SUMMARY

Objectives

1. Understand the general physical features and behavioral patterns attributed to early primates. Know their relationship to other mammals.
2. Trace primate evolutionary development through the Cenozoic era.
3. Understand the distinction between hominoid and hominid and distinguish between *Australopithecus* and *Homo.*

Key Terms

dentition
incisors
canines
premolars
molars
cusps
infant dependency
learning
Primates
arboreal
prosimians
anthropoids
New World monkeys
Old World monkeys
apes
humans
hominids
quadrupedal
bipedal
divergent
convergent
prehensile
culture
dryopiths
hominids
plasticity
australopiths
Australopithecus afarensis
A. africanus
A. robustus
A. boisei

Lecture Outline

I. The Mammalian Heritage
   A. Like all vertebrates, mammals have a nerve cord within a vertebral column and a skull with a "three-part" brain.
   B. Dentition permits the upper and lower teeth to work together—carnivores have large canines and plant eaters have "cheek teeth" with cusps.
   C. Mammals have an extended period of infant dependency and learning; there is flexibility in their responses, because the mammalian brain has a large capacity for memory and learning.

II. Primate Classification **(Fig. 42.1, p. 691; TM 181)**
   A. Primates are an order of largely arboreal mammals that includes lemurs, lorises, tarsiers, monkeys, apes, and humans.
   B. Prosimians are the oldest primates, have forward-directed eyes, and include lemurs and tarsiers.
   C. Anthropoids include monkeys, apes, and humans.

III. Trends in Primate Evolution
   A. Introduction
      1. These trends include changes in the skeleton, locomotion, hands, visual system, dentition, brain, and behavior.
   B. From Quadrapeds to Bipedal Walkers
   C. Modification of Hands
      1. Clawed paws evolved into hands.
      2. The tool-using capacity resulted because of refinements of four basic movements—divergent, convergent, prehensile, and opposable movements.
   D. Enhanced Daytime Vision
      1. The evolution of forward-directed eyes permitted an overlapping visual field and depth perception.
   E. Changes in Dentition
      1. Dentition became less specialized during evolution.
   F. Brain Expansion and Elaboration
      Ancestors of humans crossed the "cerebral Rubicon" of 750 cm^3 (the largest ape brain is 650 cm^3) to become fully human.
   G. Behavioral Evolution
      1. There are trends toward increased life spans, single births, and longer periods of infant dependency. **(Fig. 42.5, p. 696; TM 182)**
      2. Culture evolved as the behavior patterns passed between generations by learning and symbolic behavior.

IV. Primate Origins
   A. The first primates (65–54 million years ago) resembled tree shrews; they were nighttime omnivores.
   B. Some evolved into tree-living prosimian forms (54–38 millions years ago).

C. Brain size increased, the snout shortened, and the hand evolved to permit refined grasping movements.
D. Anthropoids evolved by 38–25 million years ago; reptiles on the ground encouraged their life in the trees.
E. Hominoids appeared 23–20 million years ago as major land masses moved and the climate cooled.
F. Adaptive radiation produced the dryopiths that led to gorillas, chimpanzees, and humans.

V. The Hominids **(Fig. 42.11, p. 699; TM 183)**
  A. Introduction
    1. The earliest fossils (4–2 millions years ago) in Africa show a bipedal omnivore with a larger brain.
    2. Cooler weather reduced the rain forests, and the hominids were forced out of the trees.
  B. Australopiths are the earliest known hominids.
    1. All were bipedal (they left footprints); how they are related is unknown.
  C. Stone Tools and Early *Homo*
    1. Hominids began to use stone tools 2 million years ago to get marrow out of bone and to scrape flesh from bones.
    2. They first began to make tools called choppers 1.9 million years ago at Olduvai Gorge in eastern Africa, where they invented the concept of the "home."
    3. The emergence of *Homo* was based upon plasticity, not specializations.
  D. *Homo erectus*
    1. Early humans were tested by the Pleistocene Ice Ages between 1.5 million and 300,000 years ago.
    2. *Homo erectus* made advanced stone tools and used fire as they migrated out of Africa into Asia and Europe.
  E. *Homo sapiens*
    1. Between 300,000 and 200,000 years ago, *H. sapiens* evolved from *H. erectus*.
    2. Modern humans, *H. sapiens sapiens*, evolved 40,000 years ago.
    3. Neanderthals were similar to modern humans but disappeared 35,000–40,000 years ago.
    4. Modern evolution is cultural, not biological.

Suggestions for Presenting the Material

- Depending on the region of the country and your particular institution, the discussion of human origins may, or may not, spark some controversy in your class. Whether we acknowledge it or not, this is an area where religious beliefs and scientific explanations can conflict. It has been this contributor's (LGS) experience that ridicule of a student's personal beliefs does very little to open the young mind to an alternative view. Therefore, I carefully couch my discussions to present the maximum amount of scientific evidence without deliberately being offensive.
- The origins of primates and the evolutionary trends that resulted in certain key characteristics are quite universally accepted.
- However, the exact evolutionary paths leading to modern humans are still hotly debated. This is illustrated in Figure 42.11, which presents two interpretations of the phylogenetic relationships among fossil humans.
- Be sure to distinguish "hominoid" from "hominid."
- Emphasize the fact that *Australopithecus* spp. and *Homo* spp. were contemporaries.
- Because there are several genus and species names in this chapter, each with distinguishing characteristics, perhaps you could construct a table on an overhead transparency (or printed handout) summarizing this data.

Classroom and Laboratory Enrichment

- Because of the interest in this subject, numerous videotapes and slide sets are available (check supply catalogs in your A/V center). Some may be more technical; others may be for the general public.
- If you are fortunate to have an expert in human evolution available, by all means arrange for an illustrated lecture on this complex subject.
- View museum displays depicting early human evolution and life.

Ideas for Classroom Discussion

- What characteristics define a "primate"?
- Dating of fossil remains is critical to understanding human evolutionary sequence. How is the age of fossils determined?
- What characteristics distinguish *Australopithecus* from *Homo*?
- Why do the arrangements of the phylogenetic trees in Figure 42.11 show more variation for *Australopithecus* than for *Homo*?
- Rarely are entire fossil skeletons found in one geographical site. Rather, a skeleton is a composite of many "finds." How do scientists know what bones to group together?

Term Paper Topics, Library Activities, and Special Projects

- Research the creationists' arguments against evolution of humans, and rebut each one using scientific evidence and arguments.
- Prepare a report on the methods illustrators use to reconstruct fossil humans. Assess the degree of freedom individual artists use in their work (check the "Readings" list at the end of Chapter 42).
- Investigate the controversy surrounding the skeleton called "Lucy."

Films, Filmstrips, and Videos

- *The Ascent of Man, Part I: Lower Than the Angels.* Time-Life, 52 minutes, 16 mm, color. Discusses the anatomical and physiological changes that led to the evolution of modern man.
- *Dr. Leakey and the Dawn of Man.* NGS, 1966, 51 minutes, 16 mm, color. Dr. Leakey's dramatic discoveries and their implications for the study of man's early past are examined. The film traces the development of the first theories about both the emergence of man and his place of origin and examines the application of modern technology to the study of anthropology.
- *The Great Apes: Fact Vs. Fantasy.* EBEC, 14 minutes, 16 mm, color.
- *Monkeys, Apes and Man* (2 parts). NGS, 1972, 52 minutes, 16 mm, color. From studies of primate social organization, use and manufacture of tools, and ability to learn and socialize have come unexpected insight and increased appreciation of humankind.
- *Mysteries of Mankind.* NGS, 1988, 59 minutes, 16 mm, color. Who are we and where did we come from. Mary, Louis, and Richard Leakey join Donald Johanson and other scientists who use the latest technological tools to probe the past.
- *Tobias on the Evolution of Man.* NG, 1975, 18 minutes, 16 mm, color. Phillip Tobias, a paleoanthropologist, recounts the story of human evolution with actual anthropological discoveries and shows famous hominid remains from 1 to 4 million years in age.

43

POPULATION ECOLOGY

Revision Highlights

Updated picture of human population growth, including new illustrations of average annual growth rate for 1987 in different countries, the reproductive base for various regions, and projected increases in the number of individuals age 65 and over in the United States between 1985 and 2080.

Chapter Outline

Objectives

1. Learn the language associated with the study of population ecology.
2. Understand the factors that affect population density, distribution, and dynamics.
3. Understand the meaning of the logistic growth equation and know how to calculate values for G by using the logistic growth equation. Understand the meaning of r_{max} and K.
4. Calculate a population growth rate (G); use values for natality, mortality, and number of individuals (N) that seem appropriate.
5. Use the equations for general population growth rate and the logistic growth equation.
6. Understand the significance and use of life tables; interpret survivorship curves.
7. Know the situation about the growth of human populations. Tell which factors have encouraged growth in some cultures and limited growth in others.

Key Terms

ecology
population
habitat
community
producers
consumers
herbivores
carnivores
parasites
decomposers
detritivores
ecosystem
biotic
abiotic
biosphere
population density
clumped distribution
random distribution
uniform distribution
natality
mortality
immigration
emigration
zero population growth
r, net reproduction per individual
exponential growth
J-shaped curve
K, carrying capacity
logistic growth equation
S-shaped curve
age-specific patterns
demography
life tables
cohort
survivorship
timing of reproduction
number of offspring
density-dependent factor
density-independent factor
intraspecific competition
exploitation competition
interference competition
age structure

Lecture Outline

I. Introduction
Human population growth is on a collision course with the supply of resources.

II. Ecology Defined
  A. Ecology involves the interactions of organisms with one another and with the physical and chemical environment.
  B. A population is a group of the same species that live in a specific habitat.
  C. The community encompasses all species in a habitat, including producers, consumers, decomposers, and detritivores.
  D. An ecosystem contains a community and its chemical and physical environment.
  E. The biosphere includes all living organisms on earth together with the global physical and chemical environment.

III. Population Density and Distribution
  A. Population density varies widely for each species; it is influenced by ecological relationships.
  B. Distribution in Space **(Fig. 43.2, p. 708; TM 184)**
    1. Clumped distribution occurs when offspring of intertidal organisms settle near their parents.
    2. Random distribution exists for example among spiders on a forest floor.
    3. Orchards and creosote bushes provide examples of uniform distribution (even spacing).
  C. Distribution Over Time
  Bird migrations are an example of population distribution over time.

IV. Population Dynamics
  A. Variables Affecting Population Size **(In-text art, p. 709; TM 185)**
    1. Variables are natality, mortality, immigration, and emigration.
    2. Population growth rate (G) = (births + immigration) – (deaths + emigration).
    3. Zero population growth occurs when the population is stabilized.
  B. Exponential Growth **(In-text art, p. 709; TM 185)**
    1. If G = (births per individual) – deaths per individual x (# of individuals), and if birth and death rates are constant, then r considers them together as the net reproduction per individual (*NRPI*).
    2. Thus, $G = NRPI$ x (# of individuals); or $G = rN$ => Any population shows exponential growth. **(Fig. 43.4, p. 710; TM 186a)**

C. Limits on Population Growth
   1. When an essential resource is in short supply, it becomes a limiting factor for population growth.
   2. In nature, the limiting resource may vary with time.
D. Carrying Capacity and Logistic Growth **(In-text art, pp. 710–711; TM 185)**
   1. Carrying capacity (*K*) is the population that the environment can sustain.
   2. The logistic growth equation is
      G = (Max net reproduction per individual) x (# of individuals) x (portion of unexploited resources), or
   3. $G = r_{max}N[(K - N)/K]$. **(Fig. 43.5, p. 711; TM 186b)**
   4. Study of the reindeer population on one Pribilof Island, Alaska, provides a pattern of "overshoot and crashes." **(Fig. 43.6, p. 711; TM 186c)**

V. Life History Patterns
A. Introduction
   1. Because a phenotype varies with the developmental stage, many age-specific patterns emerge.
B. Life Tables
   1. Demography shows that the probability of death changes with age. **(Fig. 43.8, p. 713; TM 187)**
C. Patterns of Survivorship and Reproduction
   1. Survivorship Curves **(Fig. 43.9, p. 714; TM 188)**
   2. Timing of Reproduction
   3. Number of Offspring
D. Evolution of Life History Patterns
   1. There may be the rapid production of many small offspring early in life or of a few large offspring later in life.
   2. Different patterns can exist within the same species. **(Fig. 43.11, p. 716; TM 189)**

VI. Limits on Population Growth
A. Density-dependent factors include the supply of nutrients.
B. Density-independent factors include rainfall levels and freezing temperatures.
C. Competition for Resources
   1. Population growth may result in limited resources.
   2. In intraspecific competition, exploitative or interference competition may occur.
D. Predation, Parasitism, and Disease
   1. Herbivores limit plant populations.
   2. Predation, parasitism, and disease increase with population density and affect population size.

VII. Human Population Growth
A. Introduction—in 1986, the population reached 5 billion while 5 to 20 million die of starvation per year.
   1. In 1988 about 86.7 million individuals were added to the world's population. **(Fig. 43.12, p. 717; TM 190)**
B. Where We Began Sidestepping Controls
   1. Humans expanded into new habitats and climatic zones. **(Fig. 43.13, p. 718; TM 191)**
   2. Agriculture increased the carrying capacity of the land.
   3. Medicine removed many limiting factors.
C. Age Structure and Reproductive Rates
   1. Rapidly growing populations have a broad age structure base. **(Figs. 43.14 and 43.15, p. 719; TM 192, 193)**

VIII. Perspective
  A. Actively growing populations have a large work force relative to the older population. **(Fig. 43.16, p. 720; TM 194)**
  B. At zero population growth, more less productive people result. **(Fig. 43.17, p. 720; TM 194)**

Suggestions for Presenting the Material

- This is the second chapter in the book to include the word *population* in its title. Chapter 36 discussed the *genetics* of populations; the present one reports on the *ecology* of populations.
- It is time to return to Figure 1.3 for the "beginning of the end" as the individual cells and organisms are integrated into a complex natural order.
- The bulk of the chapter concerns general aspects of populations: density/distribution, dynamics, survival, and limits.
- Students will perhaps be most interested in studies of population as they affect human population growth, a subject covered at the end of this chapter. Examination of the changes in worldwide patterns of population growth since the turn of the century and in the last twenty years in particular will surprise students and highlight the overwhelming need for humans to find ways to control global population growth rate. The inclusion of human population growth studies in this chapter allows students to see that we are not exempt from the rules and limitations that govern all populations.
- Exploration of population ecology also sets the stage for interesting discussion of the socioeconomic impacts of population growth and the ethical questions related to regulating population growth.

Classroom and Laboratory Enrichment

- Use flow charts on the board or on the overhead projector to show the relationships among the different components of an ecosystem.
- Select an ecosystem that consists of all or part of the campus or a nearby area. List as many of the biotic and abiotic components of that ecosystem as you can. Students could work in teams, with each team assigned to different ecosystem components.
- Choose a community, list its species, and categorize them as producers, consumers, decomposers, or detritivores.
- Graph the rates of population growth for several of the nations of the world. Discuss reasons for the differences between nations.
- Compare and contrast survivorship curves of different organisms. Ask students to first guess whether the organism you have named has a Type I, II, or III survivorship curve. Then use transparent overlays on the overhead projector to show survivorship curves for each species.
- Design and implement experiments examining the effects of resource availability, time lags, and competition on population densities of small organisms easily raised in the laboratory.
- Calculate population densities for a plant species in several small defined areas with varying environmental conditions. What pattern(s) of spatial distribution do you see? Identify factors that you believe might influence species distribution (some examples might be nutrient availability, amount of sunlight, moisture, and openings created by disturbance). Design experiments that would evaluate the role of each environmental factor in species distribution.
- Identify examples of exploitation competition and interference competition among local plant and animal communities.
- Show overhead transparencies of age structure diagrams for human populations of different nations.
- If you review Figure 1.3, you may want to insert the concept of *biomes* as contained in Chapter 46.

- If you have a collection of slides, search for a scene that will include clearly visible examples of a population, the four participating groups in a community, an ecosystem, and the biosphere. As you project each slide, ask students to identify each element.
- Use the classroom and its occupants to illustrate density and distribution:

a. Head count	a. 50 students
b. Density	b. 50 students per room
c. Clumped distribution	c. 50 students in rear half of room
d. Random distribution	d. 50 students milling around before class begins
e. Uniform distribution	e. 50 students in evenly spaced rows of chairs
f. Distribution over time	f. 50 students disperse to all parts of campus after class

- Using the equations for expressing population dynamics, work through examples of problems gleaned from an ecology text.
- Show a copy of a life table as published in a modern entomology textbook. These are especially good examples because of the differential death rates in the various stages of metamorphosis.
- The three types of survivorship curves have rather abstract designations (I, II, III). The following may aid retention:

a. Type I	Think "hI"	Signifying high survivorship during most of lifetime
b. Type II	Think "middle"	Death rate constant
c. Type III	Think "low"	"III" resembles "low" (both have three items)

- Stimulate interest in the future of human population dynamics by asking your class to vote on one of three possibilities (using Figure 43.5 as a guide):
  a. The S-curve will not "flatten out" as it reaches a carrying capacity but will continue to climb indefinitely.
  b. The curve will oscillate near the carrying capacity.
  c. The curve will plunge downward in a "crash."

Ideas for Classroom Discussion

- Discuss what would happen to an ecosystem if all of the producers disappeared. What would happen to the ecosystem if all of the consumers or decomposers or detritivores disappeared? Can you think of examples of ecosystems in which any of these events has occurred? How can an ecosystem recover from such a disturbance?
- Ask your students to think about the carrying capacity of the United States regarding its human population. Do we know what that carrying capacity is? How much can we manipulate the environment to ensure that our nation will be able to support our population? What resources might be beyond our control?
- Discuss the pros and cons of government laws regulating family size. Do you think that nations with high rates of population growth should set a legal limit on the number of children a family may have? What are some of the consequences that can occur (for example, use of of amniocentesis and abortion to ensure male children) when countries legislate family size?
- Do you think the United States should play a role in disseminating birth control information and supplies to other nations?
- What is your opinion about the stance of the Catholic church against birth control?
- How has modern medical care changed the survivorship curve for humans since the turn of the century?
- Some organisms (bamboo, cicadas) reproduce only at a single, very brief interlude during their life span but produce a large number of seeds or offspring during this short period of time. Can you think of some ways in which this pattern benefits a species?
- What is the relationship between the size of offspring and their number per reproductive event?

- Discuss the research by Reznick and Endler on the effects of different kinds of predation on the size, color, and rate of maturity among guppy populations as described in this chapter. Can you think of any parallels in the human population? What environmental factors might influence human characteristics? Can you think of any human characteristics that no longer have adaptive value in our modern society? What characteristics do you think might be useful in the future?
- Why will the growth rate of the human population of Mexico continue to soar for many years to come even if stringent birth control measures are started immediately?
- What socioeconomic challenges will the current population structure in the United States someday pose to future generations?
- How might the current trend toward delayed childbearing change the population growth in the United States? Why would an increase in parental age change the population structure?
- Distinguish between "habitat" and "niche."
- What feature distinguishes each of the following categories from the others?

  a. community b. population c. ecosystem
- What factors *should* determine the carrying capacity of the planet Earth for humans? Do these actually operate? Do others operate instead?
- What *density-independent* factors influence the size of *insect* populations? Of these, how many limit *human* populations?

Term Paper Topics, Library Activities, and Special Projects

- Examine how the widespread wave of immigrants in the late 1800s and early 1900s changed the size and rate of growth in the U.S. population.
- Examine the effects of the Black Plague on the subsequent size and growth of the human population in Europe.
- Look up information about organizations concerned with world overpopulation.
- Compare reproductive rates among different cultural and ethnic subgroups of people in the United States. Construct age structure diagrams for each subgroup.
- Look up information on population densities of several rural areas scattered randomly throughout the United States with population densities of several randomly located urban areas. Then find figures on the rates of alcoholism, crime, suicide, and divorce for each area. Is there a statistically significant correlation between population density and any one of these four types of pathology?
- Locate current data on human population growth in several countries around the world. Which countries are growing faster than the world average? Slower? Are any countries experiencing a *negative* population growth?
- Investigate the success of the world's most extensive human population limiting program—China's.
- Read several essays on human populations to learn the relationship between *numbers* of people and *rate* of consumption of nonrenewable resources. Which countries waste the most? The least?

Films, Filmstrips, and Videos

- *Factors Necessary for Life.* MGHT, 16 minutes, 16 mm, color.
- *Population Ecology.* EBF, 1964, 19 minutes, 16 mm, color. Analyzes the biological principles that are related to changes in population sizes. The population explosion of humans is examined in the light of natural enemies, food factors, and control of diseases.
- *The Riddle of the Rook.* MGHT, 25 minutes, 16 mm, color
- *Too Many Elephants.* MGHT, 25 minutes, 16 mm, color.

44

COMMUNITY INTERACTIONS

Revision Highlights

Minor changes only in this chapter.

Chapter Outline

BASIC CONCEPTS IN COMMUNITY ECOLOGY
- **Habitat and Niche**
- **Types of Species Interactions**

MUTUALISM

INTERSPECIFIC COMPETITION
- **Exploitation Competition**
- **Interference Competition**

PREDATION
- **On "Predator" Versus "Parasite"**
- **Dynamics of Predator-Prey Interactions**

PREY DEFENSES
- **Camouflage**
- **Moment-of-Truth Defenses**
- **Warning Coloration and Mimicry**

PARASITISM
- **True Parasites**
- **Parasitoids**
- **Social Parasitism**

COMMUNITY ORGANIZATION, DEVELOPMENT, AND DIVERSITY
- **Resource Partitioning**
- **Effects of Predation on Competition**
- **Species Introductions**
- **Succession**
- **Species Diversity: Island Patterns**
- **Mainland and Marine Patterns of Species Diversity**

SUMMARY

Objectives

1. Define the following ecological terms: *habitat, niche, community, symbiotic, competition, predation, parasitism*, and *mutualism*.
2. List and distinguish among the several types of species interactions.
3. Discuss the positive aspects and the negative aspects of predation on prey populations.
4. Describe how communities are organized, how they develop, and how they diversify.

Key Terms

communities
habitat
niche
commensalism
mutualism
symbiotic
interspecific competition

predation
parasitism
facultative mutualism
obligate mutualism
interspecific
intraspecific
exploitation
  competition
interference competition
competitive exclusion
predators
parasites
prey
host
functional response
numerical response
stable coexistence
oscillation
coevolution
camouflage
pheromone
mimicry
Müllerian mimics
Batesian mimics
aggressive mimicry
speed mimicry
parasitoids
social parasitism
resource partitioning
climax community
succession
primary succession
secondary succession
cyclic replacement
distance effect
area effect
species diversity

Lecture Outline

I. Introduction
   A. In the rain forests of New Guinea, there are nine species of fruit-eating pigeons.
   B. Each species has its own role in the forest.
   C. All of the organisms of the forest interact as a community.

II. Basic Concepts in Community Ecology
   A. Habitat and Niche
      1. Habitat denotes a place where a population lives, such as a tropical rain forest.
         a. It has a characteristic range of physical and chemical conditions.
         b. Species adapt to those conditions.
      2. Niche refers to a range of abiotic and biotic conditions under which a species can live and reproduce. Each species of pigeon can eat fruit of different sizes, and their seed dispersal affects where different plants grow.
   B. Types of Species Interactions
      1. Neutral: neither species directly affects the other (e.g., eagles and grass).
      2. Commensalism: one species benefits and the other is not affected (e.g., bird's nest in tree).
      3. Mutualism: there is a symbiotic relationship where both species benefit.
      4. Interspecific competition: both species are harmed by the interaction.
      5. Predation and parasitism: one species benefits while the other is harmed.

III. Mutualism
   A. The mycorrhiza is a relationship between fungal filaments and the roots of higher plants.
   B. Facultative mutualism: both species benefit but can live without the other.
   C. Obligate mutualism: neither species can live long without the other (e.g., the yucca moth and yucca plant).

IV. Interspecific Competition
   A. Introduction
      1. Interspecific competition occurs when different species compete for a limited resource.
      2. It is not as intense as intraspecific competition.
      3. Exploitative competition: all have equal access to a resource but differ in their ability to exploit that resource.
      4. Interference competition: some individuals limit others' access to the resource.
   B. Exploitation Competition
      1. Competitive exclusion suggests that complete competitors cannot coexist indefinitely. **(Fig. 44.3, p. 726; TM 195)**
      2. Field experiments can test for this by removing one species from a test plot.
   C. Interference Competition
      1. Corals kill neighboring corals of other species by poisoning them and growing over them.
      2. In most cases, one species displaces another.

V. Predation

A. On "Predator" versus "Parasite"

1. Predators get their food from prey, but they do not live on or in the prey.
2. Parasites get their food from hosts, and they live on or in the host for a good part of their life cycle; they may or may not kill the host.

B. Dynamics of Predator-Prey Interactions

1. The dynamics range from stable coexistence to erratic oscillations depending upon . . .
   a. The carrying capacity for prey in the absence of predation
   b. The reproductive rate of the prey and predator (numerical response)
   c. The functional response of the predators
2. Stable coexistence results when predators prevent prey from overshooting the carrying capacity.
3. Oscillations are likely when predators reproduce more slowly than prey.

VI. Prey Defenses

A. Introduction

1. Coevolution of predator and prey—when a new defense evolves, only those predators able to counter it will live to reproduce.

B. Camouflage

1. Prey and predators need to hide in the open—polar bears in snow provide an example.

C. Moment-of-Truth Defenses

1. When cornered, prey may startle or intimidate a predator; for example, beetles may release chemicals.

D. Warning Coloration and Mimicry

1. Toxic prey often have bright colors or bold patterns to warn predators.
2. Other prey not equipped with defenses may resemble the toxic prey by mimicry.

VII. Parasitism

A. True Parasites

1. True parasites tend to coevolve with their host; they only kill hosts without coevolved defenses.

B. Parasitoids

1. Parasitoids kill their host by eating larvae. **(Fig. 44.13, p. 736; TM 196)**
2. This provides for natural control of insect populations.

C. Social Parasitism

1. One species depends upon the social behavior of another to complete its life cycle.
2. Cowbirds lay their eggs in the nest of other birds.

VIII. Community Organization, Development, and Diversity

A. Resource Partitioning

1. Natural selection increases the differences between competing populations.
2. Similar species share the same resource in different ways. **(Fig. 44.14, p. 737; TM 197)**

B. Effects of Predation on Competition

1. Predators decrease competition among prey; removing predators increases competition.

C. Species Introductions

1. Few introductions of species into new geographic areas are without ecological consequences.

D. Succession

1. Introduction
   a. When an environment lacks life, a predictable pattern of change occurs.
   b. The first species will first flourish and then decline until a climax community is reached.
   c. Succession denotes changes in species composition that lead to a climax community.
   d. Primary succession happens in an area that was devoid of life.
   e. With secondary succession, changes disturb an inhabited area.
2. Primary succession

a. Pioneer species develop soil; these are usually small, low-growing plants with a short life cycle.
b. Perennials then appear and replace the pioneers.

3. Secondary succession
   a. Secondary succession occurs in abandoned fields and in forests where disturbances let sunlight reach the forest floor.
   b. It is similar to primary succession, but many plants grow from seeds or seedlings already present when the process begins.
4. Cyclic replacement
   a. Winds, fires, insects, and overgrazing can disturb the climax community.
   b. Community stability may require episodes of instability that permits cyclic replacement of equilibrium species and thus maintains climax community (sequoias are an example).

E. Species Diversity: Island Patterns
   1. Species numbers increase on new islands and reach a stable number.
   2. Distance effect: the farther an island is from potential colonists, the fewer species there are on it. **(Fig. 44.17, p. 743; TM 198)**
   3. Area effect: larger islands tend to support more species. **(Fig. 44.18, p. 743; TM 199)**
F. Mainland and Marine Patterns of Species Diversity
   1. The area effect holds for continents and oceans.
   2. The number of species varies with the latitude. **(Fig. 44.19, p. 744; TM 200)**

Suggestions for Presenting the Material

- Using Table 44.1 as your guide, survey the contents of this chapter:
  a. Neutralism and commensalism are mentioned but not detailed.
  b. Mutualism is briefly discussed.
  c. Interspecific competition is detailed.
  d. Predation offense and prey defense are discussed.
  e. Parasitism types are defined.
  f. Integration within the community is summarized.
- While the previous chapter covered population structure and growth, this chapter discusses the interactions among the species of a community.
- The elaborate, finely tuned species interactions described here offer another excellent opportunity to discuss coevolution. Examples such as the array of yucca species in Colorado, each pollinated exclusively by one kind of yucca moth species, emphasize the point that individuals don't evolve, populations do. Students will see many good examples of adaptive traits in this chapter.
- The coevolution of predator and prey (or would-be prey) is also another good example of the impact of one species upon the evolution of another. After reading and discussing this chapter, students should understand that communities are shaped by a complex web of many different factors.

Classroom and Laboratory Enrichment

- Design an experiment to be carried out in the lab or in the field that would test Gause's principle of competitive exclusion.
- Develop a method of graphing habitat usage and habitat overlap among two or more species living in the same habitat.
- Examine a vegetated area on campus or in an area nearby. How are the plants in the area competing for resources? Suggest some ways in which competition has shaped the plant community.
- Show 35 mm slides and/or films of examples of camouflage; ask students to distinguish the camouflaged organism.
- Select any ecosystem and look for examples of resource partitioning.

- Design and implement a study of succession. This can be done on a small scale in the lab or on a larger scale in the field. List the species and their approximate densities at the beginning of the study, then follow the changes in species composition and density as the study progresses. Students may establish a baseline study to be followed in later semesters by other students.
- Examine road cuts, construction sites, flooded river banks, plowed fields, and other places that have recently been disturbed. Can you find several plant species that you would describe as pioneer species?
- Describe patterns of succession at edges of stream beds, rivers, or coastlines.
- Ask your students to classify each of the items below as belonging to a human "habitat" or "niche":
  a. President
  b. Dorm room
  c. Secretary
  d. Lounge
  e. Counselor
  f. Student
- As each topic in the chapter is discussed, 2 x 2 transparencies should be in view because these topics are best taught and remembered by the *examples* given.
- Show the graphs (Figure 44.3) of the classic experiments of *Paramecium* growth that Gause performed. Why did *P. caudatum* decrease to near extinction in the containers but thrive in nature?
- Using the observation that grass quickly establishes itself in the cracks in the pavement of a highway in which traffic has been blocked for some time, describe what is happening using terms from the chapter.

Ideas for Classroom Discussion

- Would you expect competition between two finches of different species to be less intense or more intense than competition between two finches of the same species? Explain your answer.
- How does a shift in niche benefit competing species sharing the same environment?
- Discuss predator-prey interactions. Why are the cycles of predator and prey abundance shown in Figure 44.6 described as "idealized"? What do you think a predator would do if deprived of its primary prey item? Examine the actual diets of several predatory species; how do these diets change from one month to the next throughout the year? How can environmental disturbances such as fires, floods, climate fluctuations, and insect outbreaks influence the predator-prey cycle? What are some of the other variables that may be overlooked in predator-prey interactions?
- What is the difference between parasites and parasitoids? Between true parasitism and social parasitism?
- How do parasites help to regulate host populations in nature?
- Why is resource partitioning essential for groups of functionally similar species living together?
- What characteristics distinguish a pioneer species? Are pioneer species good competitors against later successional species? Why are pioneer species dependent on the frequent advent of open, disturbed places?
- Do you think that fires in national parks should be allowed to proceed without human intervention? Why or why not?
- What is a weed? What characteristics distinguish weeds from other plants? Are there biological features that weeds tend to share?
- In the classic graph of lynx and hare populations (Figure 44.7), what was the basis of the numerical count on the x-axis? How valid was this compared to actual field counts of these two animals?
- What do you think a "hyperparasitoid" insect is?

- The *monarch* butterfly is orange and black and tastes bad (birds eating them spit them out immediately); *viceroy* butterflies are almost indistinguishable from monarchs but taste good. Which of these is the model; which is the mimic?
- "Primary succession," according to your text, "begins with colonization of an uninhabited site by pioneer species. . . ." Are there any uninhabited places left on earth for pioneer plants and animals to colonize? What would create such a setting?
- Why do insects introduced into the United States become such pests when they were not so in their native country?
- Are certain plants "born" to be weeds? Or do they achieve that status by human condemnation?

Term Paper Topics, Library Activities, and Special Projects

- Summarize several of the classic studies involving predator-prey interactions.
- Describe several species of parasites commonly found among humans in the United States earlier in this century prior to the advent of improved hygiene and widespread medical care. What parasites are still commonly found among human populations today? What are some of the steps that can be taken to reduce parasitic infections?
- Discuss the history and results of any one of the species introductions listed in Table 44.2.
- Write a report on the effects of interspecific competition on a native species whose populations have been adversely affected by an introduced species.
- Describe succession as it has occurred following a major disaster such as a flood, debris slide, fire, or volcanic eruption.
- Discuss examples of mutualism among plants and animals.
- Discuss the role of fire in regenerating plant communities in Yellowstone National Park.
- How did disturbance by human intervention help to initiate secondary succession in American prairies?
- At what point will the growing number of plant species in a previously disturbed area stop increasing? What factors will halt the rise in species composition? Examine studies of island ecology that seek to answer this question.
- Select a group of related species for which distribution data are available, and construct a graph of patterns of species diversity corresponding to latitude.
- Discuss the discovery, history, and causes of the greenhouse effect. Summarize some of the steps recommended by scientists today to combat the greenhouse effect.
- Search for specific examples of insects that are parasitoids of other insects. Are the parasitoids "effective" controls?
- Report on the succession that has occurred since the eruption of Mount Saint Helens in Washington in 1980.
- From a book on insect pest management, report on the successful control of the cottony-cushion scale by ladybugs in California.

Films, Filmstrips, and Videos

- *Baobab: Portrait of a Tree, Parts I and II.* MGHT, 1973, 30 minutes each, 16 mm, color. A rather spectacular film that illustrates how the baobab tree provides numerous microenvironments that support the lives of a wide variety of animals. It depicts the interdependence of life in and around the tree and shows well-explained examples of animal behavior. If you only show one film about community ecology, let this be the one. Beautiful night photography.
- *Camouflage.* IFB, 1954, 10 minutes, 16 mm, color.
- *Camouflage in Nature.* CORF, 1964, 11 minutes, 16 mm.

- *Coral Jungle.* DOUBLE, 1969, 23 minutes, 16 mm, color. A Jacques Cousteau film showing the community relationships of organisms inhabiting coral reefs.
- *Life and Death in a Pond.* CORT, 1981, 15 minutes, 16 mm, color. Interaction among organisms in a fresh water ecosystem.
- *Life Between the Tides.* JF, 1977, 20 minutes, 16 mm, color. Focuses on the diversity and interdependence of life in intertidal ecosystems along the Pacific Coast, emphasizing food webs and the need for conservation.
- *Parasitism: Parasitic Flatworms.* EBF, 1963, 16 minutes, 16 mm, color. Traces the development of parasitic flatworms and shows several life cycle stages that are rarely photographed. Defines predation, mutualism, and internal and external parasitism, and gives examples of each of these relationships. Reveals the complicated life cycle of the blood flukes and lung and liver flukes.
- *Plankton.* NGS, 1976, 12 minutes, 16 mm, color. Explores the microscopic drifters of the oceans and the roles they play in the marine food webs.
- *The Temperate Deciduous Forest.* EBEC, 1962, 17 minutes, 16 mm, color. Illustrates the complex network of plant and animal relationships that make up the temperate deciduous forest community. Shows typical deciduous forest plants and animals and their adaptations to seasonal change. The full yearly cycle of spring, summer, autumn, and winter is shown through the use of time-lapse photography, microphotography, and live photography.
- *Voice of the Desert.* MGHT, 1964, 22 minutes, 16 mm, color. Shows the plants and animals of a desert community in Arizona.

45

ECOSYSTEMS

Revision Highlights

Major rewrite within same organizational framework. Ecosystems concepts in sharper focus. Better explanation of gross primary productivity, net primary productivity, net community productivity. Correction in pyramid of energy flow (Figure 45.4). New introduction to biogeochemical cycles, with clear definitions of hydrologic, atmospheric, and sedimentary cycles. New photos of Hubbard Brook watersheds; better caption for global water budget and hydrologic cycle (Figure 45.7). Rewritten text for carbon cycle. New Commentary on greenhouse effect and global warming trend, including graphs of increased concentrations of carbon dioxide, CFCs, methane, and nitrous oxide. Refined sections and illustrations on nitrogen cycle and phosphorus cycle.

Chapter Outline

STRUCTURE OF ECOSYSTEMS
- **Trophic Levels**
- **Food Webs**

ENERGY FLOW THROUGH ECOSYSTEMS
- **Primary Productivity**
- **Major Pathways of Energy Flow**
- **Pyramids Representing the Trophic Structure**

BIOGEOCHEMICAL CYCLES
- **The Hydrologic Cycle**
- **Global Movement of Carbon—An Atmospheric Cycle**
- ***Commentary:* Greenhouse Gases and a Global Warming Trend**
- **Global Movement of Nitrogen—An Atmospheric Cycle**
- **Global Movement of Phosphorus—A Sedimentary Cycle**
- **Transfer of Harmful Compounds Through Ecosystems**

SUMMARY

Objectives

1. Understand how materials and energy enter, pass through, and exit an ecosystem.
2. Describe an important study that determined the annual pattern of energy flow in an aquatic ecosystem.
3. Explain what studies in the Hubbard Brook watershed have taught us about the movement of substances (water, for example) through a forest ecosystem.
4. Understand the various trophic roles and levels.

Key Terms

system	photosynthetic autotroph	heterotrophs	open system
		ecosystem	energy input

nutrient input
energy output
nutrient output
trophic level
food chain
food web
primary producer
primary consumer
secondary consumer
tertiary consumer
primary productivity
gross primary productivity
net primary productivity
net community productivity
detritus
grazing food web
detrital food web
pyramid of numbers
pyramid of biomass
pyramid of energy
biogeochemical cycles
hydrologic cycle
atmospheric cycles
sedimentary cycle
nutrients
watershed
greenhouse effect
gaseous nitrogen, N_2
cycling processes
nitrogen fixation
ammonification
nitrification
nitrogen scarcity
denitrification
phosphorus cycle
biological magnification
ecosystem analysis

Lecture Outline

I. Introduction
   A. Each region of the earth is a system where energy is secured by autotrophs and flows to heterotrophs.
   B. Ecosystems are open systems in which there are energy inputs and outputs, nutrient recycling, and nutrient outputs.

II. Structure of Ecosystems
   A. Trophic Levels—"Who eats whom?" **(Table 45.1, p. 748; TM 201)**
   B. Food Webs
      1. Food webs are more complex than food chains because the same food resource is often part of more than one food chain.

III. Energy Flow Through Ecosystems
   A. Primary Productivity
      1. Gross productivity should be distinguished from net primary productivity.
      2. Net community productivity is affected by many factors like sunlight, rainfall, and fire.
      3. Young ecosystems have higher productivity and greater energy storage than old systems.
   B. Major Pathways of Energy Flow
      1. At equilibrium, the energy acquired by photosynthesis equals the energy lost by metabolism.
      2. Energy flows through grazing and detrital food webs.
   C. Pyramids Representing the Trophic Structure
      1. A pyramid of numbers does not consider the size of organisms.
      2. A pyramid of biomass has a large base of primary productivity.
      3. With a pyramid of energy, only 6% to 16% of the energy of one trophic level is available to the next level. **(Figs. 45.4 and 45.5, pp. 751, 752; TM 202, 203)**

IV. Biogeochemical Cycles
   A. Introduction
      1. Elements essential for life move in biogeochemical cycles.
      2. There are separate hydrologic, atmospheric, and sedimentary cycles.
   B. The Hydrologic Cycle
      1. Water is moved or stored by evaporation, precipitation, detention, and transportation.
      2. Water moves other nutrients in or out of ecosystems.
      3. A watershed funnels rain or snow into a single river.
   C. Global Movements of Carbon—An Atmospheric Cycle **(Commentary art, pp. 758, 759; TA 127, TM 204)**
   D. Global Movements of Nitrogen—An Atmospheric Cycle
      1. Nitrogen is needed for proteins and nucleic acids.
      2. It is abundant in the atmosphere but not in the earth's crust.
      3. The Cycling Processes **(Fig. 45.12, p. 761; TA 128)**
         a. In nitrogen fixation, bacteria convert N_2 to NH_3.

b. Decomposition of dead nitrogen fixers releases nitrogen-containing compounds.
c. Ammonification is when bacteria and fungi decompose dead plants and animals and release excess ammonia or ammonium ions.
d. Nitrification is a type of chemosynthesis where NH_3 or NH_4^+ is converted to NO_2^-; other nitrifying bacteria use the nitrite for energy and release NO_3^-.
e. Nitrogen Scarcity. Soil nitrogen is lost by leaching, bacterial denitrification, and farming.

E. Global Movements of Phosphorus—A Sedimentary Cycle **(Fig. 45.13, p. 762; TM 205)**
F. Transfer of Harmful Compounds Through Ecosystems
1. DDT, used to kill mosquitoes, accumulates in fatty tissues and results in biological magnification and unexpected effects.
2. Ecosystem analysis tries to predict the complex effects of a single change in an ecosystem.

Suggestions for Presenting the Material

- The material presented in this chapter will help students understand that any ecosystem is comprised of many interdependent parts. After learning about how an ecosystem can be described in terms of its food webs, its trophic levels, and its biogeochemical cycles, students will be aware of the unity that joins all organisms.
- Use as many *local* examples of ecosystems as possible in discussions, demonstrations, and lab work. While it may seem overly simplistic and sometimes inaccurate to identify and describe the different levels of an ecosystem, students should see that it is useful because it helps us to understand the functioning of an ecosystem as a whole. Such ecosystem descriptions provide a valuable baseline against which we can measure the effects of change.

Classroom and Laboratory Enrichment

- Working individually or in teams, students should select an ecosystem in a laboratory or a field setting and identify which organisms comprise each trophic level of the ecosystem.
- Devise an experiment in which one of the trophic levels of an ecosystem is removed or disrupted and the effects are measured and described.
- Set up aquatic ecosystems in the lab, and monitor them throughout the semester. Identify the trophic levels of the ecosystem, and analyze the cycling of minerals and nutrients within it. In what ways are the aquatic ecosystems in the lab similar to/different from real aquatic ecosystems?
- Discuss the primary productivities of different regions of the United States. How can human intervention change primary productivity?
- Construct a detrital food web for a typical forest or open field ecosystem. Students can list the decomposers and detritivores one might find in such areas (or identify as many of the detritivores as possible in lab) and then find out how each organism would be arranged in a food web.
- Use overhead transparencies to present the biogeochemical cycles.
- Analyze local soils to determine the mineral and organic contents.
- Using Table 45.1 as your point of departure, attempt to show several examples of each trophic level using 2 x 2 transparencies.
- The "consumers" are not labeled as primary, secondary, and so on in Figure 45.2. Project this overhead transparency and ask students to call out the level designations. Clarify the fact that the primary consumer is actually at the *second* level. What group of organisms necessary for nutrient recycling is not indicated in this figure?
- Assuming that a flat 10% of the energy in one trophic level is conserved to the next and the producer represents 100%, calculate what percent is received by the killer whale in Figure 45.2.

- Using the overhead transparencies for the major biogeochemical cycles (nitrogen, carbon, phosphorus, and water) indicate where humans are active or passive participants. How have human activities altered the "natural" cycles?

Ideas for Classroom Discussion

- Describe several trophic levels of a typical ecosystem, and ask students to arrange them in the correct order.
- What is meant by the term *nuclear winter*?
- What is the one ingredient required by all ecosystems that cannot be recycled?
- Why is the term *food chain* rarely used when describing actual ecosystems?
- What do you think were the first producers to evolve on earth?
- How (and where) do humans fit into a food web?
- Does the loss of tropical forests throughout South America affect us? How?
- Why is a pyramid of biomass a more accurate representation of an ecosystem than a pyramid of numbers?
- Why are many marine pyramids of biomass often upside down?
- Why does a pyramid of energy narrow as it goes up?
- Tropical forests are highly productive ecosystems, incorporating extremely large amounts of carbon and other nutrients into plant material. Yet, when cleared of vegetation, such areas make very poor farmlands. Why?
- Is it environmentally wise to rely on large quantities of nitrogen-rich fertilizers for crop production? What are some alternatives? Discuss the pros and cons of commercial fertilizers.
- The pesticide DDT is just one example of a substance that undergoes biological magnification as it travels through an ecosystem. Can you think of others? (One possible example is the movement of strontium 90, a by-product of nuclear testing in the 1950s, through the food web.)
- Why do most food chains have only three or four consumer links?
- Why is it nearly impossible to study a single food chain?
- Why are humans at the top of nearly every food web of which they are a part? Are they ever at any other level?
- What would the personal and ecological advantages be to humans if they were to eat "lower down" on the energy pyramid?
- Analyze each of the segments of the following expression: "bio—geo—chemical cycle."
- In each of the biogeochemical cycles indicate the route each component takes in recycling. What "invisible" component is not recyclable?

Term Paper Topics, Library Activities, and Special Projects

- Describe how a worldwide nuclear holocaust could result in a "nuclear winter." Summarize the views of those scientists who believe that such an event would occur in the wake of widespread nuclear war, and present opposing viewpoints.
- Write a report on ecosystems deep on the ocean floor at depths impenetrable by light. What kinds of organisms make up the first trophic level in such ecosystems? How are these organisms obtaining energy?
- What happens to an ecosystem if any one of its levels is removed? Find descriptions of ecosystems in which this has happened and describe the results.
- Describe the biochemical steps used by chemosynthetic autotrophs to produce energy.
- List and describe some of the primary producers in an open ocean ecosystem off the U.S. coast.

- Discuss the primary productivities of various regions around the world. Which areas have the highest and lowest primary productivities? Why?
- Look up the root/shoot ratios of several plant species, and graph these values against the latitudes of their geographic ranges.
- Describe nitrogen fixation. What kinds of organisms can perform nitrogen fixation? Discuss genetic engineering research in this area.
- Discuss farming techniques designed to minimize nitrogen loss in the soil.
- Describe the effects of strip-mining on biogeochemical cycles.
- Describe the history of the use and subsequent banning of DDT in the United States.
- Consult several geography texts to see where people from different countries of the world are located in the food webs of their areas.
- Obtain records of the yearly average temperatures and rainfall amounts from your local weather bureau. Graph the data. Are there any trends?
- The spectacular rise and fall of the most famous of all insecticides—DDT—occurred during roughly the years 1943 to 1973. Prepare a chronology of the significant events in its "life story."

Films, Filmstrips, and Videos

- *A Desert Place.* PBS, TLF, 1975, 30 minutes, 16 mm, color. *Nova Series.* Explains the adaptive features of the diverse types of organisms that must survive environmental extremes. A penetrating glimpse into the delicately balanced desert ecosystem as exemplified by the Sonoran Desert.
- *Ecology of a Hot Spring: Life at High Temperatures.* EBEC, 15 minutes, 16 mm, color. Shows examples of organisms that can survive at the high temperatures characteristic of a hot-spring environment.
- *Food Web and Energy Pyramid.* MGHT, 1967, 16 minutes, 16 mm, color.
- *Grasslands of the World.* BFA, 1978, 14 minutes, 16 mm, color.
- *High Arctic Biome.* EBEC, 1961, 23 minutes, 16 mm, color. The struggle of plant and animal life in the harsh environment of the high Arctic is shown. A fascinating ecological study of a simple, vulnerable biome.
- *Life at Salt Point (Salt Water Ecology).* CORT, INFORM, 1978, 15 minutes, 16 mm, color. Surveys open sea, wave-swept, and intertidal marine ecosystems at Salt Point, California. Above surface and underwater cinematography and photomicroscopy reveal the interaction of species and their adaptations that have enabled them to survive in diverse environments.
- *Life Between the Tides.* STF, 1982, 16 minutes, 16 mm, color. Covers the adaptations of organisms to the changing, sometimes harsh intertidal ecosystem, including the relationships among intertidal organisms and organisms in the open sea.
- *Life in a Tropical Forest.* TLF, 1971, 29 minutes, 16 mm/videotape, color. *Life Around Us Series.* Emphasizes the natural balance of three jungles and the Smithsonian Institution's natural laboratory near Panama. Based on the Life Nature Library book *The Forest.*
- *Life in Lost Creek (Fresh Water Ecology).* CORT, 1978, 15 minutes, 16 mm. Underwater and surface photography reveal habitats, organismic adaptations, niches, populations, and other ecological factors in moving and standing waters.
- *Mzima: Portrait of a Spring, Parts I and II.* MGHT, 27 minutes/26 minutes, 16 mm, color. Kenya's Mzima Springs are a microcosm of the surrounding jungle world that clearly demonstrates the interdependence of living organisms.
- *The Salt Marsh: A Question of Values.* MGHT, 1975, 21 minutes, 16 mm, color. Shows Georgian salt marshes and the work being done at the Sapelo Island marine lab investigating the energy flow and food supply interactions.
- *Succession: From Sand Dune to Forest.* EBEC, 1960, 16 minutes, 16 mm, color. Illustrates the process and general principles of ecological succession by which an area slowly and continuously changes until it becomes a stable natural community. Photographs the dunes of the southern end of Lake Michigan.

46

THE BIOSPHERE

Revision Highlights

Internal reorganization and expansion. Introduction focuses on the question of species distributions, with convergent evolution of cacti and *Euphorbia* as an example. Major section on forces shaping global patterns of climate; new illustrations of earth's atmosphere and ozone layer, global air circulation patterns, and the world's climate zones (Figures 46.3–46.5). Improved explanation of rain shadow effect. New illustration on moisture and elevation gradients and primary productivity. Biome classification system updated (e.g., sclerophyllous woodlands and shrublands and monsoon grasslands are included; forest biomes are categorized by predominant tree types and distance from the equator). Expanded coverage of water provinces. Lake ecosystems section rewritten for clarity; new illustrations on thermal stratification and human-caused eutrophication in Lake Washington. Updated picture and new table on trophic nature of lakes. Expansion of marine environments. Vertical zonation of intertidal zone described and illustrated; new photos of tide pools. New graph and text on ocean zonation; new text on upwelling. New Commentary on El Niño Southern Oscillation (ENSO) and its dramatic effects on climates and human affairs; this is a good example of the ecosystems analysis concept introduced in Chapter 45.

Chapter Outline

COMPONENTS OF THE BIOSPHERE

GLOBAL PATTERNS OF CLIMATE

- **Mediating Effects of the Earth's Atmosphere**
- **Air Currents**
- **Ocean Currents**
- **Effects of Topography**
- **Seasonal Variations in Climate**

THE WORLD'S BIOMES

- **Deserts**
- **Sclerophyllous Shrublands and Woodlands**
- **Grasslands**
- **Forests**
- **Tundra**

THE WATER PROVINCES

LAKE ECOSYSTEMS

- **Lake Zones**
- **Seasonal Changes in Lakes**
- **Trophic Nature of Lakes**

MARINE ECOSYSTEMS

- **Types of Marine Environments**
- **Estuaries**
- **Life Along the Coasts**
- **The Open Ocean**
- ***Commentary:* El Niño and Oscillations in the World's Climates**

SUMMARY

Objectives

1. Describe the ways in which climate affects the biomes of Earth and influences how organisms are shaped and how they behave.
2. Contrast life in lake ecosystems with that in oceans and estuaries.

Key Terms

biosphere
hydrosphere
atmosphere
climate
gyres
rain shadow
biogeographic realms
biome
E-horizon, zone of leaching
Regolith, bedrock
humus
laterite
desertification
sclerophyllous shrublands
sclerophyllous woodlands
shortgrass prairie
tallgrass prairie
tropical grasslands
savanna
monsoon grasslands
evergreen broadleaf forest
tropical rain forest
slash-and-burn agriculture
deciduous broadleaf forest
monsoon forests
temperate deciduous forest
evergreen coniferous forest
boreal forest
taiga
montane coniferous forest
temperate rain forest
pine barrens
arctic tundra
permafrost
alpine tundra
littoral zone
limnetic zone
plankton
profundal zone
fall overturn
spring overturn
oligotrophic
eutrophic
basin
Lake Washington
W. Edmondson
estuaries
intertidal zone
open ocean
sandy shores
muddy shores
ocean zonation
pelagic province
neritic zone
oceanic zone
benthic province
continental shelf
bathyal zone
abyssal zone
hadal zone
hydrothermal vents
upwelling
El Niño Southern Oscillation (ENSO)

Lecture Outline

I. Introduction
  A. Unrelated species in distant regions often show striking similarities.
  B. The distribution of species is related to climate, topography, and species interactions.

II. Components of the Biosphere
  A. Biosphere = earth regions where organisms live.
  B. Hydrosphere = all water on or near the earth's surface.
  C. Atmosphere = gases, particles, and water vapor enveloping the earth.

III. Global Patterns of Climate
  A. Introduction
    1. Climate includes temperature, humidity, wind velocity, cloud cover, and rainfall.
    2. The climate is shaped by the amount of solar radiation, the earth's rotational and orbital movements, the distribution of land and water, and elevation of land masses.
  B. Mediating Effects of the Earth's Atmosphere **(Fig. 46.3, p. 766; TA 129)**
    1. Ultraviolet radiation is absorbed by ozone and oxygen in the upper atmosphere.
    2. The atmosphere reflects or absorbs half the solar radiation.
    3. Heat from the sun drives the earth's weather systems.
  C. Air Currents
    1. The sun differentially heats equatorial and polar regions.
    2. Nonuniform distributions of land and water cause pressure differences.
    3. Warm air moving north or south from equatorial regions is deflected because of the earth's rotation. **(Fig. 46.4b, p. 767; TA 130)**

4. Regional differences in rainfall result—and rain influences what ecosystem can occur. **(Fig. 46.5, p. 767; TA 130)**

D. Ocean Currents **(Fig. 46.6, p. 768; TM 206)**

1. Currents form because of the earth's rotation, winds, variations in temperature, and distribution of land masses.
2. Gyres dominate each ocean and bring warm equatorial waters poleward.

E. Effects of Topography

1. The mountains of California cause the air to rise and lose moisture.

F. Seasonal Variations in Climate **(Fig. 46.8, p. 769; TM 207)**

1. The tilt in the earth's axis causes seasons.
2. Biological rhythms, such as migrations, coincide with the annual changes.

IV. The World's Biomes **(Fig. 46.11, p. 771; TA 131)**

A. Introduction

1. Climatic factors determine patterns of vegetation and why unrelated species may have similar adaptations.
2. Barriers isolate species and restrict dispersal.
3. Biomes are broad vegetational subdivisions including all animals and other organisms.
4. Biome distribution corresponds to climate and soil type. **(Fig. 46.13 left and right, pp. 772, 773; TM 208, 209)**

B. Deserts

1. Deserts are areas where evaporation exceeds rainfall.
2. More than a third of the total land area is arid or semiarid due to drought and overgrazing.

C. Sclerophyllous Shrubland and Woodlands

1. These are semiarid regions where dominant plants have tough, evergreen leaves.

D. Grasslands

1. Grasslands are dry land that is flat or rolling (they include prairies and tropical grasslands).

E. Forests

1. Evergreen Broadleaf Forests
   a. These are tropical rain forests with a great diversity of plants and animals.
2. Deciduous Broadleaf Forests
   a. This is where trees drop their leaves during a dry season (it includes monsoon and temperate deciduous forests).
3. Evergreen Coniferous Forests

F. Tundra

1. Tundra is a desolate, treeless plain with permafrost below the surface; decomposition is inhibited.

V. The Water Provinces

A. The water provinces are extensive regions that include "lentic" (e.g., lakes) and "lotic" (e.g., rivers) ecosystems, as well as oceans and seas.

VI. Lake Ecosystems

A. Lake Zones **(Fig. 46.25, p. 782; TM 210)**

1. The littoral zone extends from the shore to where rooted plants stop growing.
2. The limnetic zone goes to maximum depth of photosynthesis; it includes plankton.
3. In the profundal zone , the bacteria decompose detritus.

B. Seasonal Changes in Lakes

1. In temperate regions, deep lakes undergo changes in density and temperature.
2. In winter, most of lake is 4°C (the densest water).
3. In summer, thermal stratification occurs: epilimion, thermocline, and hypolimnion.
4. Fall and spring overturns are related to changes in primary productivity.

C. Tropic Nature of Lakes

1. Glaciers can form lakes.
2. There are oligotrophic and eutrophic lakes. **(Fig. 46.28, p. 784; TM 211)**

VII. Marine Ecosystems
A. Types of Marine Environments
1. These types include estuaries, intertidal zone, and open oceans.
B. Estuaries
1. Estuaries are partially enclosed coastal regions where fresh and salt water meet.
C. Life Along the Coasts
1. Residents here are battered by waves, changing moisture, and temperature conditions.
D. Rocky Shores
1. Erosive forces produce grazing food webs.
E. Sandy and Muddy Shores
1. These areas have unstable stretches of loose sediment.
F. The Open Ocean
1. Ocean Zonation **(Fig. 46.33, p. 788; TM 212)**
a. The pelagic province is divided into neritic and oceanic zones.
b. The benthic province includes the continental shelf of bathyl, abyssal, and hadal zones.
c. Hydrothermal vents in the abyssal zone support unique communities.
2. Upwelling
a. Upwelling occurs when currents move nutrient-rich water to the surface.

Suggestions for Presenting the Material

- Reference to Figure 1.3 should be made for one last time. You can use this figure to put a good "wrap" to the course.
- Many of the subjects discussed in this chapter can be presented very effectively with the aid of 35 mm slides, films, and videos. Many of the biomes are unfamiliar to most students, but they can be made memorable by use of 35 mm slides illustrating their different features.
- After reading about ecosystems in the previous chapter, students will be able to identify ecosystem components in the biomes and aquatic ecosystems discussed here. Use familiar examples related to local weather patterns, if possible, when discussing climate.
- Students will be interested in learning more about tropical deforestation, El Niño, the depletion of the ozone layer, and ocean pollution—subjects that have recently been in the news.
- Several aspects of human impact on the biosphere are discussed in the next chapter; you and your students may wish to wait until then before exploring some of the topics and activities suggested here.

Classroom and Laboratory Enrichment

- Use overhead transparencies to show worldwide patterns of climate distribution and biome distribution.
- Use a globe and the information you have learned about global patterns of air circulation to explain the reasons for your latest weather conditions. Explain why your region has the weather it does.
- Discuss the latest advances in meteorology. Describe the role of satellites in assisting weather prediction.
- Select an example of a high elevation mountaintop ecosystem, and examine the plant and animal communities that are found at different elevational ranges. Examine the effects of elevation on plant community composition by taking a field trip, if possible, to an area with variations in elevation.

- Prepare a diagram showing how the effects of increasing elevation are similar to those of decreasing latitude.
- View road cuts or dig trenches in different local plant communities, and examine the exposed soil profiles. Identify the soil horizons. Students can perform additional soil science experiments if desired.
- Examine changes in terrestrial plant community composition along moisture gradients.
- Study zonation in a local pond or lake. List the plants, animals, protists, and other microorganisms you find in each zone. Students in subsequent semesters could sample the lake again and describe what changes occur from season to season.
- Design and implement a study of lake stratification. Measure the temperature, dissolved oxygen, and nutrients in each layer of the lake.
- Examine zonation in a marine ecosystem. Visit a coastline area to observe shoreline species, if possible, or look at slides, films, or visit a marine aquarium.
- Construct small-scale replications of freshwater or marine ecosystems in the lab.
- Ask a weather specialist to speak to the class concerning global and local weather and trends in climate change.
- If you do not feel qualified to present the climatological information, secure a film or videotape that will explain it fully and concisely.
- If you are located near an intergrade between biomes (such as temperate deciduous forest to grassland), draw this to the students' attention. Remind them to observe the changes in plant life as they drive through the area and note some highway marker or nearby town for reference.
- To most people the American desert is viewed as a vast wasteland of sand and cactus to be bypassed as the early settlers did to reach the promise of the west coast. Gather information on how weekend revelers are destroying this fragile biome that "heals" itself very slowly.
- Prepare a table listing (a) each biome, (b) its principal location, and (c) its chief plant and animal life.

Ideas for Classroom Discussion

- What impact have human activities had on the weather in the last century?
- As each biome is discussed, ask students who have visited that biome to share their observations with the class.
- Discuss some of the ways to improve soils considered too poor for agriculture. What are some of the characteristics of desert soils and tropical soils that make them poorly suited for long-term agriculture?
- Discuss the ways in which animal morphology might differ from one biome to the next. What role does annual mean temperature play in determining animal morphology? What differences in plant morphology can you see among the different biomes?
- What is a prairie? What happened to the American prairies? Are there still patches of prairie in the United States?
- Why is species diversity so high in tropical rain forests?
- Describe the rates of nutrient cycling in the tundra and in the tropical forest. How can you explain the different rates of nutrient cycling in these two biomes?
- What is meant when a lake is described as "dead"? How does such a condition come about? How can a "dead" lake be rejuvenated?
- What is the most readily observable feature that distinguishes one biome from another?
- Relate the movement of hurricanes through the Caribbean and Gulf of Mexico to the ocean currents you see drawn in Figure 46.6.

- As a child you may have believed that if you could leave the earth's surface and fly toward the sun, you would get warmer and warmer. But of course just the opposite is true; why?
- Should modern technology change desert habitats such as Palm Springs, California, where lawns are green only because they are maintained by artificial means?
- What effect will the rising of ocean waters have on estuarine areas if the greenhouse effect melts the polar ice masses?

Term Paper Topics, Library Activities, and Special Projects

- Describe plant and animal features that are influenced by climate.
- Write a description of your local climate. List and describe all of the factors responsible for your local climate.
- Use soil science textbooks to help you write a description of local soil types.
- Write a report about desertification.
- Describe where prairie remnants are found today in the United States. Prepare plant species lists of the shortgrass prairie and the tallgrass prairie.
- Write a report about historical descriptions of the Dust Bowl.
- What changes in plant species composition would you see among deciduous forests as you traveled from the eastern United States to the western coast?
- Describe the latest efforts made by governments around the world to slow the rate of tropical deforestation.
- What happens to the arctic tundra when the permafrost becomes damaged? Discuss the effects of human habitation on the arctic tundra.
- Describe what happens when a lake undergoes eutrophication. Discuss examples of situations in which eutrophication was reversed. What changes in aquatic species composition will occur as a result of eutrophication?
- Discuss the rejuvenation of lake ecosystems that were formerly polluted.
- Describe the role of estuaries in commercial fisheries. Discuss the ecology of one of the commercially important estuaries in the United States.
- Learn more about the recent research suggesting that the ozone layer may be shrinking.
- In the Northern Hemisphere we experience warm summer temperatures when the earth is tilted *toward* the sun. In winter the tilt is just the opposite. But during which season is the earth slightly closer to the sun? Don't assume you know this one—look it up!
- Irrigation has brought "bloom to the deserts." But is there a negative side to this practice? Search for a balanced perspective.
- Prepare a colored map of the United States showing which areas of the nation provide the most food and least food; superimpose shading to show the areas of most dense consumer (human) populations.

Films, Filmstrips, and Videos

- *A Desert Place.* Time-Life, 30 minutes, 16 mm, color. Shows plant and animal life that nicely illustrate the concept of an ecosystem.
- *Life in a Tropical Forest.* Time-Life, 30 minutes, color. Shows life in the jungles of Cambodia, the upper Amazon, and an island near Panama.
- *Say Goodbye, Parts I and II.* Wolper, 1971, 51 minutes, 16 mm, color. Man's destruction of endangered wildlife populations is cataloged in this provocative and moving nature spectacular. Rod McKuen narrates.

- *The Temperate Deciduous Forest.* EBEC, 1962, 17 minutes, 16 mm, color. Illustrates the complex network of plant and animal relationships that make up the temperate deciduous forest community. Shows typical deciduous forest plants and animals and their adaptations to seasonal change. The full yearly cycle of spring, summer, autumn, and winter is shown through the use of time-lapse photography, microphotography, and live photography.
- *Voice of the Desert.* MGHT, 1964, 22 minutes, color. Shows the plants and animals of a desert community in Arizona.

47

HUMAN IMPACT ON THE BIOSPHERE

Revision Highlights

Major revision, reorganization; global rather than regional in emphasis, cast in context of human population growth. Refined definition of pollutants. Covers air pollution, water pollution, then solid waste pollution. Updated text, table, art on classes of air pollutants, acid deposition, and ozone hole. New photos of damaged forests in North America and Europe. Clear explanation of CFC effects on the ozone layer. Computer graphics of expansion of the ozone hole from 1979 through October 1987. Tighter sections on the consequences of large-scale irrigation, on maintaining water quality, and on conversion of marginal land for agriculture. New section on deforestation, including new Commentary on tropical forest destruction, plus ground-level and satellite photos of clear countries and a map of countries where the greatest destruction is occurring. New section on desertification; the Dust Bowl and current situation in Sahel are the examples. Concludes with an updated picture of limited options for energy resources and a discussion of nuclear energy and nuclear autumn and winter. Powerful perspective to get students thinking about the global picture.

Chapter Outline

ENVIRONMENTAL EFFECTS OF HUMAN POPULATION GROWTH

CHANGES IN THE ATMOSPHERE
- **Local Air Pollution**
- **Acid Deposition**
- **Damage to the Ozone Layer**

CHANGES IN THE HYDROSPHERE
- **Consequences of Large-Scale Irrigation**
- **Maintaining Water Quality**

CHANGES IN THE LAND
- **Solid Wastes**
- **Conversion of Marginal Lands for Agriculture**
- **Deforestation**
- ***Commentary:* Tropical Forests—Disappearing Biomes?**
- **Desertification**

A QUESTION OF ENERGY INPUTS
- **Fossil Fuels**
- **Nuclear Energy**

PERSPECTIVE

Objectives

1. Understand the magnitude of pollution problems in the United States.
2. Examine the effects modern agriculture has wrought on desert, grassland, and tropical rain forest ecosystems.
3. Describe how our use of fossil fuels and nuclear energy affects ecosystems.

Key Terms

greenhouse effect	dry acid deposition	tertiary treatment	net energy
pollutants	wet acid deposition	green revolution	fossil fuels
thermal inversion	Ogallala aquifer	slash-and-burn-agriculture	oil shale
industrial smog	primary treatment	desertification	meltdown
photochemical smog	secondary treatment		nuclear winter

Lecture Outline

I. Introduction
   A. Interactions between the atmosphere, oceans, and land are the engines of the biosphere.
   B. Humans have been straining these engines without appreciating that they can crack.
   C. For example, our CO_2 waste is contributing to a "greenhouse effect" (global warming).
   D. Population growth and individual demands are stressing the environment.

II. Changes in the Atmosphere
   Each day 700,000 metric tons of pollutants are dumped into the atmosphere in the United States alone.
   A. Local Air Pollution
      1. Thermal inversions can trap pollutants close to the ground. **(Fig. 47.2, p. 795; TM 213)**
      2. Industrial smog is gray air found in industrial cities that burn fossil fuel.
      3. Photochemical smog is brown air found in large cities in warm climates; for example, gases from cars form car exhaust.
   B. Acid Deposition
      1. Burning coal produces sulfur dioxides.
      2. Burning fossil fuels and fertilizers results in nitrogen oxides.
      3. They can fall to the earth as dry acid deposition or wet acid deposition—acid rain. **(Fig. 47.3, p. 795; TM 214)**
   C. Damage to the Ozone Layer
      1. Ozone in the lower stratosphere absorbs most ultraviolet radiation from the sun.
      2. The ozone layer has been thinning since 1976 and a hole appears over the Antarctic each spring. **(Fig. 47.5, p. 797; TA 132)**
      3. In response, skin cancer has increased, cataracts may increase, and phytoplankton may be affected.
      4. Chlorofluorocarbons seem to be the cause—one chlorine atom can convert 10,000 ozone molecules to oxygen.

III. Changes in the Hydrosphere
   A. Consequences of Large-Scale Irrigation **(Fig. 47.6, p. 798; TM 215)**
      1. About half of all food is raised on irrigated land.
      2. Salt buildup and waterlogging can result.
   B. Maintaining Water Quality
      1. Human waste, insecticides, herbicides, chemicals, radioactive materials, and heat can pollute water.

2. Most liquid wastes from urban populations in the United States are treated on as many as three levels.
3. Primary treatment removes and then burns sludge before it is dumped in landfills; chlorine is added to water.
4. Secondary treatment uses microbes to degrade organic matter—nitrates, viruses, toxic substances remain.
5. Tertiary treatment uses experimental methods to remove solids, phosphates, organics, etc.; it is used on only about 5% of the nation's wastewater.
6. In short, most wastewater is not properly treated.

IV. Changes in the Land
  A. Solid wastes—we face a challenge to move from a "throwaway" society to one of conservation and reuse.
  B. Conversion of Marginal Land for Agriculture **(Fig. 47.7, p. 799; TM 216)**
    1. Almost 21% of land is used for agriculture; another 28% is available but may not be worth the cost.
    2. The green revolution has increased yields 4 times but uses 100 times more energy and mineral resources.
    3. A growing human population is moving into marginal lands for increasing needs.
  C. Deforestation **(Commentary art, p. 801; TA 133)**
    1. Forests are watersheds; they control erosion, flooding, and sediment buildup in rivers and lakes.
    2. Deforestation can reduce fertility, change rainfall patterns, increase temperatures, and increase carbon dioxide.
    3. Clearing large tracts of tropical forests may have global repercussions. **(Commentary art and Fig. 47.8, pp. 801, 802; TA 133, TM 217)**
  D. Desertification
    1. Desertification is the conversion of grasslands and croplands—20 million hectares per year.

V. A Question of Energy Inputs
  A. Increases in human population and extravagant life styles increases consumption. **(Fig. 47.11, p. 804; TM 218)**
  B. Fossil Fuels
    1. Fossil fuels are a limited resource that is taking increasing energy to extract and increases atmospheric levels of carbon dioxide and sulfur dioxides.
  C. Nuclear Energy
    1. With nuclear energy, the net energy produced is low and the cost high compared with coal-burning plants.
    2. Meltdowns may release large amounts of radioactivity to the environment.
    3. Waste is so radioactive that it must be isolated for 10,000 years.
    4. A nuclear exchange and resulting nuclear winter may cause catastrophic extinctions.

VI. Perspective
  A. Other organisms in the past have changed the nature of living systems; an example would be photosynthetic organisms during the Proterozoic Era.
  B. Our capacity for accelerated change is new—where will this lead?
  C. We may avert disaster by anticipating events before they occur.

Suggestions for Presenting the Material

- This chapter is an extension of the previous one. Here the authors show how humans have altered the natural ecosystems and their functioning.

- The information in this chapter offers many possibilities for lectures, discussions, and debate. Although the concerns addressed in this chapter are worldwide in scope, students will be able to think of many examples of environmental problems that are familiar and close to home.
- Use as many local examples as possible when talking about the different aspects of our impact on the environment. As they approach the end of their study of introductory biology, students should be able to understand the enormous complexity of environmental pollution problems as well as our inability to escape from certain biological constraints. The issues discussed in this chapter can spark good discussion because they are not just biological in nature—they also involve economics, government, and personal ethics.

Classroom and Laboratory Enrichment

- Assess public understanding of environmental pollution issues by designing a brief questionnaire. Divide the public into several different groups, and compare the levels of understanding among the groups. Some possible groupings are: students in this course, past students of the course, university or college students at large, university employees, different age groups, local residents not employed by the college or university, and groups created on the basis of differences in factors such as socioeconomic status, political affiliation, age, education, and so on. How well informed are the different groups? Did they know more or less about environmental issues than you thought they would? Can you identify any myths about environmental pollution that appear to be widely held?
- Monitor the quality of air in your area. What has been the impact of humans on the air in your community?
- What happens to your trash? Trace the fate of the garbage that leaves your campus daily. What about wastes containing dangerous chemicals from research labs or medical wastes from the campus infirmary or university hospital? Does your institution have a set of rules and guidelines governing waste disposal?
- Visit a sewage treatment plant. Discuss the biological steps involved in the treatment of your local sewage. In what ways could sewage treatment be improved?
- Collect water samples from your classroom building and have the samples analyzed. Discuss the results with your class.
- Working in small groups, debate the pros and cons of different energy sources. Each group should be prepared to discuss the merits and drawbacks of one of the following energy sources: oil, coal, natural gas, hydropower, solar power, nuclear power, and other alternatives.
- Visit a nuclear power plant in your area. Invite someone who works in the nuclear power industry to speak to your class.
- Ask a representative from a local environmental group to address the class.
- Prepare a list of environmental concerns in your area and state. Ask students what *should* be done, then ask them what realistically *can* be done.
- Obtain a map of the United States showing the areas of highest cancer rates. Ask students to speculate as to why certain areas are more at risk than others.

Ideas for Classroom Discussion

- What is "low sulfur" coal?
- How have coal-burning power plants changed in the last twenty years?
- What is a "clean" energy source? Can you think of an example?
- Why is the ozone layer of the earth's atmosphere shrinking at the same time that excessive amounts of ozone at the earth's surface are present in photochemical smog?
- What is thermal pollution? What are its effects?

- Do you think our laws regarding air pollution are too lenient? What changes would you make in the laws or their enforcement?
- Why do you think relatively little has been done to eliminate the greenhouse effect? What do you think is required before governments such as our own will pass laws regulating the types of human activities responsible for phenomena such as the greenhouse effect?
- Why does normal rainwater have a pH of around 5.6?
- Where does the tap water in your classroom building come from?
- Why would it be unwise to clear tropical forests and irrigate arid lands for conversion to agriculture?
- What are some examples of renewable energy sources?
- How many students in the class recycle their aluminum cans? Newspapers? Glass? Take a hand count. Ask students for reasons why they do (or do not) recycle these materials.
- Ask students to make a list of ways in which they could modify their own life styles and behaviors to reduce environmental pollution.
- Do you think that our current environmental pollution problems reflect a fundamental shift in human values over the last fifty years? Why or why not?
- What do you think will be the most important areas of science in the next decade? In fifty years?
- How would *you* rate as a "pioneer" in a setting where there was no electricity, stores, running water, and so on?
- List the ways in which a city, especially a large one, represents one of the least stable of ecosystems.
- Your text lists several "excesses of outputs." These constitute important environmental problems, but what is the "root" problem in each case? (Answer: too many people)
- Does it seem wasteful to you that billions of gallons of pure drinking water are used to flush toilets and wash cars? Can you propose an alternative? Could you convince the city managers to adopt it?
- Your text says that 150 acres of trees are needed for one Sunday Edition of the *New York Times*. Is this necessary? When will we no longer be able to afford such extravagance? Will it be too late to reverse the ecological damage?

Term Paper Topics, Library Activities, and Special Projects

- Collect articles on the greenhouse effect from popular magazines and newspapers published in the last year. Compare and contrast the coverage of this issue provided by these publications.
- Explain the biological dangers posed by each of the classes of air pollutants listed in Table 47.1.
- Describe the actual contents of industrial smog or photochemical smog in the nearest city with an air pollution problem. Describe the ways in which the air pollution affects the health of area residents.
- Look up historical records of thermal air inversions that resulted in significant numbers of deaths due to air pollution.
- What are the biological effects of acid rain on fish populations?
- Examine the statistical link between air quality and rates of respiratory disease.
- Describe successful community recycling programs. List the features of successful programs, and describe obstacles that must be overcome to ensure success.
- Discuss the success of recycling programs in states (such as New Jersey) with laws requiring mandatory recycling.
- Write a report describing the discovery and use of new plastics that will degrade when exposed to sunlight or can be decomposed by microorganisms.
- Summarize the federal laws against pollution.

- Who are your local polluters? Find out which industries release pollutants into the atmosphere. List the pollutants and describe what steps the companies have taken to reduce emission of pollutants.
- Describe the process of illegal dumping. Why is it done? Has it been done in your area? What are the penalties for illegal dumping?
- Make a list of the contents of a toxic waste dump, and describe the potential hazards posed by such compounds.
- Discuss the environmental effects of strip-mining.
- What are the current alternatives for storing nuclear waste? What are some proposed methods that might be used to store nuclear waste in the future?
- Is a "nuclear winter" a real possibility? Compare the views of experts on this subject.
- Just a handful of states have mandatory refundable deposits on drink containers. One of the highest (ten cents per container) is in Michigan. Investigate the effect this law has had on the soft drink industry, the retailers, and of course, the environment.
- One of the highest concentrations of industrial chemical plants is along the lower Mississippi River, mainly in Louisiana. The effects of these plants on the environment are not hard to document. Why then are they allowed to continue to operate?
- Tap water in all metropolitan areas is certified as "safe," but that doesn't necessarily mean you would want to drink it. In your library find the yellow pages listings for suppliers of bottled water for these cities: New York, Chicago, Los Angeles, New Orleans, and Atlanta. Compare the number of these suppliers with the population to obtain a ratio. Did you find any surprises?

Films, Filmstrips, and Videos

- *Century III—The Gift of Life.* USIA, 1976, 28 minutes, 16 mm, color. Constructive surgery, immunology, cell biology, and genetic manipulation are four key areas that will influence the future of humans. Shows a kidney transplant operation—a gift of life from a mother to her son.
- *Countdown to Collision.* EPA, 1973, 28 minutes, 16 mm, color. Shows how man is exploiting his environment and asks where humankind is going as the volume of waste expands twice as fast as the nation's population.
- *Dawn of the Solar Age* (two parts). BBC, PBS, TLF, 1977, 57 minutes, 16 mm/videotape, color. *Nova Series.* Also see separate titles of the two parts. An examination of the sun as an alternative source of energy. Explains the NASA proposal for a giant orbiting mirror that could reflect usable light to our cities, and compares use of sunlight energy to coal and nuclear power.
- *Dawn of the Solar Age: Solar Energy.* BBC, PBS, TLF, 1977, 30 minutes, 16 mm/videotape, color. *Nova Series.* Explains the workings and utility of some of the methods for harnessing solar power, including mirrors, photoelectric cells, and space collectors.
- *Dawn of the Solar Age: Wind and Water Energy.* BBC, PBS, TLF, 1977, 25 minutes, 16 mm/videotape, color. *Nova Series.* Explores alternative sources of energy that could replace our use of dwindling supplies of coal and oil. Focuses on wind, water, and the sun. Liberal, effective use of animation and graphics.
- *Energy: Harnessing the Sun.* St. Ed. F., 1974, 19 minutes, 16 mm, color. An overview of alternative energy sources to those traditional fossil fuel types that are currently running low in various parts of the world. Animated drawings and photographs of existing experimental facilities.
- *Environment.* Wiley, 1971, 28 minutes, 16 mm, color. Probes the causes of the deterioration of the environments in modern industrial societies and examines the balance that must be struck between environmentalists and people who consume natural resources. An unusual and dramatic film that is left open-ended.
- *Food for a Modern World.* WFP, 1968, 22 minutes, 16 mm, color. Discusses current concerns about our present and future food supply and traces developments in U.S. food technology and agriculture during the past fifty years.

- *Learning About Nuclear Energy.* EBEC, 1975, 15 minutes, 16 mm, color. *Introduction to Physical Science Series.* Shows the structure of the atom through animation, models, demonstrations, and diagrams. Explains the difference between atomic and chemical energy.
- *The New Alchemists.* NFBC, 1972, 29 minutes, 16 mm, color. An engaging film that shows how a group of scientists and their families have successfully established and worked an experimental plant and fish farm near Falmouth, Massachusetts. The farm shows how organic farming techniques and alternative energy supplies are used and also shows how it is possible for such a farm to be established on a shoestring budget.
- *The Salt Marsh: A Question of Values.* MGHT, 1975, 21 minutes, 16 mm, color. Shows Georgian salt marshes and the work being done at the Sapelo Island marine lab investigating the energy flow and food supply interactions.
- *Say Goodbye, Parts I and II.* Wolper, 1971, 51 minutes, 16 mm, color. Man's destruction of endangered wildlife populations is cataloged in this provocative and moving nature spectacular. Rod McKuen narrates.
- *The Time of Man: A Natural History of Our World.* BFA, 1970, 50 minutes, 16 mm, color. Examines the relationship between twentieth-century life and the environment. Stresses the importance of understanding and preserving the environment, which sustains man and all other forms of life.

48

ANIMAL BEHAVIOR

Revision Highlights

Minor updating.

Chapter Outline

Objectives

1. Understand the components of behavior that have a genetic and/or hormonal basis.
2. Distinguish behavior that is primarily instinctive from behavior that is learned.
3. Know the aspects of behavior that have an adaptive value.

Key Terms

animal behavior
hormones
behavioral primer
seasonal photoperiodicity
song system
instinct
sign stimulus
innate releasing mechanism
learning
associative learning
conditioned reflex
extinction
latent learning
insight learning
imprinting
compass sense
navigational sense
reproductive success
adaptive behavior
selfish behavior
altruistic behavior
natural selection
sexual selection
resource-defense behavior
female-defense behavior

Lecture Outline

I. Introduction
   A. Animal behavior involves coordinated responses to external and internal stimuli.

B. Behavioral responses include integrated sensory, neural, endocrine, and effector responses; all of them have a genetic basis and are subject to natural selection.
C. Genetically based neural programs provide each new individual with a means of responding to likely situations.
D. Learning can modify neural programs.
E. Learning is adaptive when it contributes to reproductive success.

II. Mechanisms of Behavior
  A. Genetic Foundations of Behavior
    1. Bird songs have a genetic foundation.
    2. A mutation of one gene can change the frequency of the courtship song of a fruit fly.
  B. Hormonal Effects on Behavior
    1. Genes specify proteins like hormones that can profoundly affect behavior.
    2.. Environmental clues like light can affect hormone secretion—photoperiodicity in song birds is an example. That is, light can reduce melatonin secretion and affect gonad development.

III. Instinctive Behavior
  A. All behaviors have genetic and environmental contributions.
  B. Instinct is the capacity for a stereotyped response to a first-time encounter with a key stimulus.
  C. For example, male stickleback fish try to strike any red object.

IV. Learning
  A. Learning occurs when information storage results in adaptive modification of behavior.
  B. The ability to learn requires a nervous system that is genetically determined.
  C. Categories of Learning
    1. Associative Learning
      a. Associative learning is the capacity to make a connection between a new stimulus and a familiar one.
      b. Pavlov noted that dogs salivated after a meat extract was placed on their tongue; he then showed that the sound of a bell could also result in salivation.
    2. Extinction is when a behavior becomes extinguished when the reinforcing stimulus is withdrawn.
    3. Latent learning is the ability to store environmental information that can be used later.
    4. Insight learning is problem solving without trial-and-error learning.
  D. Imprinting
    1. With imprinting, the capacity to learn is time dependent.
    2. For example, young animals are primed during a short period early in life to form a learned attachment to a moving object, normally their mother.
  E. Imprinting and Migration
    1. Bird migration requires a compass sense and navigational sense.
    2. Some young birds imprint on the visual image of the night sky.
  F. Song Learning
    1. Song learning requires species-specific learning in the bird brain.

V. The Adaptive Value of Behavior
  A. Foundations of Behavioral Evolution
    1. Reproductive success refers to the survival and production of offspring.
    2. Adaptive behavior promotes reproductive success.
    3. Selfish behavior occurs when an individual increases its chances of producing offspring.
    4. Altruistic behavior is self-sacrificing behavior that helps others and decreases the individual's own chance to reproduce.
    5. Natural selection is thought to select individuals, not groups.
  B. Adaptation and Feeding Behavior

1. Two populations of garter snakes show different feeding preferences from birth. **(Fig. 48.8, p. 815; TM 219)**

C. Anti-Predator Behavior

1. Animals compromise foraging efficiency to improve their chances of avoiding predation.
2. For example, woodchucks forage near burrows even if the location does not have the most lush vegetation.

D. Reproductive Behavior

1. Because of individual selection, in sexually reproducing animals, other members of a species create obstacles to reproductive success.
2. This intraspecific pressure is sexual selection.
3. A male's reproductive success may be measured by the number of descendants.
4. A female's success is measured by how many eggs she can produce or how many offspring she can care for—females are concerned with the quality of a mate, not the number of matings.
5. Environmental factors affecting the distribution of females are central to male competition for access to mates.
6. If the resources sought by females are clumped in space, then resource-defense behavior should evolve.
7. As a defense against predators, females may live together; this is female-defense behavior.
8. Females benefit by choosing superior genes or superior material benefits. For example, the female hangingfly mates with mates that offer them the largest fly or moth.

Suggestions for Presenting the Material

- This chapter and the next deal with topics that are not *strictly* biological but rather encompass the realm of social interaction.
- Although the chapters are separated for convenience, you may wish to integrate the content to customize your presentations.
- One approach to presenting the material is the inclusion of representative examples (with slides) that illustrate each principle of learning.
- Stress the "interpretative" aspect of studying behavior; that is, rarely do two investigators give exactly the same interpretation of the behavior they see.
- Many of the principles of animal behavior are difficult to demonstrate in the lab, although behavior textbooks and lab manuals contain possibilities for classroom lab experiments that can be used to accompany this chapter or the following chapter on social behavior.
- Films of such aspects of animal behavior as instinctive behaviors and learned behaviors make an excellent addition to the classroom. Students can also read and report on animal behavior experiments performed by others or the careers of famous animal behaviorists.
- The topic of animal behavior can serve as an opening for discussing the challenges of experimental design, especially with regard to experiments conducted in the field.

Classroom and Laboratory Enrichment

- Design an animal behavior experiment that can be performed outdoors in your campus environment. Begin by listing animal species found on campus, and then design methods of observing their behaviors.
- Learn more about experiments investigating instinctive behaviors in newly hatched chicks.
- Design experiments to test the ability of rats to learn mazes.
- Can zoos or aquaria be used as laboratories for animal behavior research? Visit a nearby zoo or aquarium to observe animal behavior. What types of behaviors do you see that are a product of captivity?

- Develop lab experiments investigating territoriality among fishes or small vertebrates.
- The most reliable sources of visual enhancement for these lectures are films and videotapes (the Disney series is excellent) because they are edited to show the critical behavior that may have occurred only after patient waiting.
- Although this chapter focuses on behavior in general using a variety of animals, you may want to prepare a listing of human examples for each topic and ask students to evaluate your list and supplement it.
- Survey the class opinion on whether each of the topics below is controlled more by *heredity* or *environment*, or equally:
  a. Intelligence
  b. Body size
  c. Beauty
  d. Speech patterns
  e. Health

Ideas for Classroom Discussion

- What are some of the difficulties in designing research experiments in animal behavior? Discuss problems that must be overcome when performing animal behavior research.
- Can avoidance behavior in *Paramecium* be considered learning?
- What is the link between environment and behavior?
- What are the characteristics of innate behaviors? What are the characteristics of learned behaviors?
- Can you think of some reasons why displays of aggression between males of the same species rarely result in actual bloodshed?
- Distinguish among associative learning, latent learning, and insight learning.
- Give an example of a sign stimulus.
- How much of the animal behavior we see and interpret is the result of what "we want to see"?
- What part has natural selection played in the behaviors we observe in animals today?
- What kinds of methods are used to alter animal behavior (for example, training dogs)? Are these the same methods as used for altering behavior in children?
- In what ways is human behavior altered by prison incarceration?

Term Paper Topics, Library Activities, and Special Projects

- Describe the role of environmental cues in determining migratory behaviors in birds. Do biologists understand the physiology underlying the navigational senses of birds?
- Discuss the annual migration of salmon to spawning grounds.
- How are green sea turtles able to find their way from their home in ocean waters off Brazil to their breeding grounds hundreds of miles away on Ascension Island? Describe the latest research efforts attempting to answer this question.
- Describe the mood-altering effects of the steroid drugs sometimes taken by athletes to increase muscle mass.
- Examine the effects of any one of the so-called recreational drugs (alcohol, nicotine, marijuana, cocaine, LSD, amphetamines, barbiturates) on human behavior.
- Describe the role of melatonin in vertebrate behavior.
- Examine the role of sex in determining human behavior. Are there some behaviors that occur with significantly greater frequency among individuals of one sex than among individuals of the opposite sex?

- Station yourself in a variety of busy spots on campus. Record the usual and unusual behaviors you see. Make note of the day of the week, time of day, gender of participants, race, weather conditions, and so on. Compare your observations with others in the class.
- What is unusual about monarch butterfly migrations compared to bird migrations? (Hint: Investigate the migration of the offspring.)

Films, Filmstrips, and Videos

- *Animal Communication, Parts I and II.* Time-Life, 1971, 30 minutes, 16 mm, color. Shows several types of animal communication, from insects to primates.
- *Animal Migration.* BFA, 1977, 12 minutes, 16 mm, color. Examines animal migration, using several species of birds, insects, fish, and mammals. Includes theories of motivation, orientation, and navigation techniques.
- *Army Ants: A Study in Social Behavior.* EBEC, 1966, 19 minutes, 16 mm, color. Shows the development of ant behavioral patterns and examines the social structure of an ant society.
- *Baboon Behavior.* IU, 1961, 31 minutes, 16 mm, color. Shows baboons in their native habitat in Kenya and compares their behavior with that of counterparts in human development.
- *Behavioral Biology.* CORT, 1981, 15 minutes, 16 mm, color. *Biological Sciences Series.* Compares the roles of heredity with learning and environment as determinants of behavior. Provides examples in protists, invertebrates, and vertebrates.
- *Biological Clocks.* GRACUR, 25 minutes, 16 mm, color. Discusses how animals and humans regulate their lives according to natural rhythms.
- *Bird Brain: The Mystery of Bird Navigation.* BBC, TLF, 1976, 27 minutes, 16 mm/videocassette, color. *Nova Series.* Probes the mysteries of bird migration and navigation, revealing that birds use several means to navigate. Internal clocks, navigation, and the importance of migration are explained generally too.
- *Do Animals Reason?* NGS, 1975, 14 minutes, 16 mm, color. Shows various examples of animal behavior using triggerfish, dolphins, starlings,and others to suggest that animals can and do learn by reasoning.
- *The Fruit Fly: A Look at Behavioral Biology (Drosophila).* CRMP, 1973, 21 minutes, 16 mm, color. The relationship of genes to behavior is examined, using the fruit fly as an example. Linkage of genes into linear arrays on the chromosomes, crossover, and sex determination are shown to occur in the fruit fly. The work of Morgan, Beadle, and Sturtevant is discussed, and the recent work occurring in Benzer's laboratory is examined in some detail.
- *Konrad Lorenz: Science of Animal Behavior.* NG, 1975, 14 minutes, 16 mm, color. Shows imprinting in young goslings and discusses its relationship to pecking orders.
- *Miss Goodall and the Wild Chimpanzees.* NGS, 1966, 51 minutes, 16 mm, color. British zoologist Jane Goodall's study of chimpanzees in East Africa is recorded in a fascinating close-up. Tool usage is closely examined.
- *Protist Behavior.* MLA-Wards, 1977, 11 minutes, 16 mm, color. Excellent photography shows basic behavioral mechanisms occurring in protists. Stereotyped aggregations, avoidance, orientation to light, and foraging behavior are all shown.
- *The Social Insects: The Honeybee.* EBEC, 1960, 24 minutes, 16 mm, color. Illustrates that honeybees are social insects (societies) that are divided into castes that contribute to the life of the colony. Shows how the various castes adapt to reproduction, food gathering, and how they differ during development. Shows the development of a honeybee from egg to adult. Also shows and explains how bees communicate.
- *The Tool Users.* NG, 1975, 14 minutes, 16 mm, color. Shows several examples (ants, finches, chimpanzees) of animals that use tools to accomplish tasks that their own body structures alone cannot perform.

49

SOCIAL BEHAVIOR

Revision Highlights

Minor updating.

Chapter Outline

Objectives

1. Describe how forms of communication organize social behavior.
2. List the costs and benefits of social life.
3. Explain the roles of self-sacrifice and altruism in social life.

Key Terms

social behavior	bioluminescent	altruistic behavior	self-sacrificing behavior
communication signal	dominance hierarchy	cooperation	kin selection

Lecture Outline

I. Communication and Social Behavior
   A. Consider the Termite
      1. A colony may have a million members working together, so communication is necessary.

2. Social behavior is the tendency of animals to enter into cooperative and interdependent relationships

B. Communication Defined

1. Termite soldiers can release alarm scents or pheromones to achieve communication.
2. A communication signal is a stimulus produced by one animal that changes the behavior of another such that they both benefit.
3. Natural selection has favored the evolution of information exchanges.

II. Channels of Communication

A. Visual Signals

1. Visual signals are used by animals that are active during the day.
2. Male baboons may threaten a rival with a "yawn."
3. Courtship in birds often involves visual displays.
4. Fireflies use bioluminescent signals.

B. Chemical Signals

1. Insects and mammals may release pheromones to attract a mate.

C. Tactile Signals

1. Honeybees "dance" in the dark beehive to tell others the location of food.

D. Acoustical Signals

1. These include bird songs and frog calls.

III. Costs and Benefits of Social Life

A. The degrees of sociality are different adaptations to different environmental challenges.

B. Social animals may be better able to hunt but more likely to deplete local food supplies or contract a contagious disease.

IV. Predation and Sociality

A. Dilution Effect

1. In a large group, a prey animal is less likely to become a victim.

B. Improved Vigilance

1. In large groups, each animal has more time to feed and yet is better able to be warned of prey.

C. Group Defense

1. Social animals are better able to repeal a predator.

D. The Selfish Herd

1. Animals in the center of a herd are the least likely to become prey.

V. Social Life and Self-Sacrifice

A. Self-sacrifice exists if it enhances the continuity of the genetic line of the altruistic individual.

B. Parental Behavior

1. Parents use time and energy that might otherwise improve their own chances to live and reproduce again.
2. Because of the genetic similarity between a parent and its offspring, it is possible for a parent to give up some of its own chances for survival and still have its genes spread through the population.

C. Cooperative Societies

1. A wolf can share its kill with members of the pack that are not its offspring.

D. Dominance Hierarchies

1. Dominance hierarchies minimize aggression within the society.
2. Dominant members leave more offspring than subordinates.

VI. Evolution of Altruism

A. Introduction

1. With altruistic behavior, the "helper" reduces its own reproductive potential while the "helped" has increased its reproductive success.
2. With cooperation, all parties have increased numbers of surviving offspring.

3. Self-sacrificing behavior reduces the probability of survival but may increase reproductive success.
4. In kin selection, the direct form is aid to offspring, while the indirect form is aid to relatives.

B. Kin Selection
1. Altruism (with self-sacrifice) can be genetically advantageous if the beneficiaries of the sacrifice are relatives.

C. Suicide, Sterility, and Social Insects
1. In insect societies like bees, sterile guards may protect the queen by stinging an intruder and committing suicide.
2. By sacrificing their own reproductive chances, they increase the number of genetically similar offspring produced.

D. Human Social Behavior
1. Some behavior is universal; for example, smiling is innate and helps to tie infants to parents.
2. Many human emotions are linked to specific facial movements—pleasure, anger, surprise, and rage are examples.
3. Genes code for proteins, not behavior—genes give us a capacity to behave.
4. Adaptive does not mean the same thing as moral; it means valuable in the transmission of an individual's genes.

Suggestions for Presenting the Material

- This chapter continues the discussion of behavior begun in the previous chapter by delving into some interesting aspects of *social* behavior.
- If you feel capable of commenting on the relationship between behavior as viewed by a biologist and as viewed by a sociologist, your students would no doubt appreciate this "bridging" of the disciplines.
- Emphasize the evolutionary value of each of the social behaviors described in the text. Students have seen many examples of adaptive traits by this point in their study of introductory biology and will be fascinated to think of behaviors as yet another example.
- Students will ask questions about aspects of human behavior that will serve as the focal points of some interesting discussions.
- Because many social behaviors are difficult to describe or demonstrate in class, films can be effectively used to present many of the more complex examples of social behavior, such as communication among honeybees.

Classroom and Laboratory Enrichment

- Design and implement a lab experiment involving schooling behavior among tropical fish species. How many fish must swim together before a lone fish will recognize them as a school and join them? Will fish school selectively with members of their own species?
- Look up information about body language and visual cues used by humans. Design experiments to be performed on campus that will allow you to observe these and other forms of nonverbal human communication. Is nonverbal communication different in males and females? Do students use different types of body language in different campus locations (classrooms, library, cafeteria, dormitories)?
- Design and perform a lab experiment investigating maternal behaviors in white mice. Is a mother able to quickly retrieve her pups if they are removed from the nest? Why is this an adaptive trait? What signals does she use to find her pups?

- What kinds of acoustical signaling can you hear in your area on a spring or summer evening? What kinds of acoustical signaling can you hear during the daytime hours? Determine the species responsible for the sounds you hear, and analyze the meaning of their calls.
- The best example of highly social insect behavior is the waggle dance of the honeybee. There are several sources of videotape and film; check your biological supply house catalogs.
- If the opportunity is available, consider a demonstration of firefly signaling. This may be overly ambitious but might be of interest to an honors group as a special project.
- To stimulate interest in this material, try to include a human behavior example for each of the topics discussed in the text.

Ideas for Classroom Discussion

- Discuss the ways in which elaborate mating displays benefit a species.
- In what ways do elements of sociality such as schooling benefit a species? What are the drawbacks of sociality?
- How do dominance hierarchies benefit those species that use them?
- Explain the benefits that accrue to the young vertebrate "helpers" who forego reproduction to assist in the rearing of their siblings.
- What is meant by the term *selfish herd*?
- How much of our human behavior is determined by our culture? What behaviors seem to be universal among all cultures? Can you think of unusual behaviors that seem to be found only among one culture?
- Can you think of some selection pressures that are responsible for the evolution of many human traits? (Some examples include selection for capturing large prey, coping with carnivorous competitors, and caring for infants.)
- Is child abuse a human behavior of modern times in Western cultures? Can you think of some reasons why this might be so?
- Why do you think true social behavior has evolved in only a few insect species?
- What aspects of bee behavior are "learned"; which are innate?
- In what respects are countries that describe themselves as "socialist" similar to truly social insects? How do they differ?

Term Paper Topics, Library Activities, and Special Projects

- Describe the chemical nature, distribution, and role of pheromones in animal behavior.
- Describe the visual communication system of fireflies.
- Make a list of altruistic behaviors among humans.
- Describe behavioral studies of human infants.
- How can human behaviors change as a result of brain surgery (such as a lobotomy) or injury? Describe the relationship between the different areas of the brain and human behavior.
- Discuss research in primate behavior. What have such experiments taught us about human behavior?
- What examples of nonverbal communication (for example, certain gestures and facial displays) seem to be universally understood? What are some examples of nonverbal communication that exist only within a particular culture or have different meanings among different cultures?
- Examine the growing commercial use of pheromones to control insect pests. How do pheromones work? How are they produced?

- Why do humans universally regard infants as "cute"? Are there facial features present only among infants that elicit a characteristic loving, nurturing response from adults?
- Discuss behaviorists' views on war as a human behavior. Why do humans fight wars with one another?
- Read about Diane Fosse's research on the societies of primates in Africa.
- E. O. Wilson is well known for his synthesis of biology and sociology as presented in his book on "sociobiology." Prepare a brief synopsis of his ideas.

Films, Filmstrips, and Videos

- *Army Ants: A Study in Social Behavior.* EBEC, 1966, 19 minutes, 16 mm, color. Shows the development of ant behavioral patterns and examines the social structure of an ant society.
- *Baboon Behavior.* IU, 1961, 31 minutes, 16 mm, color. Shows baboons in their native habitat in Kenya and compares their behavior with that of counterparts in human development.

Appendix I

WRITING ESSAYS AND TERM PAPERS

A term paper is really just a long essay, its greater length reflecting more extensive treatment of a broader issue. Both assignments present critical evaluations of what you have read. In preparing an essay, you synthesize information, explore relationships, analyze, compare, contrast, evaluate, and organize your own arguments clearly, logically, and persuasively, gradually leading up to an assessment of your own. A good term paper or short essay is a creative work; you must interpret thoughtfully what you have read and come up with something that goes beyond what is presented in any single article or book consulted.

Getting Started

You must first decide on a general subject of interest. Often your instructor will suggest topics that former students have successfully exploited. Use these suggestions as guides, but do not feel compelled to select one of these topics unless so instructed. Be sure to choose or develop a subject that interests you. It is much easier to write successfully about something of interest than about something that bores you.

All you need for getting started is a general subject, not a specific topic. Stay flexible. As you research your selected subject, you usually will find that you must narrow your focus to a particular topic because you encounter an unmanageable number of references pertinent to your original idea. You cannot, for instance, write about the entire field of primate behavior because the field has many different facets, each associated with a large and growing literature. In such a case, you will find a smaller topic, such as the social significance of primate grooming behavior, to be more appropriate; as you continue your literature search, you may even find it necessary to restrict your attention to a few primate species.

Alternatively, you may find that the topic originally selected is too narrow and that you cannot find enough information on which to base a substantial paper. You must then broaden your topic, or switch topics entirely, so that you will end up with something to discuss. Don't be afraid to discard a topic on which you cannot find much information.

Choose a topic you can understand fully. You can't possibly write clearly and convincingly on something beyond your grasp. Don't set out to impress your instructor with complexity; instead, dazzle your instructor with clarity and understanding. Simple topics often make the best ones for essays.

Researching Your Topic

Begin by carefully reading the appropriate section of your textbook to get an overview of the general subject of which your topic is a part. It is usually wise to then consult one or two additional textbooks before venturing into the recent literature; a solid construction requires a firm foundation. Your instructor may have placed a number of pertinent textbooks on reserve in your college library. Alternatively, you can consult your librarian, or the library card file or computer system, looking for books listed under the topic you have chosen to investigate.

Plagiarism and Note-Taking

The essay or term paper you submit for evaluation must be original work: yours. Submitting anyone else's work under your name is plagiarism and can get you expelled from college. Presenting someone else's ideas as your own is also plagiarism. Consider the following two paragraphs.

> Smith (1981) suggests that this discrepancy in feeding rates may reflect differences in light levels used in the two different experiments. Jones (1984), however, found that light level did not influence the feeding rates of these animals and suggested that the rate differences reflect differences in the density at which the animals were held during the two experiments.

> This discrepancy in feeding rates might reflect differences in light levels. Jones (1984), however, found that light level did not influence feeding rates. Perhaps the difference in rates reflects differences in the density at which the animals were held during the two experiments.

The first example is fine. In the second example, however, the writer takes credit for the ideas of Smith and Jones; the writer has plagiarized.

Plagiarism sometimes occurs unintentionally, through faulty note-taking. Photocopying an article or book chapter does not constitute note-taking; neither does copying a passage by hand, occasionally substituting a synonym for a word used by the source's author. Take notes using your own words; you must get away from being awed by other people's words and move toward building confidence in your own thoughts and phrasings. Note-taking involves critical evaluation; as you read, you must decide either that particular facts or ideas are relevant to your topic or that they are irrelevant. As Sylvan Barnet says in *A Short Guide to Writing About Art* (1981. Little, Brown and Company, second edition, p. 142), "You are not doing stenography; rather, you are assimilating knowledge and you are thinking, and so for the most part your source should be digested rather than engorged whole." If an idea is relevant, you should jot down a summary using your own words. Avoid writing complete sentences as you take notes; this will help prevent unintentional plagiarism later and will encourage you to see through to the essence of a statement while note-taking.

Sometimes the authors' words seem so perfect that you cannot see how they might be revised to best advantage for your paper. In this case, you may wish to copy a phrase or a sentence or two verbatim, but be sure to enclose this material in quotation marks as you write, and clearly indicate the source and page number from which the quotation derives. If you modify the original wording slightly as you take notes, you should indicate this as well, perhaps by using modified quotation marks: "......". If your notes on a particular passage are in your own words, you should also indicate this as you write. I precede such notes, reflecting my own ideas or my own choice of words, with the word *Me* and a colon; my wife, who is also a biologist, uses her initials. If you take notes in this manner you will avoid the unintentional plagiarism that occurs when you later forget who is actually responsible for the wording of your notes or who is really responsible for the origin of an idea.

You probably cannot take notes in your own words if you do not understand what you are reading. Similarly, it is also difficult to be selective in your note-taking until you have achieved a general understanding of the material. I suggest that you first consult at least one general reference text and read the material carefully, as recommended earlier. Once you have located a particularly promising scientific article, read the entire paper through at least once without taking any notes. Resist the (strong) temptation to annotate and take notes during this first reading, even though you may feel that without a pen in your hand you are accomplishing nothing. Put your pencils, pens, and notecards or paper away and read. Read slowly and with care. Read to understand. Study the illustrations, figure captions, tables, and graphs carefully, and try to develop your own interpretations before reading those of the author(s). Don't be frustrated by not understanding the paper at the first reading; understanding scientific literature takes time and patience.

By the time you have completed your first reading of the paper, you may find that the article is not really relevant to your topic after all or is of little help in developing your theme. If so, the preliminary read-through will have saved you from wasted note-taking.

Some people suggest taking notes on index cards, with one idea per card so that the notes can be sorted readily into categories at a later stage of the paper's development. If you prefer to take notes on full-sized paper, beginning a separate page for each new source and writing on only one side of each page will facilitate sorting later.

As you take notes, be sure to make a complete record of each source used: author(s), year of publication, volume and page numbers (if consulting a scientific journal), title of article or book, publisher, and total number of pages (if consulting a book). It is not always easy to relocate a source once returned to the library stacks; the source you forget to record completely is always the one that vanishes as soon as you realize that you need it again. Also, before you finish with a source, it is good practice to read the source through one last time to be sure that your notes accurately reflect the content of what you have read.

Writing the Paper

Begin by reading all your notes. Again, do this without pen or pencil in hand. Having completed a reading of your notes to get an overview of what you have accomplished, reread them, this time with the intention of sorting your ideas into categories. Notes taken on index cards are particularly easy to sort, provided that you have not written many different ideas on a single card; one idea per card is a good rule to follow. To arrange notes written on full-sized sheets of paper, some people suggest annotating the notes with pens of different colors or using a variety of symbols, with each color or symbol representing a particular aspect of the topic. Still other people simply use scissors to snip out sections of the notes, and then group the resulting scraps of paper into piles of related ideas. You should experiment to find a system that works well for you.

At this point you must eliminate those notes that are irrelevant to the specific topic you have finally decided to write about. No matter how interesting a fact or idea is, it has no place in your paper unless it clearly relates to the rest of the paper and therefore helps you develop your argument. Some of the notes you took early on in your exploration of the literature are especially likely to be irrelevant to your essay, since these notes were taken before you had developed a firm focus. Put these irrelevant notes in a safe place for later use; don't let them coax their way into your paper.

You must next decide how best to arrange your categorized notes, so that your essay or term paper progresses toward some conclusion. The direction your paper will take should be clearly and specifically indicated in the opening paragraph, as in the following example written by Student *A*:

> Most shelled molluscs, including clams, oysters, muscles, snails, and chitons, are sedentary; they live either attached to hard substrate (like rock) or in soft-substrate burrows. A few bivalve species, however, can actually swim, by expelling water from their mantle cavities. One such swimming mollusc is the scallop *Pecten maximus*. This paper will describe the morphological features that make swimming

possible in *P. maximus* and will consider some of the evolutionary pressures that might have been selected for these adaptations.

The nature of the problem being addressed is clearly indicated in this first paragraph, and Student *A* tells us clearly why the problem is of interest: (1) the typical bivalve doesn't move and certainly doesn't swim; (2) a few bivalves can swim; (3) so what is there about these exceptional species that enables them to do what other species can't; (4) and why might this swimming ability have evolved? Note that use of the pronoun *I* is now perfectly acceptable in scientific writing.

The first paragraph of your paper must state clearly what you are setting out to accomplish and why. Every paragraph that follows the first paragraph should advance your argument clearly and logically toward the stated goal.

State your case, and build it carefully. Use your information and ideas to build an argument, to develop a point, to synthesize. Avoid the tendency to simply summarize papers one by one: They did this, then they did that, and then they suggested the following explanation. Instead, set out to compare, to contrast, to illustrate, to discuss.

In referring to specific experiments, don't simply state that a particular experiment supports some particular hypothesis; describe the relevant parts of the experiment and explain how the results relate to the hypothesis under question.

In all writing, avoid quotations unless they are absolutely necessary; use your own words whenever possible. At the end of your essay, summarize the problem addressed and the major points you have made so that the reader will remember the key elements of your paper.

Never introduce any new information in your summary paragraph.

Citing Sources

Unless you are told otherwise, do not footnote. Instead, cite references directly in the text by author and date of publication. For example: Landscapes can be classified according to the dominant plant species (Slobodkin, 1988). Jones (1981), for example, refers to white oak forests.

At the end of your paper, include a section entitled Literature Cited, listing all references you have referred to in your paper. Do not include any references you have not actually read. Each reference listed must give author(s), date of publication, title of article, title of journal, and volume and page numbers. If the reference is a book, the citation must include the publisher, place of publication, and total number of pages in the book, or the page numbers pertinent to the citation. Your instructor may specify a particular format for preparing this section of your paper.

Creating a Title

By the time you have finished writing, you should be ready to title your creation. Give the essay or term paper a title that is appropriate and interesting, one that conveys significant information about the specific topic of your paper.

Good title: Behavioral and Chemical Defense Mechanisms of Gastropods and Bivalves

Poor title: Molluscan Defenses

Good title: The Effects of Spilled Fuel Oil on the Breeding of Shorebirds

Poor title: Pollutants and Birds

The following are good sources of information for developing essays and term papers:

General biology textbooks

Specialized textbooks, such as general texts on human physiology, invertebrate zoology, marine biology, and ecology.

Science section of major newspapers, such as the *Boston Globe*, the *New York Times*, and the *Los Angeles Times*. Most major daily newspapers have a science section once each week.

BioScience

The New England Journal of Medicine

Oceans

Science News

Scientific American

Sea Frontiers

Appendix II

ANNOTATED LIST OF COMPUTER SOFTWARE

This appendix lists some software programs that can be helpful for your students. Each item listed gives information about the software in the following order: title, publisher, source, computer, comments, copyright date, cost, and type of software. The items are listed according to the major units of the text.

Unit One: Unifying Concepts in Biology

- *Characteristics of a Scientist.* Cygnus Software / Cambridge; Educational Images / Apple / Puzzles, quizzes, tricks, and observations challenge the user to think and solve problems / 1983 / $44.95 / Interactive tutorial
- *CLASSIFY (A Classification Key Program).* DEE / Diversified Educational Enterprises / Apple II; TRS-80 / Requires the accompanying documentation / 1983 / $60.95 / Simulation
- *Interpreting Graphs, Escape (Game).* Conduit / Conduit / Apple / Twenty examples of graphs are presented to give the student a concept of the relevancy of graphs to events. A graphing game (Escape!) is included to reinforce graph reading and interpretive skills / 1983 / $50.00 / Drill and practice; simulation
- *Mathematics for Science Series 1-4.* Merlan Scientific, Ltd. / Cambridge / Apple; Commodore / A very good program for strengthening math concepts for science students. Lessons include measurement, scales, basic math skills, scientific notation, graphs, metrics / 1983 / $216.95 / Drill and practice tutorial
- *The Scientific Method.* Cygnus Software / Cambridge; Educational Images / Apple / By using brain teasers, puzzles, and problems, this program helps students organize their thinking by implementing the scientific method / 1983 / $44.95 / Interactive tutorial
- *Tribbles (An Introduction to the Scientific Method).* Conduit / Carolina Biological / Apple; IBM / The user forms hypotheses to explain growth patterns of alien life forms. A very good simulation for developing thinking skills / 1976 / $40.00 / Interactive simulation

Unit Two: The Cellular Basis of Life

- *Acid-Base Titrations.* Conduit / Conduit / Apple II / A very good simulation of laboratory titrations / 1984 / $75.00 / Simulation
- *Biochemistry Unit (Carbohydrates).* Bio Learning Systems, Inc. / Educational Images; Carolina Biological; Cambridge / Apple / An excellent tutorial that demonstrates the basics of carbohydrate structure and function / 1983 / $60.00 / Tutorial simulation
- *Biochemistry Unit (Lipids).* Bio Learning Systems, Inc. / Educational Images; Carolina Biological; Cambridge / Apple / Excellent tutorial outlining the basic properties of lipids. Includes dehydration synthesis, hydrolysis, phospholipid structure, and saturated and unsaturated fats / 1983 / $60.00 / Tutorial

- *Biochemistry Unit (Nucleic Acids).* Bio Learning Systems, Inc. / Educational Images; Carolina Biological; Cambridge / Apple / An excellent tutorial highlighting the unique features of nucleic acids / 1983 / $60.00 / Tutorial
- *Biochemistry Unit (Proteins).* Bio Learning Systems, Inc. / Educational Images; Carolina Biological; Cambridge / Apple / Introduces the basic properties of proteins, including structure, function, polymers, denaturation, and hydrolysis / 1983 / $60.00 / Tutorial
- *Biochemistry Unit Test Disk.* Bio Learning Systems, Inc. / Educational Images; Carolina Biological; Cambridge / Apple / Drill and practice of the contents of the Biochemistry Unit (Proteins, Lipids, Nucleic Acids, and Carbohydrates) / 1983 / $60.00 / Drill and practice
- *Biology: Energy and Life.* Britannica / Encyclopaedia Britannica / Apple / Good review or introduction covering molecules, energy storage, energy transfer, and metabolism / 1983 / $125.00 / Tutorial
- *Biology: The Cell.* Britannica / Encyclopaedia Britannica / Apple / A good introduction to or review of basic cell theory / 1983 / $125.00 / Drill and practice
- *Cell Respiration—Advanced Unit.* Bio Learning Systems, Inc. / Cambridge; Carolina Biological; Educational Images / Apple / A high-level tutorial on the chemical pathways for aerobic and anaerobic respiration / 1984 / $250.00 / Tutorial
- *Cell Respiration—Basic Unit.* Bio Learning Systems, Inc. / Cambridge; Carolina Biological / Apple / An introductory presentation of redox reactions, ATP-ADP transformations, carrier molecules, glycolysis, and the Krebs cycle / 1984 / $129.95 / Tutorial
- *Cell Structure.* Educational Images, Ltd. / Educational Images; Cambridge / Apple; IBM / A good overview of the structures in an animal cell / 1984 / $54.95 / Tutorial
- *Chemicals of Life I: The Structure of Matter.* IBM / IBM / A good tutorial simulation covering atomic structure, bonds, ions, and molecular theory / 1985 / $60.00 / Tutorial
- *Chemistry for Biologists.* Educational Images, Ltd. / Cambridge; Educational Images / Apple; IBM / Outlines the chemical basis of living systems. Ionic, covalent, and hydrogen bonds, as well as their biological importance; are presented / 1984 / $54.95 / Tutorial
- *ENZKIN (Unit on Enzyme Kinetics).* Conduit / Conduit / Apple / A simulation of enzyme experiments using six different enzymes and eleven values each of pH, substrate volume, enzyme volume, incubation time (in minutes), and temperature / 1979 / $50.00 / Simulation
- *ENZYME (A Simulation of the Lock and Key Method).* DEE / Diversified Educational Enterprises / Apple II; TRS 80 / A good simulation that requires the accompanying documentation / 1983 / $69.95 / Simulation
- *Experiments in Metabolism.* Educational Images, Ltd. / Educational Images; Cambridge / Apple; IBM / With color graphics and animation, the student determines the basic metabolic rate of a simulated laboratory mouse / 1984 / $54.95 / Tutorial simulation
- *General Chemistry—Atomic Weights.* COMPress / Carolina Biological; Cambridge / Apple; IBM / Excellent for introduction or review of atomic weight concepts / 1983 / $70.00 / Tutorial simulation
- *General Chemistry—pH: Acids and Bases in Water.* COMPress / Carolina Biological; Cambridge / Apple; IBM / Excellent for introduction or review of H+ :: OH- concepts / 1984 / $70.00 / Tutorial simulation
- *Osmotic Pressure.* Conduit / Carolina Biological / Apple / A good simulation of the effects of differing solutes, solvents, and molecular weights on osmotic pressure / 1983 / $50.00 / Simulation
- *Photosynthesis Advanced Unit.* Bio Learning Systems, Inc. / Cambridge; Educational Images; Carolina Biological / Apple / This five-part program presents the fundamentals of photosystems I and II, the Calvin cycle, C4 pathways, and the structure and functions of both chloroplasts and chlorophyll / 1984 / $250.00 / Tutorial
- *Photosynthesis—Basic Unit.* Bio Learning Systems, Inc. / Cambridge; Carolina Biological / Apple / Beginning with a presentation of chloroplast structure and function, this program presents the pathways of the light and dark reactions, including the splitting of water, the Calvin cycle, and phosphorylation / 1984 / $130.00 / Tutorial

- *Solar Food: An Exploration of Photosynthesis.* HRM Software / HRM / Apple / Tutorials and simulated experiments with plants introduce photosynthesis and the effects of light, carbon dioxide, and temperature / 1985 / $69.00 / Tutorial simulation
- *StarCell.* Merlan Scientific, Ltd. / Merlan Scientific, Ltd. / Apple; Commodore / A "spaceship" based review of cytological concepts, including mitosis, respiration, phagocytosis, excretion, protein synthesis, and lysosomes / 1986 / Game simulation

Unit Three: The Ongoing Flow of Life

- *Birdbreed.* EduTech, Inc. / Cambridge / Apple / The genetic concepts of simple dominance, linkage, independent assortment, multiple alleles, and gene interaction are presented on four levels of difficulty / 1982 / $94.95 / Simulation
- *CatLab.* Conduit / Carolina Biological; Cambridge; Conduit / Apple; IBM / A relevant simulation of the genetics of the domestic cat that demonstrates Mendelian genetics and the scientific method / 1982 / $75.00 / Simulation
- *DICROSS.* DEE / Educational Images / Apple II / A very good simulation of dihybrid crosses showing independent assortment and linkage / 1983 / $69.95 / Simulation
- *DNA: The Genetic Code.* Prentice-Hall Media, Inc. / Prentice-Hall Media / Apple II / A very good presentation of DNA structure, DNA replication, and protein synthesis / 1984 / $119.00 / Tutorial
- *Gene Structure and Function, Disk 1, DNA.* EduTech, Inc. / Cambridge / Apple / Describes nucleotide structure and bonding to form DNA. Replication, including complementary bases and hydrogen bonding, is demonstrated with the formation of two daughter strands / 1984 / $44.00 / Tutorial simulation
- *Gene Structure and Function, Disk 2, RNA.* EduTech, Inc. / Cambridge / Apple / Animated simulations show RNA synthesis as well as the functions of ribosomal, messenger, and transfer RNA / 1984 / $44.00 / Tutorial
- *Gene Structure and Function, Disk 3, Protein Synthesis.* EduTech, Inc. / Cambridge / Apple / Structure of amino acids and the formation of peptide bonds are highlighted along with codons for the twenty amino acids / 1984 / $44.00 / Tutorial
- *Gene Structure and Function, Disk 4, Animations and Quizzes.* EduTech, Inc. / Cambridge / Apple / Animations and quizzes from the other three disks in this outstanding series / 1984 / $44.00 / Drill and practice; simulations
- *Genetic Engineering.* Helix Educational Software, Inc. / Educational Images; Carolina Biological; Cambridge / Apple; IBM / A very good tutorial simulation on the basics and applications of genetic engineering / 1985 / $95.00 / Tutorial simulation
- *Heredity Dog.* HRM Software / HRM; Cambridge / Apple; Commodore / A very good simulation of the inheritance of one and two genes in domestic dogs / 1983 / $69.00 / Simulation
- *Human Genetic Disorders.* HRM Software / Carolina Biological; Cambridge / Apple / A very good simulation of the four basic patterns of inheritance using twenty-four human disorders. Crosses may be performed or genotypes may be inferred from pedigrees / 1985 / $69.00 / Simulation
- *LINKOVER (Unit on Genetic Mapping).* Conduit / Cambridge; Carolina Biological / Apple / A good simulation of genetic mapping experiments used to determine the sequence of genes in a linkage group on a single chromosome / 1975 / $50.00 / Simulation
- *Meiosis.* EME / Carolina Biological; Cambridge / Apple; IBM / An excellent review or introduction to the basic principles of meiosis / 1984 / $54.95 / Interactive simulation
- *Mendelian Genetics (A Problem-Solving Approach).* COMPress / Carolina Biological; Cambridge; Educational Images / Apple / A thorough treatment of the basic principles of Mendelian genetics, offering simulated laboratory experiments to demonstrate these concepts / 1979 / $70.00 / Tutorial simulation

- *Mitosis.* Educational Images, Ltd. / Educational Images; Cambridge / Apple; IBM / The cell structures and processes involved in mitosis are presented in tutorial form using many graphics and simulations / 1984 / $54.95 / Tutorial; drill and practice
- *MONOCROSS (A Monohybrid Cross Simulation).* DEE / Educational Images; Carolina Biological / Apple II / A good simulation of monohybrid crosses using peas, four-o-clocks, and Drosophila / 1982 / $69.95 / Simulation
- *The Gene Machine.* HRM Software / Carolina Biological; Educational Images; Cambridge / IBM; Commodore 64 / A very good simulation using clear graphics to demonstrate the fundamental concepts of DNA and RNA structure and function / 1983 / $69.00 / Tutorial simulation

Unit Four: Plant Systems and Their Control

- *COMPETE (Unit on Plant Competition).* Conduit / Educational Images; Carolina Biological / Apple / A simulation of the various causes and effects of competition in plants / 1979 / $50.00 / Simulation
- *Exploring the Amazing Food Factory: The Leaf.* Thoroughbred / Queue; Carolina Biological / Apple; IBM / A generally good tutorial on leaf structure and function. Included in this program are sections covering stomate action, transport in xylem and phloem, and photosynthesis / 1984 / $49.95 / Tutorial
- *How Plants Grow: The Inside Story.* Thoroughbred / Queue; Carolina Biological; Cambridge / Apple; IBM / This program presents material on plant growth, including sections on meristems and plant hormones / 1984 / $49.95 / Drill and practice
- *Leaf: Structure and Physiology.* IBM / IBM / IBM / A generally good tutorial on leaf structure and function. Included in this program are sections covering stomate action, transport in xylem and phloem, and photosynthesis / 1985 / $60.00 / Tutorial
- *PLANT (A Simulation of Plant Growth).* DEE / Diversified Educational Enterprises / Apple II / A good simulation that requires the accompanying documentation / 1983 / $69.95 / Simulation
- *Plants: Growth and Specialization.* IBM / IBM / IBM / This program presents material on plant growth, including sections on meristems and plant hormones / 1985 / $60.00 / Tutorial
- *Pollination and Fertilization: Seeds, Fruits, and Embryos.* IBM / IBM / IBM / An excellent module on flower structure and function. Topics covered include pollination, fertilization, gametogenesis, and germination / 1986 / $60.00 / Tutorial

Unit Five: Animal Systems and Their Control

- *All About Circulation. Part 2, Circulation System.* Micro Power & Light Co. / Micro Power and Light Co. / Apple / Good tutorial on the human circulatory system. Coverage includes the key organs as well as the heart, blood, and the pulmonary and systemic circulatory systems / 1980 / $39.95 / Tutorial
- *Biochemistry of the Immune System.* Helix Educational Software / Educational Images; Carolina Biological; Cambridge / Apple / This "user friendly" program presents the modern biology of the immune system while stressing biochemistry and genetics / 1985 / $95.00 / Tutorial
- *Enzyme Investigations.* HRM Software / HRM; Carolina Biological / Apple II 48K / A tutorial covering human digestive enzymes leads to a simulation of a trip through the alimentary canal. Simulated enzyme experiments may be performed by varying pH, temperature, and substrate concentration / 1985 / $69.00 / Tutorial simulation
- *Exercises in Muscle Contraction.* Educational Images, Inc. / Educational Images; Carolina Biological / Apple; IBM / With animation and graphics, four types of stimuli and their effect on the phases of motor unit activity are presented / 1985 / $64.95 / Simulation

- *Human Anatomy.* Britannica / Encyclopaedia Britannica / Apple / Excellent review or introduction to seven systems of the human body / 1980 / $99.00 / Drill and practice
- *Kidney Structure and Function: A Directed Lab Study.* Bio Learning Systems / Carolina Biological; Educational Images / Apple / Through a variety of simulations the relationship between the structure of the kidney and its function is presented / 1985 / $69.95 / Tutorial
- *Microbe.* Synergistic Software / Educational Images; Cambridge / Apple / A high-level game that simulates a trip through the human body to attend to several medical emergencies / 1982 / $49.95 / Game simulation
- *Neuromuscular Concepts.* Biosource Software / Queue / Apple II / An excellent tutorial on muscle contractions / $49.95 / Tutorial
- *Nutrient Analysis System 2.* University of Central Florida / Queue / Apple; IBM / An excellent program for showing the user the total nutritional value of specific foods, entire meals, or daily diets / $289.95 / Computer analysis
- *Skeletal Muscle Anatomy and Physiology.* Biosource Software / Queue / Apple II / A very thorough treatment of several concepts of skeletal muscle / $49.95 / Tutorial
- *The Heart Simulator.* Focus Media, Inc. / Focus Media; Carolina Biological; Cambridge / Apple II / High resolution graphics and animation depict systemic, pulmonary, and coronary circulation. Interactive drill and practice identify the different parts of the heart. Excellent / 1984 / $55.00 / Drill and practice; simulation
- *The Skeletal System.* Brain Bank, Inc. / Cambridge; Carolina Biological / Apple / An interactive program that identifies bone structure, major bones, joints, ligaments, and cartilage. Thorough and entertaining / 1983 / $70.00 / Tutorial; drill and practice
- *Your Body: Series I and II.* Focus Media, Inc. / Focus Media; Carolina Biological / Apple; TRS-80; Commodore 64 / Good questions on the human body are integrated with various games / 1984 / $258.00 / Drill and practice

Unit Six: Evolution

- *EVOLUT.* Conduit / Educational Images / Apple / Demonstrates the effect of selection on gene frequency in wild populations / 1980 / $50.00 / Simulation
- *Moths (The Evolution of the Pepper Moth).* DEE / Educational Images / Apple; IBM / User should generate a graph on paper (data supplied by program). Simulation of the classic application of the Hardy-Weinberg principle in industrial England / 1983 / $69.95 / Simulation
- *POPGEN (Population Genetics Simulation).* DEE / Diversified Educational Enterprises / Apple II / Requires the accompanying documentation / 1983 / $59.95 / Simulation

Unit Seven: Diversity: Evolutionary Force, Evolutionary Product

- *The Worm: Invertebrates Anatomy.* Ventura Educational Systems / Educational Images; Carolina Biological / Apple / Using very clear graphics, this program covers systems and individual structures common to many annelids/ 1984 / $39.95 / Tutorial simulation

Unit Eight: Ecology and Behavior

- *BALANCE (A Simulation of Predator-Prey Relationship).* DEE / Cambridge; Carolina Biological / Apple; IBM / Graphs and tables show the results of changing variables on predator-prey relationships / 1983 / $54.95 / Simulation
- *Oh, Deer.* Minnesota Educational Computing Corporation / Cambridge / Apple / The user has the opportunity to manage a simulated herd of white-tail deer in a residential area / 1983 / $44.95 / Simulation

- *POLLUTE (A Simulation of Water Pollution).* DEE / Cambridge / Apple; IBM/ Graphs and tables indicate the impact of various kinds of pollution on bodies of water and the survival of the fish living there / 1983 / $54.95 / Simulation
- *Population Concepts.* Educational Materials and Equipment Co. / Educational Materials and Equipment Co. / Apple / Presents the basics of population dynamics, including birth and death rates, migration, biotic potential, exponential growth, carrying capacity, and breeding density. Experiments are also possible / 1986 / $32.95 / Simulation
- *Population Growth.* Conduit / Conduit / Apple / A good explanation of growth curves including the concepts of growth rate, starting population, carrying capacity, rate constant, and doubling time / 1984 / $50.00 / Tutorial simulation
- *Population Growth.* COMPress / Carolina Biological; Educational Images / Apple / A computer simulation with sections covering exponential growth, density-dependent growth, variation in carrying capacity, and delays in regulatory response / 1981 / $70.00 / Simulation

Names and Addresses of Producers and Distributors of Computer Software

- Cambridge Development Laboratory, Inc.
  42 Fourth Avenue
  Waltham, MA 02154
- Carolina Biological Supply
  Burlington, NC 27215
- Conduit
  The University of Iowa
  Oakdale Campus
  Iowa City, IA 52242
- Diversified Educational Enterprises, Inc.
  725 Main Street
  Lafayette, IN 47901
- Educational Images, Ltd.
  P.O. Box 3456, West Side
  Elmira, NY 14905
- Educational Materials and Equipment Company
  P.O. Box 2805
  Danbury, CT 06813
- Encyclopaedia Britannica Educational Corporation
  425 North Michigan Avenue
  Chicago, IL 60611
- Focus Media, Inc.
  839 Stewart Avenue
  Garden City, NY 11530
- HRM Software
  Room Z 51
  175 Tompkins Avenue
  Pleasantville, NY 10570-9973
- Learning Corporation of America
  1350 Avenue of the Americas
  New York, NY 10019
- Merlan Scientific, Ltd.
  247 Armstrong Avenue
  Georgetown, Ontario
  Canada

- Micro Power and Light Co.
  12820 Hillcrest Road
  Dallas, TX 75230
- Prentice-Hall Media
  150 White Plains Road
  Tarrytown, NY 10591
- Queue
  Room S
  562 Boston Avenue
  Bridgeport, CT 06610

Appendix III

SOURCES FOR FILMS, FILMSTRIPS, AND VIDEOS

BBC	British Broadcasting Company–TV 630 Fifth Avenue New York, NY 10020
BFA	Phoenix/BFA Films and Video, Inc. 470 Park Avenue South New York, NY 10016
BFI	British Film Institute 72 Dean Street London W1V 5HB, England
BOUH	Boulton-Hawker Films Hadleigh, Ipswich Suffolk, England
CORF, CORT	Coronet Instructional Films (Distributed by Simon & Schuster) 108 Wilmot Road Deerfield, IL 60015
CRMP	McGraw-Hill Training Systems (formerly CRM Educational Films) P.O. Box 641 Del Mar, CA 90214
DOUBLE	Doubleday Media 1370 Reynolds Avenue Santa Ana, CA 92705
EBEC, EBF	Encyclopedia Britannica Educational Corporation 425 North Michigan Avenue Chicago, IL 60611
EMCORP	Educational Media Corporation Educational Materials Media Division 180 East Sixth Street P.O. Box 21311 Minneapolis, MN 55421
FPSERV, FPSR	Film Production Service 523 East Main Street Richmond, VA 23219

GRACUR	Graphic Curriculum, Inc. P.O. Box 565 Lenox Hill Station New York, NY 10021
HRAW	Holt, Rinehart and Winston 383 Madison Avenue New York, NY 10017
IFB	Internatonal Film Bureau 332 South Michigan Avenue Chicago, IL 60604
IU	Indiana University Audio-Visual Center Bloomington, IN 47405
JF	Journal Films 930 Pitner Avenue Evanston, IL 60202
Lilly	Eli Lilly & Co. 307 East McCarty Street Indianapolis, IN 46285
MCGH	CRM/McGraw-Hill Educational Films P.O. Box 641 110 Fifteenth Street Del Mar, CA 92014
MFIORH	Montefiore Hospital for Chronic Diseases 111 East 210th Street Bronx, NY 10467
MG	Media Guild 118 South Acacia, Box 881 Solana Beach, CA 92075
MGHT, MGHF	McGraw-Hill Films 674 Via De La Valle P.O. Box 641 Del Mar, CA 92014
MIFE	Milner-Fenwick, Inc. 2125 Greenspring Drive Timonium, MD 21093
MLA	Modern Learning Aids (see Wards) Divison of Ward's Natural Science Establishment P.O. Box 1712 Rochester, NY 14603
NFBC	National Film Board of Canada 1251 Avenue of the Americas, 16th Floor New York, NY 10020

NG, NGS	National Geographic Society 17th and M Streets, NW Washington, DC 20036
PBS	Public Broadcasting Service 1320 Braddock Place Alexandria, VA 22314
STF	Stanton Films 2417 Artesia Boulevard Redondo Beach, CA 90278
Time-Life, TLF	Time-Life Films and Video 100 Eisenhower Drive P.O. Box 644 Paramus, NJ 07652
UCEMC	University of California Extension Media Center 2223 Fulton Street Berkeley, CA 94720
USIA	U.S. Information Agency 1750 Pennsylvania Avenue, Northwest Washington, DC 20547
Wards	Ward's Natural Science Establishment 5100 West Henrietta Road P.O. Box 92912 Rochester, NY 14692
WFP	Wexler Film Productions, Inc. 801 North Seward Street Los Angeles, CA 90038
Wiley	John Wiley and Sons, Inc. 605 Third Avenue New York, NY 10016
Wolper	Wolper Productions 8489 West Third Street Los Angeles, CA 90048